8|3|-

RADIATION AND PROPAGATION OF ELECTROMAGNETIC WAVES

ELECTRICAL SCIENCE

A Series of Monographs and Texts

Edited by

Henry G. Booker
UNIVERSITY OF CALIFORNIA AT SAN DIEGO
LA JOLLA, CALIFORNIA

Nicholas DeClaris
UNIVERSITY OF MARYLAND
COLLEGE PARK, MARYLAND

JOSEPH E. ROWE. Nonlinear Electron-Wave Interaction Phenomena. 1965

MAX J. O. STRUTT. Semiconductor Devices: Volume I. Semiconductors and Semiconductor Diodes. 1966

AUSTIN BLAQUIERE. Nonlinear System Analysis. 1966

VICTOR RUMSEY. Frequency Independent Antennas. 1966

CHARLES K. BIRDSALL AND WILLIAM B. BRIDGES. Electron Dynamics of Diode Regions. 1966

A. D. KUZ'MIN AND A. E. SALOMONOVICH. Radioastronomical Methods of Antenna Measurements. 1966

CHARLES COOK AND MARVIN BERNFELD. Radar Signals: An Introduction to Theory and Application. 1967

J. W. CRISPIN, JR., AND K. M. SIEGEL (eds.). Methods of Radar Cross Section Analysis. 1968

GIUSEPPE BIORCI (ed.). Network and Switching Theory. 1968

ERNEST C. OKRESS (ed.). Microwave Power Engineering:
Volume 1. Generation, Transmission, Rectification. 1968
Volume 2. Applications. 1968

T. R. BASHKOW (ed.). Engineering Applications of Digital Computers. 1968

JULIUS T. TOU (ed.). Applied Automata Theory. 1968

ROBERT LYON-CAEN. Diodes, Transistors, and Integrated Circuits for Switching Systems. 1969

M. RONALD WOHLERS. Lumped and Distributed Passive Networks. 1969

MICHEL CUENOD AND ALLEN E. DURLING. A Discrete-Time Approach for System Analysis. 1969

K. KUROKAWA. An Introduction to the Theory of Microwave Circuits. 1969

H. K. MESSERLE. Energy Conversion Statics. 1969

GEORGE TYRAS. Radiation and Propagation of Electromagnetic Waves. 1969

GEORGES METZGER AND JEAN-PAUL VABRE. Transmission Lines with Pulse Excitation. 1969

C. L. SHENG. Threshold Logic. 1969

RADIATION AND PROPAGATION OF ELECTROMAGNETIC WAVES

George Tyras

Department of Electrical Engineering
Cullen College of Engineering
University of Houston
Houston, Texas

ACADEMIC PRESS New York and London 1969

ACADEMIC PRESS, INC.
111 Fifth Avenue, New York, New York 10003

United Kingdom Edition published by
ACADEMIC PRESS, INC. (LONDON) LTD.
Berkeley Square House, London W1X 6BA

LIBRARY OF CONGRESS CATALOG CARD NUMBER: 69-18341

PRINTED IN THE UNITED STATES OF AMERICA

To my children
Wanda Jean and Daryl George

CONTENTS

3. Plane Waves in Homogeneous Anisotropic Media

4. Plane Waves in Inhomogeneous Stratified Media

5. Spectral Representation of Elementary Sources

6. Field of a Dipole in a Stratified Medium

7. Radiation in Anisotropic Plasma

8. Green's Function Method of Solution

9. Axial Currents and Cylindrical Boundaries

10. Diffraction by Cylindrical Structures

11. Apertures on Cylindrical Structures

PREFACE

The period of the last twenty years has witnessed a great amount of research activity and many notable accomplishments both in the scope and in the understanding of the field of electromagnetics. Through the application of more modern and better mathematical methods it became possible to achieve a clearer exposition of the subject as well as a more systematic and logical treatment of various boundary value problems.

Unfortunately, not many of these accomplishments have found their way into the graduate teaching of the subject of electromagnetics in most schools. This phenomenon is due primarily to the fact that the development of modern textbooks on this subject has not, generally, kept pace with the research advances. Specialized monographs, devoted either to a single topic or to a special class of problems, have appeared from time to time. Such monographs are generally difficult to use in a classroom, especially if several of them are necessary in a single course.

The underlying philosophy of this graduate text is to narrow the existing gap between graduate teaching and modern research in electromagnetics. It is designed for a two-semester course in electrical engineering or electrophysics for students with only undergraduate preparation in electromagnetic theory. A practicing engineer should find this book useful for self-study as well as for a reference toward a better understanding of the current literature.

This book represents the author's viewpoint concerning the subject of an electrical engineer's graduate training in electromagnetic fields. A reader will find in this book a number of basic electromagnetic boundary value problems discussed in sufficient depth to enable subsequent serious work in more specialized branches of electromagnetic engineering, such as antennas, radio-wave propagation, microwaves, and electrooptics. Among the topics covered here are the following: plane waves in anisotropic media, plane waves in inhomogeneous media, spectral representation of elementary sources, field of a dipole in a stratified medium, radiation in anisotropic plasma, axial currents and cylindrical boundaries, diffraction by cylindrical structures, and apertures on cylindrical structures.

To treat the subject in sufficient depth, it has been necessary to introduce, as the material demanded, certain mathematical techniques which, even though well known to the researchers in the field, are ordinarily not available in a proper form for classroom presentation. These techniques, which certainly are not beyond the grasp of an average graduate student of today, usually represent a major obstacle to the understanding of modern literature for those who have been exposed only to the "conventional" treatment of the subject as it is still being presented in many graduate schools today. The mathematical techniques used in the treatment of the various topics include the Fourier transform technique, Green's function and its spectral representations, contour integration, asymptotic expansions via the method of the steepest descents, the WKB method including the expansion about a turning point, the Lebedev–Kontorovich transform, and the Wiener–Hopf technique.

Clarity of presentation was the guiding thought in the writing of this book and originality was considered to be only of secondary importance. Consequently, some ideas have been borrowed from other books, monographs, and published papers. A list of references which most closely resemble the treatment of a particular topic is included with each chapter. A sufficient number of problems is also included at the end of each chapter. These problems are intended to serve the purpose of additional drill as well as wider applications of the mathematical techniques presented.

No special attempt has been made to make this book self-sufficient in its mathematical aspects. Because of the prevailing trends in graduate engineering education, it is reasonable to assume that either prior to or concurrently with taking the course for which this book is intended, the student has acquired mathematical competence at the level of, e.g., G. Arfken: "Mathematical Methods for Physicists" (Academic Press, 1966). In addition, the readily available "Handbook of Mathematical Functions," edited by M. Abramowitz and I. A. Stegun (Natl. Bur. Std. Appl. Math. Ser. 55), contains almost all of the mathematical formulas referred to in this text. In exceptional cases, when not readily available, the necessary mathematical formulas are developed as they become needed.

GEORGE TYRAS

Houston, Texas
August 1969

ACKNOWLEDGMENTS

Many contributors throughout the years have influenced my thinking concerning the material presented in the various sections of this book. In addition, many of my graduate students, former and present, who have struggled with me through the early editions of this material, by their enthusiasm and constructive criticism, have helped me to achieve a clearer exposition of the subject. To all of these people I wish to express my sincere thanks. Special thanks are due to Professor Leopold B. Felsen of the Polytechnic Institute of Brooklyn for his reading of the portions of the manuscript and his constructive criticism. To Professor Henry G. Booker of the University of California at San Diego, I wish to express my sincere appreciation for his advice and encouragement during the writing of the final draft. To Dean C. V. Kirkpatrick and Professor H. S. Hayre, Cullen College of Engineering, University of Houston, I wish to express my appreciation for lending me support and encouragement in the preparation of the final draft. The help of Mr. Richard G. Schell, who read the galley proof and corrected many errors, is gratefully acknowledged. Finally, very special thanks are due to my wife, Rosemary Louise, without whose help, understanding, and forbearance throughout the years, this book would never materialize.

RADIATION AND PROPAGATION OF ELECTROMAGNETIC WAVES

CHAPTER 1

ELECTROMAGNETIC THEORY

In this chapter we shall review the basic laws of electromagnetic theory, which were formulated in the early nineteenth century and combined by Maxwell in 1864 into a consistent set of equations. The reader is assumed to be familiar with the basic experiments that lead logically to the formulation of Maxwell's equations. To the best of our knowledge, Maxwell's equations correctly describe the large-scale (macroscopic) electromagnetic phenomena that occur in nature. Ample experimental evidence is available to support this view, including the classical experiments of Coulomb, Ampère, and Faraday, which provided the original motivation for Maxwell to postulate the concept of the "electromagnetic theory."

In what follows, we shall review the basic equations and derive a number of general formulas without specific reference to any physical problem. In the succeeding chapters, we shall undertake the task of solving specific boundary value problems in which these formulas will prove useful.

1.1 Maxwell's Equations

The electromagnetic fields arise from stationary or moving charges called currents. The electromagnetic theory is concerned with the average behavior of the moving charges, the effect of which can be measured. Thus, we are not concerned with the behavior of an individual charge, and when such a term is used in this book, it should be understood to mean an "average" behavior of an ensemble of charges.

The *current density* is defined by

$$\mathbf{J} = \rho \mathbf{v} \tag{1.1}$$

where ρ is the average charge density and $\mathbf{v}$ is the average velocity. The direction of the current density vector will coincide with the direction of the charge velocity if the current is produced by the motion of the positive charges, and it will be opposite to the velocity vector if the current is produced by the motion of electrons. The scalar quantity I, the *current*, is defined by

$$I = \int_S \mathbf{J} \cdot \mathbf{n} \, ds \tag{1.2}$$

where $\mathbf{n}$ is a unit vector normal to the surface S.

The electric field intensity vector $\mathbf{E}$ is defined as the force $\mathbf{F}$ acting on a stationary test charge $\mathbf{q}$ in the limit as the magnitude of the test charge becomes infinitesimally small so that it does not perturb the field; i.e.,

$$\mathbf{E} = \lim_{q \to 0} \mathbf{F}/q. \tag{1.3}$$

When a charge is in motion, there may be an additional force acting on it, even though the electric field $\mathbf{E} = 0$. This force is written as

$$\mathbf{F} = q\mathbf{v} \times \mathbf{B}. \tag{1.4}$$

Equation (1.4) may be considered as defining a new quantity, the *magnetic flux density* $\mathbf{B}$.

In addition to the two field vectors defined above, it will be convenient to define two more vectors, $\mathbf{D}$ and $\mathbf{H}$, as follows. The field vector $\mathbf{D}$, the *electric flux density*, will be thought of as a measure of the number of lines of force produced by a charge. The vector $\mathbf{H}$, the *magnetic intensity*, when measured on a closed contour, will relate the magnitude of the total current enclosed.

The basic postulate relating to the four defined vector field quantities is that they satisfy Maxwell's equations

$$\nabla \times \mathbf{E} = -\frac{\partial \mathbf{B}}{\partial t}, \tag{1.5a}$$

$$\nabla \times \mathbf{H} = \frac{\partial \mathbf{D}}{\partial t} + \mathbf{J}, \tag{1.5b}$$

$$\nabla \cdot \mathbf{D} = \rho, \tag{1.5c}$$

$$\nabla \cdot \mathbf{B} = 0, \tag{1.5d}$$

where all the necessary conditions for the continuity and existence of the derivatives at all points in space are tacitly implied.

The *equation of continuity*

$$\nabla \cdot \mathbf{J} + \frac{\partial \rho}{\partial t} = 0 \tag{1.6}$$

is implied by Maxwell's equations. It can be derived by taking a divergence of (1.5b) and subsequently using (1.5c).

Maxwell's equations can also be written in an integral form. Using

Stokes's theorem

$$\int_S (\nabla \times \mathbf{F}) \cdot \mathbf{n}\, ds = \oint_C \mathbf{F} \cdot \boldsymbol{\tau}\, dc \tag{1.7}$$

where $\mathbf{n}$ is a unit vector normal to the area element ds of the surface S and $\boldsymbol{\tau}$ is a unit vector tangent to the element dc of the curve C, as shown in Fig. 1.1,

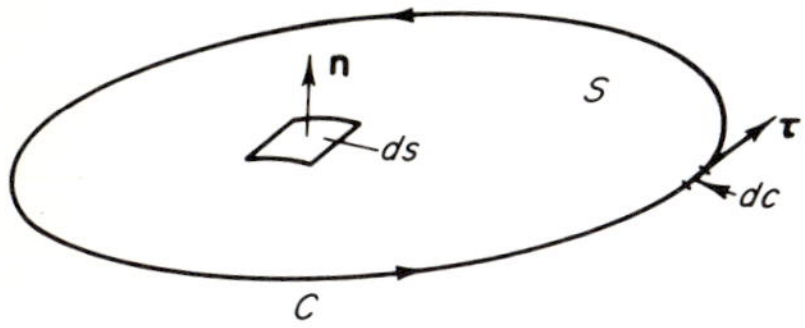

Fig. 1.1. Relevant relations for the application of Stokes's theorem.

the two curl equations in (1.5) can be written

$$\oint_C \mathbf{E} \cdot \boldsymbol{\tau}\, dc = -\frac{\partial}{\partial t} \int_S \mathbf{B} \cdot \mathbf{n}\, ds, \tag{1.8a}$$

$$\oint_C \mathbf{H} \cdot \boldsymbol{\tau}\, dc = I + \frac{\partial}{\partial t} \int_S \mathbf{D} \cdot \mathbf{n}\, ds \tag{1.8b}$$

where we also used the definition of the current in (1.2). The time derivative is taken outside the integral sign with the understanding that the surface S does not vary with time. Moreover, using the divergence (Gauss) theorem

$$\int_V \nabla \cdot \mathbf{F}\, dv = \oint_S \mathbf{F} \cdot \mathbf{n}\, ds \tag{1.9}$$

where S denotes a closed surface enclosing the volume V and $\mathbf{n}$ is a unit outward normal, as shown in Fig. 1.2, we obtain for the remaining two

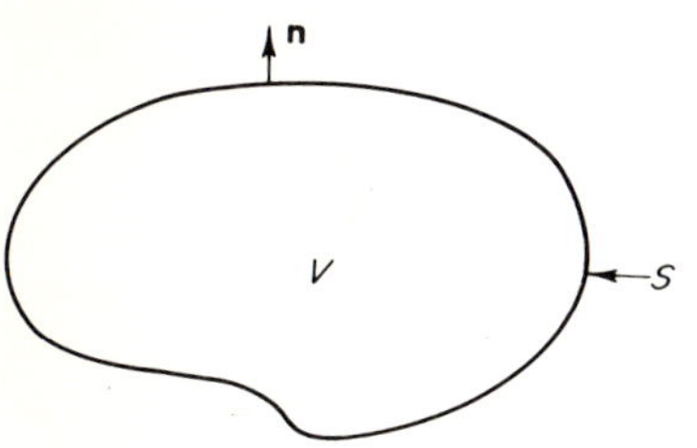

Fig. 1.2. Relevant relations for the application of the divergence (Gauss) theorem.

equations in (1.5) and (1.6)

$$\oint_S \mathbf{D} \cdot \mathbf{n}\, ds = \int_V \rho\, dv, \tag{1.10a}$$

$$\oint_S \mathbf{B} \cdot \mathbf{n}\, ds = 0, \tag{1.10b}$$

$$\oint_S \mathbf{J} \cdot \mathbf{n}\, ds = -\frac{\partial}{\partial t} \int_V \rho\, dv. \tag{1.10c}$$

The integral form of Maxwell's equations will be useful in the derivation of the boundary conditions at interfaces.

1.2 Macroscopic Properties of Matter

1.2.1 Electric Permittivity and Magnetic Permeability

The four equations in (1.5) contain twelve unknown field components to be determined if the currents and charges are specified. In order to render the problem determinate, additional relations between the field vectors must be specified. These relations can be obtained by considering the macroscopic properties of matter. To simplify the discussion we distinguish the following three groups.

Free Space (Vacuum)

In vacuum we have

$$\mathbf{D} = \varepsilon_0 \mathbf{E}, \qquad \mathbf{B} = \mu_0 \mathbf{H} \tag{1.11}$$

where

$$\varepsilon_0 = (1/36\pi) \times 10^{-9} \quad \text{farad/meter},$$
$$\mu_0 = 4\pi \times 10^{-7} \quad \text{henry/meter},$$
$$c = (\mu_0 \varepsilon_0)^{-1/2} \sim 3 \times 10^8 \quad \text{meters/second}.$$

Isotropic Media

Media whose physical properties are independent of their orientation with respect to the electric and magnetic fields are called *isotropic*. In this case we have

$$\mathbf{D} = \varepsilon_0 \varepsilon \mathbf{E}, \qquad \mathbf{B} = \mu_0 \mu \mathbf{H} \tag{1.12}$$

where ε is the scalar electric permittivity and μ is the scalar magnetic permeability.

If ε and μ are constants independent of the spatial coordinates, the medium is called *spatially homogeneous and isotropic*.

A medium may be spatially homogeneous but it may vary in time. We shall refer to such a medium as *temporally inhomogeneous and isotropic*.

If a medium is isotropic but varies in space as well as in time, it will be termed *isotropic, spatially, and temporally inhomogeneous*.

Anisotropic Media

Media whose physical properties are dependent on their orientation with respect to either the electric or the magnetic field are called *anisotropic.* In this case we write

$$\begin{pmatrix} D_x \\ D_y \\ D_z \end{pmatrix} = \varepsilon_0 \begin{pmatrix} \varepsilon_{11} & \varepsilon_{12} & \varepsilon_{13} \\ \varepsilon_{21} & \varepsilon_{22} & \varepsilon_{23} \\ \varepsilon_{31} & \varepsilon_{32} & \varepsilon_{33} \end{pmatrix} \begin{pmatrix} E_x \\ E_y \\ E_z \end{pmatrix} \tag{1.13a}$$

and

$$\begin{pmatrix} B_x \\ B_y \\ B_z \end{pmatrix} = \mu_0 \begin{pmatrix} \mu_{11} & \mu_{12} & \mu_{13} \\ \mu_{21} & \mu_{22} & \mu_{23} \\ \mu_{31} & \mu_{32} & \mu_{33} \end{pmatrix} \begin{pmatrix} H_x \\ H_y \\ H_z \end{pmatrix}, \tag{1.13b}$$

or in short notation

$$\mathbf{D} = \varepsilon_0 \boldsymbol{\varepsilon} \mathbf{E}, \qquad \mathbf{B} = \mu_0 \boldsymbol{\mu} \mathbf{H} \tag{1.14}$$

where $\boldsymbol{\varepsilon}$ and $\boldsymbol{\mu}$ are called the *electric permittivity* and *magnetic permeability tensors*, respectively.

In addition to being anisotropic, a medium may be space or time dependent or both. Such media, although not unusual in nature, are extremely difficult to deal with from the standpoint of being able to solve even the simplest boundary value problem.

1.2.2. Electric and Magnetic Polarization, Conductivity

In order to obtain the governing relations connecting the vectors **D** and **E** and **B** and **H**, it will prove useful to define two additional vectors and write **D** and **B** as follows.

$$\mathbf{D} = \varepsilon_0 \mathbf{E} + \mathbf{P}, \tag{1.15a}$$

$$\mathbf{B} = \mu_0 (\mathbf{H} + \mathbf{M}), \tag{1.15b}$$

where **P** and **M** are the electric and magnetic polarization vectors, respectively. It follows from the definitions above that the polarization vectors arise only in material media and vanish in free space.

Taking the divergence of each side of (1.15a) and remembering that $\nabla \cdot \mathbf{D} = \rho_{\text{true}}$ denotes the true charge density, whereas $\nabla \cdot \mathbf{E} = \rho_{\text{tot}}$ denotes the total charge density and $-\nabla \cdot \mathbf{P} = \rho_{\text{pol}}$ denotes the polarization charge density, we obtain

$$\rho_{\text{tot}} = (\varepsilon_0)^{-1} (\rho_{\text{true}} + \rho_{\text{pol}}). \tag{1.16}$$

Similarly, taking the divergence of (1.15b), we obtain

$$\nabla \cdot \mathbf{H} = -\nabla \cdot \mathbf{M} = \rho^{(m)} \tag{1.17}$$

since the divergence of **B** vanishes indentically.

The electric polarization charge density ρ_{pol} and the magnetic charge density $\rho^{(m)}$ can exist only in regions of matter where **P** or **M**, respectively, varies from point to point or at sharp discontinuities (boundaries) of the material.

In some media that possess a crystalline structure, the conduction electrons are relatively free to move through the crystal lattice. Under the influence of an applied electric field, the conduction electrons are forced to move along the lines of force, but their motion is retarded because of the collisions with the lattice. The proportionality factor between the conduction current density and the electric field is called the *conductivity* σ, and we write

$$\mathbf{J}_c = \sigma \mathbf{E}. \tag{1.18}$$

The current density will also depend on the applied magnetic field (the *Hall effect*) but this effect is usually very small in relation to the effect produced by the electric field and, hence, it is usually neglected.

1.3 Energy and Power

In an electric field, the time-averaged energy stored in a volume V is given by

$$\langle W_e \rangle = \tfrac{1}{4} \operatorname{Re} \int_V \mathbf{D} \cdot \mathbf{E}^* \, dv, \tag{1.19}$$

where the asterisk (*) denotes the complex conjugate.

The time-averaged energy stored in a magnetic field is given equivalently by

$$\langle W_m \rangle = \tfrac{1}{4} \operatorname{Re} \int_V \mathbf{B} \cdot \mathbf{H}^* \, dv. \tag{1.20}$$

In (1.19) and (1.20) the peak values of the field vectors are implied. The numerical factor $\frac{1}{4}$ arises from the fact that the time averaging is done over one period.

The *complex Poynting vector* is defined by

$$\langle \mathbf{S} \rangle = \tfrac{1}{2} \mathbf{E} \times \mathbf{H}^* \tag{1.21}$$

and the complex power is given by

$$P = \oint_s \langle \mathbf{S} \rangle \cdot \mathbf{n} \, ds \tag{1.22}$$

where the integration must be carried over a closed surface. The real part of P represents the real power and the imaginary part represents the power stored in the system.

1.4 Boundary Conditions

The Maxwell equations properly determine the behavior of electromagnetic fields in regions in which the physical properties of the medium vary continuously. If the medium's parameters vary discontinuously across some surface, we may expect corresponding changes in the field vectors. In order to properly continue the solution to Maxwell's equations from one region to another in such a way that the resulting solution be unique and valid everywhere, it is necessary to establish the conditions to be imposed on the field vectors at the interfaces. These *boundary conditions* can be readily derived by using the integral form of Maxwell's equations:

$$\oint_C \mathbf{E} \cdot \boldsymbol{\tau}\, dc = -\int_S \frac{\partial \mathbf{B}}{\partial t} \cdot \mathbf{n}\, ds; \tag{1.23a}$$

$$\oint_C \mathbf{H} \cdot \boldsymbol{\tau}\, dc = I + \int_S \frac{\partial \mathbf{D}}{\partial t} \cdot \mathbf{n}\, ds; \tag{1.23b}$$

$$\oint_S \mathbf{B} \cdot \mathbf{n}\, ds = 0; \tag{1.23c}$$

$$\oint_S \mathbf{D} \cdot \mathbf{n}\, ds = q; \tag{1.23d}$$

where I is the true current across the surface S as defined in (1.2).

Consider the interface separating medium 1 with parameters ε_1, μ_1, σ_1 from medium 2 with parameters ε_2, μ_2, ρ_2. If the integral (1.23c) is taken over the closed surface S of the small cylinder in Fig. 1.3 that is normal to

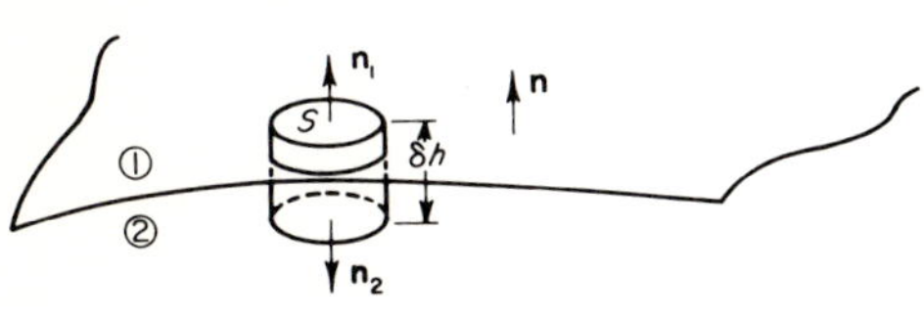

Fig. 1.3. Surface S pertinent to the derivation of the boundary conditions for the normal components of fields.

the interface, then for a sufficiently small cross section of the cylinder the field may be considered constant over each face. If the height of the cylinder is allowed to shrink to zero, the contribution from the curved surface will vanish and according to (1.23c) we obtain

$$(\mathbf{B}_1 - \mathbf{B}_2) \cdot \mathbf{n} = 0, \tag{1.24}$$

denoting the fact that *the normal component of* **B** *is continuous across an interface.*

In a similar fashion we can apply the integral (1.23d) to the cylinder in Fig. 1.3. We obtain in this case

$$(\mathbf{D}_1 - \mathbf{D}_2) \cdot \mathbf{n} = \lim_{\delta h \to 0} \rho\, \delta h = \rho_s \tag{1.25}$$

where ρ_s is called the *surface charge density.* Equation (1.25) denotes the fact that *the normal component of* **D** *has a jump discontinuity across an interface equal to the magnitude of the surface charge density.*

Now consider the integral expression (1.23a) as related to Fig. 1.4. The

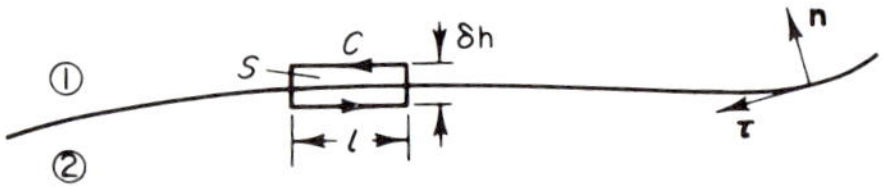

Fig. 1.4. Contour C pertinent to the derivation of the boundary conditions for the tangential components of fields.

line integral of the electric field taken around the closed contour C must be equal to the surface integral of the negative rate of change of the magnetic flux density. If the loop is sufficiently small, the fields can be considered constant. Moreover, in the limit as the height of the loop δh approaches zero, the surface integral vanishes and only the horizontal parts of the contour C contribute. Thus, we obtain

$$(\mathbf{E}_1 - \mathbf{E}_2) \cdot \boldsymbol{\tau} = 0. \tag{1.26}$$

Alternately, (1.26) can be expressed in terms of the unit normal as

$$\mathbf{n} \times (\mathbf{E}_1 - \mathbf{E}_2) = 0. \tag{1.27}$$

Equation (1.26) or (1.27) expresses the fact that *the tangential components of* **E** *are continuous across an interface of two media.*

In a similar way it can be shown that (1.23b) will lead to the relation

$$\mathbf{n} \times (\mathbf{H}_1 - \mathbf{H}_2) = \lim_{\delta h \to 0} \mathbf{J}\, \delta h = \mathbf{J}_s \tag{1.28a}$$

where $\mathbf{J}_s$ is called the *surface current density.*

If the current density is bounded, as it must be in any medium of finite conductivity, then the limit on the right in (1.28a) is zero and hence we have

$$\mathbf{n} \times (\mathbf{H}_1 - \mathbf{H}_2) = 0, \tag{1.28b}$$

denoting the fact that *the tangential components of* **H** *are continuous across an interface of two media.*

As a final remark, it is worthwhile to mention that it is sufficient to require that the tangential components of the electric and magnetic fields be continuous across an interface. Since all of the fields must satisfy Maxwell's equations, this automatically ensures that the normal component of **B** must be continuous (Problem 1.1).

1.5 The Wave Equation

The central problem in this book is the solution of Maxwell's equations subject to some known boundary conditions when the sources of the fields (currents and charges) are specified and the electric permittivity and magnetic permeability are known everywhere. It is then desirable to have a single differential equation expressing a relation between the known sources and the fields they produce. To this end it will be convenient to consider separately the following cases.

1.5.1 Spatially and Temporally Homogeneous Conducting Isotropic Medium

Performing the curl operation on (1.5a) and substituting from (1.5b) while using (1.12) and (1.18), we obtain

$$\nabla \times \nabla \times \mathbf{E} + \mu_0 \mu \left(\sigma \frac{\partial \mathbf{E}}{\partial t} + \varepsilon_0 \varepsilon \frac{\partial^2 \mathbf{E}}{\partial t^2} \right) = -\mu_0 \mu \frac{\partial \mathbf{J}}{\partial t} \tag{1.29}$$

where **J** in this case denotes the *impressed current density*. Using the identity

$$\nabla \times \nabla \times \; = \nabla \nabla \cdot \; - \nabla^2 \tag{1.30}$$

along with (1.5c) and (1.12), we obtain

$$\nabla^2 \mathbf{E} - \mu_0 \mu \left(\sigma \frac{\partial \mathbf{E}}{\partial t} + \varepsilon_0 \varepsilon \frac{\partial^2 \mathbf{E}}{\partial t^2} \right) = \mu_0 \mu \frac{\partial \mathbf{J}}{\partial t} + \frac{\nabla \rho}{\varepsilon_0 \varepsilon}. \tag{1.31}$$

Alternately, we may use the continuity equation (1.6) to express ρ in terms of **J** in (1.31).

In a similar way we can arrive at a differential equation for the magnetic field

$$\nabla^2 \mathbf{H} - \mu_0 \mu \left(\sigma \frac{\partial \mathbf{H}}{\partial t} + \varepsilon_0 \varepsilon \frac{\partial^2 \mathbf{H}}{\partial t^2} \right) = -\nabla \times \mathbf{J} \tag{1.32}$$

where, again, **J** is the impressed current density.

1.5.2 Spatially Inhomogeneous Isotropic Dielectric Medium

Consider a medium whose properties do not vary in time but whose electric permittivity is a function of the spatial coordinates. Assuming that the magnetic permeability is constant and the medium is nonconducting, we obtain (1.29) as before. In this case, however,

$$\nabla \cdot \mathbf{E} = \frac{\rho}{\varepsilon_0 \varepsilon} - \mathbf{E} \cdot \frac{\nabla \varepsilon}{\varepsilon}. \tag{1.33}$$

Hence, the wave equation becomes

$$\nabla^2 \mathbf{E} + \nabla\left(\mathbf{E} \cdot \frac{\nabla \varepsilon}{\varepsilon}\right) - \mu_0 \varepsilon_0 \mu\varepsilon \frac{\partial^2 \mathbf{E}}{\partial t^2} = \mu_0 \mu \frac{\partial \mathbf{J}}{\partial t} + \nabla\left(\frac{\rho}{\varepsilon_0 \varepsilon}\right). \tag{1.34}$$

It is left as an exercise to show that the corresponding wave equation for the magnetic field vector is in this case

$$\nabla^2 \mathbf{H} + \frac{\nabla \varepsilon}{\varepsilon} \times \nabla \times \mathbf{H} - \mu_0 \varepsilon_0 \mu\varepsilon \frac{\partial^2 \mathbf{H}}{\partial t^2} = -\nabla \times \mathbf{J} + \frac{\nabla \varepsilon}{\varepsilon} \times \mathbf{J}. \tag{1.35}$$

1.5.3 Temporally Inhomogeneous Isotropic Dielectric Medium

If a medium is spatially homogeneous and isotropic, but the electric permittivity varies with time while the magnetic permeability is constant in both space and time, then the following wave equations are obtained.

$$\nabla^2 \mathbf{E} - \mu_0 \varepsilon_0 \mu\left(2 \frac{\partial \varepsilon}{\partial t} \frac{\partial \mathbf{E}}{\partial t} + \varepsilon \frac{\partial^2 \mathbf{E}}{\partial t^2} + \frac{\partial^2 \varepsilon}{\partial t^2} \mathbf{E}\right) = \mu_0 \mu \frac{\partial \mathbf{J}}{\partial t} + \frac{\nabla \rho}{\varepsilon_0 \varepsilon} \tag{1.36}$$

and

$$\nabla^2 \mathbf{H} - \mu_0 \varepsilon_0 \mu\left(\frac{\partial \varepsilon}{\partial t} \frac{\partial \mathbf{H}}{\partial t} + \varepsilon \frac{\partial^2 \mathbf{H}}{\partial t^2}\right) = -\nabla \times \mathbf{J}. \tag{1.37}$$

1.5.4 Electrically Anisotropic Homogeneous Medium

If a medium is homogeneous in both space and time and its electric permittivity is in the form of a tensor $\boldsymbol{\varepsilon}$ while the magnetic permeability is a scalar, then Maxwell's equations can be written

$$\nabla \times \mathbf{E} = -\mu_0 \mu \frac{\partial \mathbf{H}}{\partial t}, \tag{1.38a}$$

$$\nabla \times \mathbf{H} = \varepsilon_0 \boldsymbol{\varepsilon} \frac{\partial \mathbf{E}}{\partial t} + \mathbf{J}, \tag{1.38b}$$

$$\nabla \cdot \boldsymbol{\varepsilon}\mathbf{E} = \frac{\rho}{\varepsilon_0}, \tag{1.38c}$$

$$\nabla \cdot \mathbf{H} = 0. \tag{1.38d}$$

A wave equation for the electric field vector can be formed by operating first with $\nabla \times$ on the first equation and then substituting from the second. We obtain in this case

$$\nabla \times \nabla \times \mathbf{E} + \mu_0 \mu \varepsilon_0 \boldsymbol{\varepsilon} \frac{\partial^2 \mathbf{E}}{\partial t^2} = -\mu_0 \mu \frac{\partial \mathbf{J}}{\partial t}. \tag{1.39}$$

The splitting of the double curl operator in the form of (1.30) is not convenient in this case because of the rather involved nature of the third equation in (1.38). Instead, it may be more convenient to express all of the vectors and vector operators in the form of matrices and use the rules of matrix multiplication. To this end we shall write all vectors as column vectors, e.g.,

$$\mathbf{E} = \begin{pmatrix} E_x \\ E_y \\ E_z \end{pmatrix}. \tag{1.40}$$

The curl operator can then be written

$$\nabla \times = \begin{pmatrix} 0 & -\partial_z & \partial_y \\ \partial_z & 0 & -\partial_x \\ -\partial_y & \partial_x & 0 \end{pmatrix}. \tag{1.41}$$

The double curl operator $\nabla \times \nabla \times$ is simply the square of the matrix in (1.41). We find

$$\nabla \times \nabla \times = \begin{pmatrix} -(\partial_y^{\,2} + \partial_z^{\,2}) & \partial_x \partial_y & \partial_x \partial_z \\ \partial_x \partial_y & -(\partial_x^{\,2} + \partial_z^{\,2}) & \partial_y \partial_z \\ \partial_x \partial_z & \partial_y \partial_z & -(\partial_x^{\,2} + \partial_y^{\,2}) \end{pmatrix}. \tag{1.42}$$

Finally, to be consistent, the divergence operator $\nabla \cdot$ must be written as a row vector, i.e.,

$$\nabla \cdot = (\partial_x \; \partial_y \; \partial_z) \tag{1.43}$$

so that when operating on a vector, say $\mathbf{H}$, it will correctly give

$$\nabla \cdot \mathbf{H} = (\partial_x \; \partial_y \; \partial_z) \begin{pmatrix} H_x \\ H_y \\ H_z \end{pmatrix} = \partial_x H_x + \partial_y H_y + \partial_z H_z. \tag{1.44}$$

To form a differential equation for the magnetic field vector it is expedient first to multiply (1.38b) by the inverse of $\boldsymbol{\varepsilon}$, i.e., $\boldsymbol{\varepsilon}^{-1}$, and then operate with the curl and subsequently substitute from (1.38a). We obtain in this case

$$\nabla \times \boldsymbol{\varepsilon}^{-1} \nabla \times \mathbf{H} + \mu_0 \varepsilon_0 \mu \frac{\partial^2 \mathbf{H}}{\partial t^2} = \nabla \times \boldsymbol{\varepsilon}^{-1} \mathbf{J}. \tag{1.45}$$

Again, in order to pursue the problem further, it is convenient to resort to matrix algebra.

1.5.5 Magnetically Anisotropic Homogeneous Medium

If a medium is characterized by a magnetic permeability tensor $\boldsymbol{\mu}$ while the electric permittivity is a scalar and the medium is homogeneous in both space and time, then it will follow that the magnetic field vector will satisfy the wave equation

$$\nabla \times \nabla \times \mathbf{H} + \mu_0 \varepsilon_0 \varepsilon \boldsymbol{\mu} \frac{\partial^2 \mathbf{H}}{\partial t^2} = \nabla \times \mathbf{J} \tag{1.46}$$

while the electric field vector will satisfy

$$\nabla \times \boldsymbol{\mu}^{-1} \nabla \times \mathbf{E} + \mu_0 \varepsilon_0 \varepsilon \frac{\partial^2 \mathbf{E}}{\partial t^2} = -\mu_0 \frac{\partial \mathbf{J}}{\partial t}. \tag{1.47}$$

1.6 Electromagnetic Potentials in Homogeneous Isotropic Media

1.6.1 Hertzian Vectors

We have previously introduced the electric and magnetic polarization vectors as

$$\mathbf{D} = \varepsilon_0 \mathbf{E} + \mathbf{P}, \tag{1.48a}$$

$$\mathbf{B} = \mu_0 (\mathbf{H} + \mathbf{M}). \tag{1.48b}$$

In what follows it will be convenient to distinguish the part of the polarization vector that arises from the applied field and the part that is due to external, impressed forces. We shall denote, respectively, by $\mathbf{P}_0$ and $\mathbf{M}_0$ the parts of the electric and magnetic polarizations that are due to electric and magnetic dipole distribution that is controlled solely by external sources. The parts of the polarizations that are induced by the existing fields we shall denote by $\mathbf{P}_1$ and $\mathbf{M}_1$. Thus, we write

$$\mathbf{D} = \varepsilon_0 \mathbf{E} + \mathbf{P}_1 + \mathbf{P}_0 = \varepsilon_0 \varepsilon \mathbf{E} + \mathbf{P}_0, \tag{1.49a}$$

$$\mathbf{B} = \mu_0(\mathbf{H} + \mathbf{M}_1) + \mu_0\mathbf{M}_0 = \mu_0 \mu\mathbf{H} + \mu_0\mathbf{M}_0, \tag{1.49b}$$

where, effectively, we have absorbed the effect of the induced polarizations in the coefficients ε and μ. In addition, if the current density $\mathbf{J}$ is the one due to conduction only, $\mathbf{J}_c = \sigma\mathbf{E}$, and the time dependence is of the form $\exp(-i\omega t)$, Maxwell's equations become

$$\nabla \times \mathbf{E} = i\omega\mu_0 \mu\mathbf{H} + i\omega\mu_0\mathbf{M}_0, \tag{1.50a}$$

$$\nabla \times \mathbf{H} = -\frac{i\omega\varepsilon_0 n^2}{\mu}\mathbf{E} - i\omega\mathbf{P}_0, \tag{1.50b}$$

$$\varepsilon_0 \varepsilon\nabla \cdot \mathbf{E} + \nabla \cdot \mathbf{P}_0 = \rho, \tag{1.50c}$$

$$\mu\nabla \cdot \mathbf{H} + \nabla \cdot \mathbf{M}_0 = 0, \tag{1.50d}$$

where

$$n^2 = \mu\varepsilon\left(1 + \frac{i\sigma}{\omega\varepsilon_0 \varepsilon}\right) \tag{1.51}$$

is the *complex refractive index*.

Consider a homogeneous isotropic medium that contains no magnetic dipoles, the only source of the fields being an electric dipole distribution $\mathbf{P}_0$. In this case $\nabla \cdot \mathbf{H} = 0$ and consequently we can define

$$\mathbf{H} = -\frac{i\omega\varepsilon_0 n^2}{\mu}\nabla \times \mathbf{\Pi}^{(e)}, \tag{1.52}$$

where $\mathbf{\Pi}^{(e)}$ is called the *Hertzian vector of the electric type*. Substituting (1.52) into (1.50b) and (1.50a) gives

$$\nabla^2\mathbf{\Pi}^{(e)} - \nabla(\nabla \cdot \mathbf{\Pi}^{(e)}) + \mathbf{E} = -\frac{\mu}{\varepsilon_0 n^2}\mathbf{P}_0 \tag{1.53}$$

and

$$\nabla \times (\mathbf{E} - k_0{}^2n^2\mathbf{\Pi}^{(e)}) = 0 = \nabla \times \nabla\varphi, \tag{1.54}$$

where

$$k_0 = \frac{\omega}{c} = \frac{2\pi}{\lambda_0} \tag{1.55}$$

and φ is any arbitrary scalar function of position. From (1.54) it will follow that

$$\mathbf{E} = \nabla\varphi + k_0{}^2n^2\mathbf{\Pi}^{(e)}. \tag{1.56}$$

Substituting (1.56) into (1.53) yields

$$(\nabla^2 + k^2)\mathbf{\Pi}^{(e)} = -\frac{\mu \mathbf{P}_0}{\varepsilon_0 n^2}, \tag{1.57}$$

where $k = k_0 n$ and $\varphi = \nabla \cdot \mathbf{\Pi}^{(e)}$ (Lorentz condition). Hence, we have obtained a self-consistent set of equations.

$$(\nabla^2 + k^2)\mathbf{\Pi}^{(e)} = -\frac{\mu}{\varepsilon_0 n^2} \mathbf{P}_0, \tag{1.58a}$$

$$\mathbf{H} = -\frac{i\omega\varepsilon_0 n^2}{\mu} \nabla \times \mathbf{\Pi}^{(e)}, \tag{1.58b}$$

$$\mathbf{E} = k^2 \mathbf{\Pi}^{(e)} + \nabla(\nabla \cdot \mathbf{\Pi}^{(e)}). \tag{1.58c}$$

If the medium is free of electric dipoles and charges, and the electromagnetic fields arise solely from a magnetic dipole distribution, then $\nabla \cdot \mathbf{E} = 0$. We can in this case define the *Hertzian vector of the magnetic type* $\mathbf{\Pi}^{(m)}$ and find a self-consistent set of equations as follows.

$$(\nabla^2 + k^2)\mathbf{\Pi}^{(m)} = -\frac{\mathbf{M}_0}{\mu}, \tag{1.59a}$$

$$\mathbf{E} = i\omega\mu_0 \mu \nabla \times \mathbf{\Pi}^{(m)}, \tag{1.59b}$$

$$\mathbf{H} = k^2 \mathbf{\Pi}^{(m)} + \nabla(\nabla \cdot \mathbf{\Pi}^{(m)}). \tag{1.59c}$$

Finally, we remark that if the medium contains both the electric and magnetic current distribution, then the total field can be obtained by superposing the partial fields of (1.58) and (1.59).

Consider a finite region of space to which the "forced" polarization $\mathbf{P}_0$ is confined. Since $\mathbf{P}_0 = \rho\boldsymbol{\xi}$ where ρ is the charge density and $\boldsymbol{\xi}$ is the charge displacement due to external forces from the equilibrium position, then the current density can be written

$$\mathbf{J}^{(e)} = \rho\mathbf{v} = \frac{1}{V}\sum_{n=1}^{N} q_n \dot{\boldsymbol{\xi}}_n = \frac{1}{V}\sum_{n=1}^{N} \dot{\mathbf{p}}_n = \frac{\partial \mathbf{P}_0}{\partial t}, \tag{1.60}$$

where $\mathbf{J}^{(e)}$ denotes the electric polarization current density. The corresponding relation between the magnetic polarization and the magnetic current density is

$$\mathbf{J}^{(m)} = \frac{\partial \mathbf{M}_0}{\partial t}. \tag{1.61}$$

Hence (1.58a) and (1.59a) can be written in terms of polarization current densities as

$$(\nabla^2 + k^2)\mathbf{\Pi}^{(e)} = -\frac{i\mu}{\varepsilon_0 n^2 \omega}\mathbf{J}^{(e)} \tag{1.62}$$

and

$$(\nabla^2 + k^2)\mathbf{\Pi}^{(m)} = -\frac{i}{\omega\mu}\mathbf{J}^{(m)}. \tag{1.63}$$

1.6.2 Representation of the Field Vectors in Terms of Two Scalars

Consider a homogeneous isotropic region that is free of electric and magnetic dipole distribution so that the right side of both (1.58a) and (1.59a) is identically zero. Moreover, consider a single cartesian component of each of the Hertzian vectors, say the z component, so that

$$\begin{aligned} \mathbf{\Pi}^{(e)} &= \Pi^{(e)}\mathbf{1}_z, \\ \mathbf{\Pi}^{(m)} &= \Pi^{(m)}\mathbf{1}_z. \end{aligned} \tag{1.64}$$

Hence, in this case, we have two scalar wave equations:

$$(\nabla^2 + k^2)\Pi^{(e,\,m)} = 0. \tag{1.65}$$

With each scalar function $\Pi^{(e)}$ and $\Pi^{(m)}$ there are associated definite components of the electromagnetic fields as defined in (1.58) and (1.59). These fields, arising from either $\Pi^{(e)}$ or $\Pi^{(m)}$, can be thought of as partial fields and since the problem is linear, we can use superposition and write

$$\mathbf{E} = \mathbf{E}^{(e)} + \mathbf{E}^{(m)}, \tag{1.66a}$$

$$\mathbf{H} = \mathbf{H}^{(e)} + \mathbf{H}^{(m)}, \tag{1.66b}$$

where

$$\mathbf{H}^{(e)} = -\frac{i\omega\varepsilon_0 n^2}{\mu}\nabla \times \mathbf{1}_z \Pi^{(e)}, \tag{1.67a}$$

$$\mathbf{E}^{(e)} = \left(\mathbf{1}_z k^2 + \nabla\frac{\partial}{\partial z}\right)\Pi^{(e)} \tag{1.67b}$$

and

$$\mathbf{E}^{(m)} = i\omega\mu_0 \mu\nabla \times \mathbf{1}_z \Pi^{(m)}, \tag{1.68a}$$

$$\mathbf{H}^{(m)} = \left(\mathbf{1}_z k^2 + \nabla\frac{\partial}{\partial z}\right)\Pi^{(m)}. \tag{1.68b}$$

Substituting (1.67) and (1.68) into (1.66), we obtain

$$\begin{aligned}\mathbf{H} = \mathbf{1}_x&\left(-\frac{i\omega\varepsilon_0 n^2}{\mu}\frac{\partial\Pi^{(e)}}{\partial y} + \frac{\partial^2\Pi^{(m)}}{\partial x\,\partial z}\right)\\ + \mathbf{1}_y&\left(\frac{i\omega\varepsilon_0 n^2}{\mu}\frac{\partial\Pi^{(e)}}{\partial x} + \frac{\partial^2\Pi^{(m)}}{\partial y\,\partial z}\right)\\ + \mathbf{1}_z&\left(k^2\Pi^{(m)} + \frac{\partial^2\Pi^{(m)}}{\partial z^2}\right)\end{aligned} \tag{1.69}$$

and

$$\begin{aligned}\mathbf{E} = \mathbf{1}_x&\left(i\omega\mu_0\,\mu\frac{\partial\Pi^{(m)}}{\partial y} + \frac{\partial^2\Pi^{(e)}}{\partial x\,\partial z}\right)\\ + \mathbf{1}_y&\left(-i\omega\mu_0\,\mu\frac{\partial\Pi^{(m)}}{\partial x} + \frac{\partial^2\Pi^{(e)}}{\partial y\,\partial z}\right)\\ + \mathbf{1}_z&\left(k^2\Pi^{(e)} + \frac{\partial^2\Pi^{(e)}}{\partial z^2}\right),\end{aligned} \tag{1.70}$$

which show the representation of all electromagnetic field components.

Inasmuch as the direction of the Hertzian vector chosen was completely arbitrary, it will follow that any electromagnetic field can be represented in terms of two scalars in a region that is homogeneous, isotropic, and free of impressed electric and magnetic polarizations.

1.7 Uniqueness Theorem for Electromagnetic Boundary Value Problems

In what follows it will often be necessary to make assumptions concerning the validity of certain mathematical manipulations which should have no effect on the final result. Clearly, what is needed in this case is a uniqueness theorem that would quarantee that the solution obtained is the only one that is physically admissible. We shall demonstrate the proof of uniqueness of solution when the medium is isotropic, linear, and time independent but whose properties may vary spatially.

Suppose that there exist two solutions (subscripts 1 and 2) that satisfy the time-harmonic Maxwell equations

$$\begin{aligned}\nabla\times\mathbf{E}_{1,2} &= i\omega\mu_0\,\mu\mathbf{H}_{1,2},\\ \nabla\times\mathbf{H}_{1,2} &= (\sigma - i\omega\varepsilon_0\,\varepsilon)\mathbf{E}_{1,2} + \mathbf{J},\\ \nabla\cdot\varepsilon\mathbf{E}_{1,2} &= \rho/\varepsilon_0,\\ \nabla\cdot\mathbf{H}_{1,2} &= 0.\end{aligned} \tag{1.71}$$

If we denote $\mathbf{E} = \mathbf{E}_2 - \mathbf{E}_1$ and $\mathbf{H} = \mathbf{H}_2 - \mathbf{H}_1$, then it will follow that the difference of the two fields must satisfy the source-free Maxwell equations

$$\nabla \times \mathbf{E} = i\omega\mu_0 \mu \mathbf{H}, \tag{1.72a}$$

$$\nabla \times \mathbf{H} = (\sigma - i\omega\varepsilon_0 \varepsilon)\mathbf{E}, \tag{1.72b}$$

$$\nabla \cdot \varepsilon\mathbf{E} = \nabla \cdot \mathbf{H} = 0. \tag{1.72c}$$

If we multiply (1.72a) by $\mathbf{H}^*$ and the conjugate of (1.72b) by $\mathbf{E}$ and subtract one from the other, we obtain

$$\nabla \cdot (\mathbf{E} \times \mathbf{H}^*) = i\omega(\mu_0 \mu |\mathbf{H}|^2 - \varepsilon_0 \varepsilon |\mathbf{E}|^2) - \sigma |\mathbf{E}|^2 \tag{1.73}$$

since

$$\nabla \cdot (\mathbf{E} \times \mathbf{H}^*) = \mathbf{H}^* \cdot \nabla \times \mathbf{E} - \mathbf{E} \cdot \nabla \times \mathbf{H}^*. \tag{1.74}$$

We integrate (1.73) over a volume V bounded by a surface S and obtain

$$\oint_S \mathbf{E} \times \mathbf{H}^* \cdot \mathbf{n}\, da = -i\omega \int_V (\mu_0 \mu |\mathbf{H}|^2 - \varepsilon_0 \varepsilon |\mathbf{E}|^2)\, dv + \sigma \int_V |\mathbf{E}|^2\, dv \tag{1.75}$$

where $\mathbf{n}$ is the unit normal pointing into the volume V.

The only components of $\mathbf{E}$ and $\mathbf{H}$ that contribute to the value of the surface integral on the left in (1.75) are the tangential components. Thus, if the tangential components of $\mathbf{E}_1$ and $\mathbf{E}_2$ or $\mathbf{H}_1$ and $\mathbf{H}_2$ are equal over the surface S, then the surface integral vanishes identically. Under these conditions, the right side of (1.75) must vanish also where the real and the imaginary parts must vanish independently. Thus, it will follow that $\mathbf{E} = 0$ due to the second integral on the right and, hence, also $\mathbf{H} = 0$ everywhere within the volume V.

The extension of the uniqueness theorem to anisotropic media is not a simple matter, although it has been asserted by Stratton (1941) that such an extension is possible.

1.8 The Dirac Delta Function

Consider the concept of a point charge and the corresponding charge density. Evidently, we must visualize the charge density to be zero everywhere in space and become infinite at the location of the point charge $\mathbf{r}$ in such a way that

$$\int_{V'} \rho(\mathbf{r}, \mathbf{r}')\, d\mathbf{r}' = \begin{cases} q, & \mathbf{r} \text{ in } V', \\ 0, & \mathbf{r} \text{ not in } V'. \end{cases} \tag{1.76}$$

What is needed in this case is a functional description of the charge density to satisfy the stated requirements. To this end, we shall use the *Dirac delta function*

$$\delta(\mathbf{r} - \mathbf{r}') = \delta(x - x')\,\delta(y - y')\,\delta(z - z) \tag{1.77}$$

with the integral property

$$\int_{V'} f(\mathbf{r}')\,\delta(\mathbf{r} - \mathbf{r}')\,d\mathbf{r}' = f(\mathbf{r}) \tag{1.78}$$

if $\mathbf{r}$ is included in the volume V', and zero otherwise, and providing that $f(\mathbf{r}')$ is continuous at $\mathbf{r}' = \mathbf{r}$. Thus, in the case of a single charge, the corresponding charge density will be written

$$\rho(\mathbf{r}, \mathbf{r}') = q\,\delta(\mathbf{r} - \mathbf{r}') \tag{1.79}$$

so that (1.76) is satisfied.

Equation (1.78) will be looked on as one defining the delta function. In this sense, the delta function serves as an instrument for integration. In reality it is not a function but a *distribution* or *generalized function*. It is important to note that because of its singular character, the delta function cannot be the end result of the calculation. The delta function has meaning only as long as a subsequent integration over its argument is carried out.

The delta function has a number of interesting properties, the most important of which, perhaps, is its symmetry. Thus, for instance, in one dimension

$$f(x) = \int_{-\infty}^{\infty} f(x')\,\delta(x - x')\,dx' \tag{1.80}$$

and it will follow that

$$\delta(x - x') = \delta(x' - x) \tag{1.81}$$

for all functions $f(x')$ continuous at $x' = x$. Some of the additional properties are

$$\delta'(x) = -\delta'(-x), \tag{1.82a}$$

$$x\,\delta(x) = 0, \tag{1.82b}$$

$$x\,\delta'(x) = -\delta(x), \tag{1.82c}$$

$$\delta(ax) = |a|^{-1}\,\delta(x), \tag{1.82d}$$

$$\int_{-\infty}^{\infty} \delta(x' - x)\,\delta(x - x'')\,dx = \delta(x' - x''), \tag{1.82e}$$

$$f(x)\,\delta(x - x') = f(x')\,\delta(x - x'). \tag{1.82f}$$

We note that in circular cylindrical coordinates

$$\delta(\mathbf{r} - \mathbf{r}') = \frac{\delta(\rho - \rho')\,\delta(\varphi - \varphi')\,\delta(z - z')}{\rho}, \tag{1.83}$$

and in spherical coordinates

$$\delta(\mathbf{r} - \mathbf{r}') = \frac{\delta(r - r')\,\delta(\theta - \theta')\,\delta(\varphi - \varphi')}{r^2 \sin\theta}. \tag{1.84}$$

The validity of the last two representations can be readily demonstrated by multiplying by the corresponding volume element and integrating.

The concept of the delta function is inherent to any known integral transform. Consider for instance the Fourier series

$$f(x) = \sum_{-\infty}^{\infty} a_n e^{inx}, \tag{1.85a}$$

$$a_n = (2\pi)^{-1} \int_{-\pi}^{\pi} f(x')e^{-inx'}\, dx' \tag{1.85b}$$

where $f(x)$ is any periodic, piecewise continuous function satisfying the Dirichlet condition. If we substitute from (1.85b) into (1.85a) and rearrange, then

$$f(x) = \int_{-\pi}^{\pi} f(x')\left\{(2\pi)^{-1} \sum_{-\infty}^{\infty} \exp[in(x - x')]\right\} dx'. \tag{1.86}$$

In the sense of the definition of the delta function in (1.80) it follows that in this case

$$\begin{aligned}\delta(x - x') &= (2\pi)^{-1} \sum_{-\infty}^{\infty} \exp[in(x - x')] \\ &= (2\pi)^{-1} \sum_{0}^{\infty} \varepsilon_n \cos n(x - x')\end{aligned} \tag{1.87}$$

where ε_n is the *Neumann factor*

$$\varepsilon_n = \begin{cases} 1, & n = 0, \\ 2, & n \neq 0. \end{cases} \tag{1.88}$$

As a second example, consider the Fourier transform pair

$$\tilde{f}(\alpha) = (2\pi)^{-1/2} \int_{-\infty}^{\infty} f(x) \exp(-i\alpha x)\, dx, \tag{1.89a}$$

$$f(x) = (2\pi)^{-1/2} \int_{-\infty}^{\infty} \tilde{f}(\alpha) \exp(i\alpha x)\, d\alpha. \tag{1.89b}$$

Substituting (1.89b) into (1.89a) and rearranging gives

$$\tilde{f}(\alpha) = \int_{-\infty}^{\infty} \tilde{f}(\alpha')\left\{(2\pi)^{-1} \int_{-\infty}^{\infty} \exp[-ix(\alpha - \alpha')]\, dx\right\} d\alpha' \tag{1.90}$$

where, again, the quantity in curly braces on the right is identified as a delta function, that is

$$\begin{aligned}\delta(\alpha - \alpha') &= (2\pi)^{-1} \int_{-\infty}^{\infty} \exp[-ix(\alpha - \alpha')]\, dx \\ &= (\pi)^{-1} \int_{0}^{\infty} \cos x(\alpha - \alpha')\, dx.\end{aligned} \tag{1.91}$$

Alternately, it will also follow from (1.89) that

$$\begin{aligned}\delta(x - x') &= (2\pi)^{-1} \int_{-\infty}^{\infty} \exp[i\alpha(x - x')]\, d\alpha \\ &= (\pi)^{-1} \int_{0}^{\infty} \cos \alpha(x - x')\, d\alpha.\end{aligned} \tag{1.92}$$

Many other often useful representations of the delta function can be found by using the known integral transforms. We list below two of them for future reference.

$$\frac{\delta(\rho - \rho')}{\rho'} = \frac{\delta(\rho' - \rho)}{\rho} = \int_{0}^{\infty} \lambda J_\nu(\lambda\rho) J_\nu(\lambda\rho')\, d\lambda, \tag{1.93a}$$

$$\frac{\delta(r - r')}{r'^2} = \frac{\delta(r' - r)}{r^2} = \frac{2}{\pi} \int_{0}^{\infty} k^2 j_\nu(kr) j_\nu(kr')\, dk. \tag{1.93b}$$

These can be obtained in a straightforward manner from the Fourier–Bessel transform for cylindrical and spherical Bessel functions, respectively.

Problems

1.1 Show that, except possibly for a static case, the continuity of the tangential components of **E** and **H** at an interface implies the continuity of the normal components of **D** and **B**. *Hint:* Use cartesian coordinates.

1.2 Verify Eq. (1.35).

1.3 Derive Eqs. (1.46) and (1.47).

1.4 Derive Eqs. (1.59a–c).

1.5 Demonstrate the properties of the delta function listed in (1.82a–f).

1.6 Using the Fourier–Bessel transform pair

$$f(\rho) = \int_0^\infty \tilde{f}(\lambda) J_\nu(\lambda\rho)\lambda \, d\lambda,$$

$$\tilde{f}(\lambda) = \int_0^\infty f(\rho) J_\nu(\lambda\rho)\rho \, d\rho$$

and

$$f(r) = \left(\frac{2}{\pi}\right)^{1/2} \int_0^\infty \tilde{f}(k) j_\nu(kr) k^2 \, dk,$$

$$\tilde{f}(k) = \left(\frac{2}{\pi}\right)^{1/2} \int_0^\infty f(r) j_\nu(kr) r^2 \, dr,$$

derive (1.93a) and (1.93b), respectively.

1.7 Demonstrate the validity of the following approximate representations of the delta function.

(a) $$\delta(x) = \lim_{\varepsilon \to 0} (\varepsilon\sqrt{\pi})^{-1} \exp(-x^2/\varepsilon^2);$$

(b) $$\delta(x) = \lim_{N \to \infty} \frac{\sin Nx}{\pi x};$$

(c) $$\delta(x) = \lim_{y \to 0} \frac{y}{\pi(x^2 + y^2)}.$$

References

Electromagnetic Theory

Elliot, R. S. (1966). "Electromagnetics." McGraw-Hill, New York.

Jones, D. S. (1964). "The Theory of Electromagnetism." Pergamon, Oxford.

Panofsky, W. K. H., and Phillips, M. (1962). "Classical Electricity and Magnetism," 2nd ed. Addison-Wesley, Reading, Massachusetts.

Stratton, J. A. (1941). "Electromagnetic Theory," p. 488. McGraw-Hill, New York.

Applied Mathematics

Arfken, G. (1966). "Mathematical Methods for Physicists." Academic Press, New York.

Duff, G. F. D., and Naylor, D. (1966). "Differential Equations of Applied Mathematics." Wiley, New York.

Friedman, B. (1959). "Principles and Techniques of Applied Mathematics." Wiley, New York.

Morse, P. M., and Feshbach, H. (1953). "Methods of Theoretical Physics," Vols. I and II. McGraw-Hill, New York.

Papoulis, A. (1962). "The Fourier Integral and Its Applications." McGraw-Hill, New York.

CHAPTER 2

PLANE WAVES IN HOMOGENEOUS ISOTROPIC MEDIA

In this chapter we shall study the simplest and most fundamental solution of Maxwell's equations, which results in so-called plane waves that depend on time and a single space variable only. It will turn out that the solutions to some very complex boundary value problems can often be expressed as a spectrum of these elementary plane waves. It is, therefore, instructive to study the associated phenomena in their simplest possible forms, devoid of the mathematical complexities that will follow.

2.1 General Properties of Plane Waves

Let us seek a solution to Maxwell's equations in a spatially and temporally homogeneous region of infinite extent containing no free currents or charges. Let the desired solution depend only on time and distance measured along a single axis in space. Without loss of generality, we can assume that the wave varies only along the z' axis of the wave coordinate system x', y', z'. On this system we superpose the "laboratory" or the observer's coordinate system in such a manner that the two systems have a common origin as shown in Fig. 2.1.

Let $\mathbf{n}$ be a unit vector along the z' axis, then its projections n_x, n_y, n_z on the x, y, z axes are the corresponding directional cosines. The distance along the z' axis can be expressed in terms of the unprimed laboratory coordinate system as follows.

$$z' = n_x x + n_y y + n_z z = \mathbf{n} \cdot \mathbf{r}. \tag{2.1}$$

Using the chain rule of differentiation, we convert the partial derivatives of the unprimed system to the partial derivatives with respect to the x' axis. Thus, since

$$\frac{\partial}{\partial x_i} = \frac{\partial z'}{\partial x_i} \cdot \frac{\partial}{\partial z'} \tag{2.2}$$

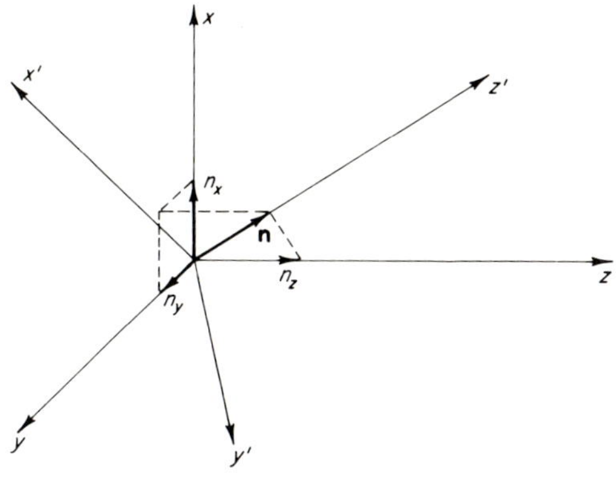

Fig. 2.1. Relation between the wave and laboratory coordinate systems.

and $\partial z'/\partial x_i = n_i$, we obtain

$$\mathbf{\nabla} = \mathbf{n}\,\frac{\partial}{\partial z'}. \tag{2.3}$$

It is a straightforward matter to show that, in the primed coordinates, Maxwell's equations take on the form

$$\mathbf{n} \times \frac{\partial \mathbf{E}}{\partial z'} = -\mu_0\,\mu\,\frac{\partial \mathbf{H}}{\partial t}, \tag{2.4a}$$

$$\mathbf{n} \times \frac{\partial \mathbf{H}}{\partial z'} = \varepsilon_0\,\varepsilon\,\frac{\partial \mathbf{E}}{\partial t} + \sigma\mathbf{E}, \tag{2.4b}$$

$$\mathbf{n} \cdot \frac{\partial \mathbf{H}}{\partial z'} = 0, \tag{2.4c}$$

$$\mathbf{n} \cdot \frac{\partial \mathbf{E}}{\partial z'} = 0. \tag{2.4d}$$

The foregoing set of equations can be solved simultaneously by performing the $\mathbf{n}\,\times$ operation on either the first or the second, differentiating once more with respect to z' and substituting from one into the other. The results are

$$\frac{\partial^2 \mathbf{E}}{\partial z'^2} - \mu_0\,\varepsilon_0\,\mu\varepsilon\,\frac{\partial^2 \mathbf{E}}{\partial t^2} - \mu_0\,\mu\sigma\,\frac{\partial \mathbf{E}}{\partial t} = 0, \tag{2.5a}$$

$$\frac{\partial^2 \mathbf{H}}{\partial z'^2} - \mu_0\,\varepsilon_0\,\mu\varepsilon\,\frac{\partial^2 \mathbf{H}}{\partial t^2} - \mu_0\,\mu\sigma\,\frac{\partial \mathbf{H}}{\partial t} = 0. \tag{2.5b}$$

Now let us return for a moment to (2.4a). Operating with $\mathbf{n}\cdot$ we note that

$$-\mu_0\mu\left(\mathbf{n}\cdot\frac{\partial\mathbf{H}}{\partial t}\right)=\mathbf{n}\cdot\left(\mathbf{n}\times\frac{\partial\mathbf{E}}{\partial z'}\right)=\frac{\partial\mathbf{E}}{\partial z'}\cdot(\mathbf{n}\times\mathbf{n})=0, \tag{2.6}$$

from which it follows that $\mathbf{n}\cdot\partial\mathbf{H}/\partial t=0$. Multiplying this by dt and adding to (2.4c) multiplied by dz', we obtain $\mathbf{n}\cdot[(\partial\mathbf{H}/\partial t)\,dt+(\partial\mathbf{H}/\partial z')\,dz']=0$, or equivalently

$$dH_{z'}=0. \tag{2.7}$$

It follows that $\mathbf{H}_{z'}$ cannot vary along either the z' coordinate or with time. The only possibility is a static magnetic field in the z' direction, but since we are concerned with time-varying fields, we must conclude that $H_{z'}=0$.

Similarly, multiplying (2.4b) by $\mathbf{n}\cdot$, we obtain

$$\mathbf{n}\cdot\left(\frac{\partial\mathbf{E}}{\partial t}+\frac{\sigma}{\varepsilon_0\,\varepsilon}\mathbf{E}\right)=0. \tag{2.8}$$

Multiplying (2.8) by dt and adding to (2.4d) multiplied by dz' gives

$$dE_{z'}+\frac{\sigma}{\varepsilon_0\,\varepsilon}E_{z'}\,dt=0, \tag{2.9}$$

from which it follows that $E=A(z')\exp(-(\sigma/\varepsilon_0\,\varepsilon)t)$. In the case of non-vanishing conductivity, $\sigma\neq 0$, the longitudinal component of the electric field vanishes exponentially with time and in the case of vanishing conductivity, $\sigma=0$, only the static longitudinal component of the electric field may exist.

The foregoing analysis can now be concluded with the statement that the *vectors* **E** *and* **H** *of any electromagnetic field that is a function of time and one space variable only, i.e.,* $\mathbf{E}(z',t)$ *and* $\mathbf{H}(z',t)$, *lie in the plane normal to the axis of the coordinate* z'.

Let us now return to the two equations (2.5a, b). Since we have just shown that the longitudinal fields cannot exist, we are concerned only with the transverse fields. Let the scalar function $E(z',t)$ denote either $E_{x'}$ or $E_{y'}$. We shall seek a solution by the method of separation of variables. Let us put

$$E(z',t)=f(z')g(t); \tag{2.10}$$

then $f(z')$ and $g(t)$ must satisfy the ordinary differential equations

$$\left(\frac{d^2}{dz'^2}+k^2\right)f=0, \tag{2.11a}$$

$$\left(\frac{d^2}{dt^2} + \frac{\sigma}{\varepsilon_0 \varepsilon}\frac{d}{dt} + \frac{k^2}{\mu_0 \varepsilon_0 \mu\varepsilon}\right)g = 0, \tag{2.11b}$$

where k in this case is the separation constant. The solution to the differential equations above are well known. Thus, we write

$$\mathbf{E}(z', t) = (\mathbf{E}_1 e^{ikz'} + \mathbf{E}_2 e^{-ikz'})e^{-pt}, \tag{2.12a}$$

$$\mathbf{H}(z', t) = (\mathbf{H}_1 e^{ikz'} + \mathbf{H}_2 e^{-ikz'})e^{-pt}, \tag{2.12b}$$

where

$$p = \frac{\sigma}{2\varepsilon_0 \varepsilon} \pm \left[\left(\frac{\sigma}{2\varepsilon\varepsilon_0}\right)^2 - \frac{k^2}{\mu_0 \varepsilon_0 \mu\varepsilon}\right]^{1/2} \tag{2.13}$$

and $\mathbf{E}_{1,2}$ and $\mathbf{H}_{1,2}$ are constant vectors, orthogonal to the z' axis. Using (2.4a) together with (2.12), we obtain

$$\mathbf{n} \times (\mathbf{E}_1 e^{ikz'} - \mathbf{E}_2 e^{-ikz'}) = \frac{\mu_0 \mu p}{ik}(\mathbf{H}_1 e^{ikz'} + \mathbf{H}_2 e^{-ikz'}). \tag{2.14}$$

Equations (2.14) must hold for all z' and this will be possible only if the coefficients of the exponentials vanish independently. The following must then be true.

$$\mathbf{H}_1 = \frac{ik}{\mu_0 \mu p}\mathbf{n} \times \mathbf{E}_1, \tag{2.15a}$$

$$\mathbf{H}_2 = -\frac{ik}{\mu_0 \mu p}\mathbf{n} \times \mathbf{E}_2, \tag{2.15b}$$

which express the orthogonality of the pairs $\mathbf{H}_1$, $\mathbf{E}_1$ and $\mathbf{H}_2$, $\mathbf{E}_2$. Thus, *the electric and magnetic fields that vary only with respect to time and one space variable are orthogonal to the axis of the said space variable and to each other.*

2.2 Spectral Decomposition

A convenient method of plane wave analysis is that of spectral decomposition. When ε, μ, and σ are linear, the principle of superposition holds and the analysis of the behavior of any wave form can be reduced to the analysis of the simplest harmonic waves by means of the spectral method.

Consider the differential equation (2.5a) or (2.5b) for the particular case when the medium is nonconducting, i.e., $\sigma = 0$. Let F denote any cartesian component of either the electric or magnetic field, then

$$\left(v\frac{\partial}{\partial z'} - \frac{\partial}{\partial t}\right)\left(v\frac{\partial}{\partial z'} + \frac{\partial}{\partial t}\right)F = 0 \tag{2.16}$$

where $v = (\mu_0 \varepsilon_0 \mu\varepsilon)^{-1/2}$. If we put $\xi = z'/v - t$ and $\eta = z'/v + t$, use the chain rule for differentiation

$$\begin{aligned}\frac{\partial}{\partial \xi} &= \frac{\partial z'}{\partial \xi}\cdot\frac{\partial}{\partial z'} + \frac{\partial t}{\partial \xi}\frac{\partial}{\partial t},\\ \frac{\partial}{\partial \eta} &= \frac{\partial z'}{\partial \eta}\cdot\frac{\partial}{\partial z'} + \frac{\partial t}{\partial \eta}\frac{\partial}{\partial t},\end{aligned} \tag{2.17}$$

and note that

$$z' = \frac{v}{2}(\xi + \eta), \qquad t = \tfrac{1}{2}(\eta - \xi), \tag{2.18}$$

we find that (2.16) can be written

$$\frac{\partial^2 F}{\partial \xi\, \partial \eta} = 0. \tag{2.19}$$

The solution to (2.19) is a linear combination of the functions $F_1(\xi)$ and $F_2(\eta)$. Thus, in the original notation,

$$F(z', t) = F_1\left(\frac{z'}{v} - t\right) + F_2\left(\frac{z'}{v} + t\right). \tag{2.20}$$

Suppose that $F_2(z'/v + t) = 0$ so that $F(z', t) = F_1(z'/v - t)$. In each plane z' = constant, the field changes with time and at each given moment the field is different for a different value of z'. The field has the same value for the coordinate z' and time t satisfying the relation $z' = v(T + t)$. This implies that if at some time $t = 0$ the field at a certain point along the z' coordinate had some definite value, then after an interval T, the field would have the same value at a distance vT from the original place. We can say that all the values of the electromagnetic field are propagated in space along the z' axis with a velocity equal to v.

Thus, $F_1(z'/v - t)$ represents a plane wave moving in the positive z' direction. Employing an analogous argument, it can be shown that $F_2(z'/v + t)$ represents a plane wave moving in the opposite z' direction.

Let us now attempt to represent the function $F_1(\xi)$ by the real part of the integral:

$$F_1(\xi) = \mathrm{Re}\int_0^\infty \tilde{F}_1(\omega)e^{i\omega\xi}\, d\omega \tag{2.21a}$$

or alternately

$$F_1(\xi) = \frac{1}{2}\left\{\int_0^\infty \tilde{F}_1(\omega)e^{i\omega\xi}\,d\omega + \int_0^\infty \tilde{F}_1{}^*(\omega)e^{-i\omega\xi}\,d\omega\right\} \tag{2.21b}$$

since the real part of any complex number can always be written as one half of that number plus its conjugate.

To find $\tilde{F}_1(\omega)$ in terms of $F_1(\xi)$, we multiply (2.21b) by $\exp(-i\omega'\xi)$ and integrate over ξ from minus to plus infinity. Thus,

$$\int_{-\infty}^{\infty} F_1(\xi)\exp(-i\omega'\xi)\,d\xi = \frac{1}{2}\int_0^\infty d\omega\left\{\tilde{F}_1(\omega)\int_{-\infty}^{\infty}\exp[i(\omega-\omega')\xi]\,d\xi\right.$$

$$\left.+\ \tilde{F}_1{}^*(\omega)\int_{-\infty}^{\infty}\exp[-i(\omega+\omega')\xi]\,d\xi\right\}. \tag{2.22}$$

The integrals in the curly braces on the right can be identified as $2\pi\,\delta(\omega-\omega')$ and $2\pi\,\delta(\omega+\omega')$, respectively. Thus, it will follow that the second integral in the curly braces will be zero because the integration is carried over positive ω only. Hence, we find

$$\tilde{F}_1(\omega) = \frac{1}{\pi}\int_{-\infty}^{\infty} F_1(\xi)e^{-i\omega\xi}\,d\xi, \tag{2.23a}$$

$$F_1(\xi) = \mathrm{Re}\int_0^\infty \tilde{F}_1(\omega)e^{i\omega\xi}\,d\omega. \tag{2.23b}$$

We note that $F_1(\xi)$ is a result of summation over all positive frequencies. If the wave is monochromatic, we can write

$$f_1(\omega,\xi) = \tilde{f}_1(\omega)e^{i\omega\xi} \tag{2.24a}$$

or, in the original notation

$$f_1(\omega, z', t) = \tilde{f}_1(\omega)\exp[i(n_x x + n_y y + n_z z - vt)\omega/v]. \tag{2.24b}$$

Let us define

$$k = \frac{\omega}{v} = \frac{2\pi}{\lambda}, \tag{2.25a}$$

$$k_j = kn_j \tag{2.25b}$$

and noting that $k^2 = k_x{}^2 + k_y{}^2 + k_z{}^2$, we can think of the propagation vector $\mathbf{k}$ having magnitude defined in (2.25a) and components k_x, k_y, and k_z along the axes of the cartesian coordinate system. Equation (2.24) becomes

$$f_1(\omega, \mathbf{r}, t) = A_1(\omega)\exp[i(\mathbf{k}\cdot\mathbf{r} - \omega t + \theta_1(\omega))] \tag{2.26}$$

where $A_1(\omega)$ is the amplitude and $\theta_1(\omega)$ is the phase factor of $\tilde{f}_1(\omega)$. It is clear that any physical meaning is attached to the real part of (2.26).

It is easy to show that $f_1(\omega, \mathbf{r}, t)$ satisfies the differential equation

$$(\nabla^2 + k^2)f_1(\omega, \mathbf{r}, t) = 0, \tag{2.27}$$

which we shall henceforth call the reduced wave equation. Note that (2.27) could be obtained from (2.16) by assuming a solution of the form $F(z't) = f(z') \exp(-i\omega t)$.

2.3 Inhomogeneous Plane Waves

We noted that the real numbers k_j can be considered as components of the propagation vector $\mathbf{k}$ whose magnitude squared is given by

$$k^2 = \left(\frac{\omega}{v}\right)^2 = k_x{}^2 + k_y{}^2 + k_z{}^2. \tag{2.28}$$

We now investigate the possibility of satisfying (2.28) with a triplet of complex numbers

$$k_j = k_j' + ik_j'' \tag{2.29}$$

in such a way that $k = \omega/v$ will remain real and the reduced wave equation (2.27) will be satisfied as before. Evidently, the solution to (2.27) can now be written

$$f_1(\omega, \mathbf{r}, t) = \tilde{f}_1(\omega) \exp[i(\mathbf{k}' \cdot \mathbf{r} - \omega t)] \exp(-\mathbf{k}'' \cdot \mathbf{r}), \tag{2.30}$$

where

$$\mathbf{k}' = \mathbf{l}_x k_x' + \mathbf{l}_y k_y' + \mathbf{l}_z k_z', \tag{2.31a}$$

$$\mathbf{k}'' = \mathbf{l}_x k_x'' + \mathbf{l}_y k_y'' + \mathbf{l}_z k_z''. \tag{2.31b}$$

Equation (2.30) describes a wave of varying amplitude or an *inhomogeneous plane wave*. It will follow that the planes of constant amplitude are given by $\mathbf{k}'' \cdot \mathbf{r} =$ constant while the planes of constant phase are determined by $\mathbf{k}' \cdot \mathbf{r} =$ constant. Moreover, since in this case

$$k^2 = |\mathbf{k}'|^2 - |\mathbf{k}''|^2 \tag{2.32a}$$

and

$$\mathbf{k}' \cdot \mathbf{k}'' = 0, \tag{2.32b}$$

it will follow that the planes of constant phase and constant amplitude are orthogonal to each other. Thus, the wave propagates in the direction of the vector $\mathbf{k}'$ and its amplitude decreases in the direction of the vector $\mathbf{k}''$, perpendicular to $\mathbf{k}'$.

The inhomogeneous plane waves are characteristic of wave propagation in conducting media. In this case, $\sigma \neq 0$ and the wave equation is

$$\left(\frac{\partial^2}{\partial z'^2} - \mu_0 \varepsilon_0 \mu\varepsilon \frac{\partial^2}{\partial t^2} - \mu_0 \mu\sigma \frac{\partial}{\partial t}\right)F = 0, \tag{2.33}$$

where F stands for any cartesian component of either the electric or magnetic field. A time-harmonic solution can be found by making the substitution $F(z', t) = f(z') \exp(-i\omega t)$ with the understanding that the real part of this expression must be taken. Equation (2.33) then becomes

$$\left(\frac{d^2}{dz'^2} + k^2\right)f = 0, \tag{2.34}$$

where now

$$k = k_0\left[\mu\varepsilon\left(1 + \frac{i\sigma}{\omega\varepsilon_0 \varepsilon}\right)\right]^{1/2}, \tag{2.35a}$$

$$k_0 = \frac{\omega}{c} = \frac{2\pi}{\lambda_0}, \tag{2.35b}$$

and k itself is complex. A positive sign was chosen for the square root in (2.35a) so that $k \to k_0$ as $\sigma \to 0$. It will also follow that $\mathrm{Im}(k) \geq 0$ or

$$k = k' + ik'', \qquad k'' > 0. \tag{2.36}$$

A complete solution to (2.33) can be written

$$F(z', t) = [A_1(\omega) \exp(i\theta_1(\omega)) \exp(ikz') + A_2(\omega) \exp(i\theta_2(\omega)) \times \exp(-ikz')] \exp(-i\omega t), \tag{2.37}$$

where $A_{1,2}(\omega)$ is a real amplitude at $z' = 0$ and $\theta_{1,2}(\omega)$ is a phase factor. Equation (2.37) in connection with (2.36) implies that the first term in the square brackets describes a wave traveling in the positive z' direction and the second term describes a wave traveling in the negative z' direction.

Equation (2.37) can be written in a general way as

$$F(\mathbf{r}, t) = A(\omega) \exp(i\theta(\omega)) \exp[i(\mathbf{k} \cdot \mathbf{r} - \omega t)], \tag{2.38}$$

where $\mathbf{k}$ is a complex vector. Now since

$$k^2 = k_x{}^2 + k_y{}^2 + k_z{}^2 = \omega^2 \mu_0 \varepsilon_0 \mu\varepsilon + i\sigma\omega\mu_0 \mu, \tag{2.39}$$

then, upon separation of real and imaginary parts, it will follow that in this case

$$|\mathbf{k}'|^2 - |\mathbf{k}''|^2 = \omega^2 \mu_0 \varepsilon_0 \mu\varepsilon, \tag{2.40a}$$

$$\mathbf{k}' \cdot \mathbf{k}'' = \tfrac{1}{2}\sigma\omega\mu_0 \mu. \tag{2.40b}$$

Again, the planes of constant amplitude are determined from $\mathbf{k}'' \cdot \mathbf{r} =$ constant and the planes of constant phase from $\mathbf{k}' \cdot \mathbf{r} =$ constant. Note, however, that unlike in the previous case, the two planes are not perpendicular, inasmuch as $\mathbf{k}' \cdot \mathbf{k}'' \neq 0$.

2.4 Reflection and Refraction

Consider a monochromatic plane wave of angular frequency ω. The associated time dependence can be written in the form $\exp(-i\omega t)$ where the real (or imaginary) part of this expression is implied. With this assumed time variation, the electric and the magnetic field components will satisfy the reduced wave equation

$$(\nabla^2 + k^2)\begin{pmatrix}\mathbf{E}\\ \mathbf{H}\end{pmatrix} = 0. \tag{2.41}$$

Since Maxwell's equations relate the electric and magnetic field components, only one of these equations needs to be solved. The solution is of the form

$$\mathbf{E}(\mathbf{r}, t) = \mathbf{E}_0 \exp[i(\mathbf{k}\cdot\mathbf{r} - \omega t)], \tag{2.42}$$

where $\mathbf{E}_0$ is a constant vector, $k = k_0 n$, and $n^2 = \mu\varepsilon(1 + i\sigma/\omega\varepsilon_0 \varepsilon)$ is the complex index of refraction. The magnetic field $\mathbf{H}$ is found from the appropriate curl equation. Noting that $\nabla = i\mathbf{k}$, we obtain

$$\mathbf{H} = (\omega\mu_0 \mu)^{-1}\mathbf{k} \times \mathbf{E}. \tag{2.43}$$

Consider a plane interface of two adjoining media, both of which are homogeneous and isotropic but otherwise arbitrary. The geometry of the problem is shown in Fig. 2.2. A plane wave polarized in the direction perpendicular to the plane of incidence is incident from above on the interface, the x-y plane, at an angle θ_1 with respect to the normal. Part of this wave will be reflected at the interface and part of it will be transmitted into the lower region. Our task will be to find the amplitudes of the reflected and transmitted waves. Since the wave is assumed not to vary in the y direction, the appropriate solutions for the fields in the upper region are

$$E_{1y} = E_0[\exp(-ik_{1z} z) + R_\perp \exp(ik_{1z} z)] \exp(ik_{1x} x), \tag{2.44a}$$

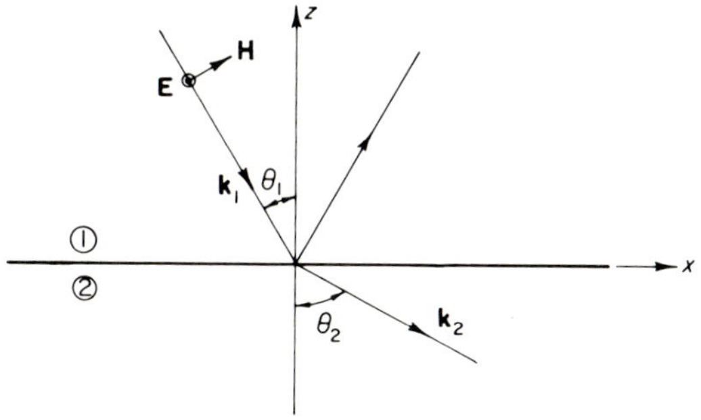

Fig. 2.2. Reflection and refraction at a plane interface (perpendicular polarization).

$$H_{1x} = \frac{k_{1z} E_0}{\omega \mu_0 \mu_1} [\exp(-ik_{1z} z) - R_\perp \exp(ik_{1z} z)] \exp(ik_{1x} x). \tag{2.44b}$$

Note that the x dependence must be identical for both the downward and upward waves since both of these waves propagate in the positive x direction. Moreover, since $k_x^2 + k_z^2 = k^2$ it will follow that k_z must also be the same except possibly for a change in sign. Indeed, we must choose the negative sign of the z dependence of the downward wave and the positive sign for the upward wave to ensure that the fields remain finite everywhere for $\text{Im}(k) \geq 0$.

In the lower region no upward waves are admissible; hence we must choose

$$E_{2y} = E_0 T_\perp \exp[i(k_{2x} x - k_{2z} z)], \tag{2.45a}$$

$$H_{2x} = \frac{k_{2z}}{\omega \mu_0 \mu_2} E_0 T_\perp \exp[i(k_{2x} x - k_{2z} z)], \tag{2.45b}$$

where $R_\perp$ and $T_\perp$ are the reflection and transmission coefficients to be determined. The continuity of the tangential electric and magnetic fields at $z = 0$, for all x implies the following.

$$k_{1x} = k_{2x}, \tag{2.46a}$$

$$1 + R_\perp = T_\perp, \tag{2.46b}$$

$$1 - R_\perp = \frac{k_{2z} \mu_1}{k_{1z} \mu_2} T_\perp. \tag{2.46c}$$

The last two equations can be solved simultaneously to give

$$R_\perp = \frac{\mu_2 k_{1z} - \mu_1 k_{2z}}{\mu_2 k_{1z} + \mu_1 k_{2z}}, \tag{2.47a}$$

$$T_\perp = \frac{2\mu_2 k_{1z}}{\mu_2 k_{1z} + \mu_1 k_{2z}}. \tag{2.47b}$$

Alternately, noting that $k_{jx} = k_j \sin\theta_j$ and $k_{jz} = k_j \cos\theta_j$, the reflection and transmission coefficients become

$$R_\perp = \frac{\mu_{21} \cos\theta_1 - n_{21}\cos\theta_2}{\mu_{21}\cos\theta_1 + n_{21}\cos\theta_2}, \tag{2.48a}$$

$$T_\perp = \frac{2\mu_{21}\cos\theta_1}{\mu_{21}\cos\theta_1 + n_{21}\cos\theta_2}, \tag{2.48b}$$

where $\mu_{21} = \mu_2/\mu_1$ and $n_{21} = n_2/n_1$. These coefficients can also be expressed in terms of a single angle θ_1 or θ_2 by noting (2.46a), which gives

$$\cos\theta_1 = (1 - n_{21}^2 \sin^2\theta_2)^{1/2}, \tag{2.49a}$$

$$\cos\theta_2 = (1 - n_{12}^2 \sin^2\theta_1)^{1/2}. \tag{2.49b}$$

One detail of the preceding manipulations deserves special attention. In the lower region the fields depend on the vertical distance from the interface through the factor $\exp(-ik_{2z}z)$ and in order that they remain finite as $z \to -\infty$ we must demand that

$$\operatorname{Im}\,({n_2}^2 - {n_1}^2 \sin^2\theta_1)^{1/2} \geq 0. \tag{2.50}$$

In the corresponding case of parallel polarization (electric field vector in the plane of incidence) it can be shown that the reflection and transmission coefficients are given by

$$R_\parallel = \frac{\cos\theta_2 - n_{21}\mu_{12}\cos\theta_1}{\cos\theta_2 + n_{21}\mu_{12}\cos\theta_1}, \tag{2.51a}$$

$$T_\parallel = 1 + R_\parallel. \tag{2.51b}$$

It should be noted that in each of the preceding cases we have defined the reflection and transmission coefficients as the appropriate ratios of the electric fields. A difference in sign in the reflection coefficient will result when the latter is defined as the ratio of the corresponding magnetic field components.

In general, the reflection coefficient $R_\perp$ and $R_\parallel$ are complex quantities since the refractive index is complex. This means that not only the amplitude of the wave changes upon reflection, but also its phase.

2.5 The Zenneck Surface Wave

Consider a special case of reflection when the electric field vector is in the plane of incidence, the upper medium is air, $n_1 = 1$, and the permeabilities of both media are the same, $\mu_1 = \mu_2 = 1$. The reflection coefficient takes on a simpler form:

$$R_{\parallel} = \frac{k_{2z} - n_2{}^2 k_{1z}}{k_{2z} + n_2{}^2 k_{1z}}, \tag{2.52a}$$

$$n_2 = \left(\varepsilon_2 + \frac{i\sigma_2}{\omega\varepsilon_0}\right)^{1/2}. \tag{2.52b}$$

We wish to investigate the consequences that follow in the case when there is no reflection, $R_{\parallel} = 0$, satisfied by the so-called Brewster's angle. The tangential components of the electric field in the upper and lower regions then become

$$E_{1x} = E_0 \exp[i(k_{1x}x - k_{1z}z)], \tag{2.53a}$$

$$E_{2x} = E_0 \exp[i(k_{1x}x - k_{2z}z)]. \tag{2.53b}$$

Moreover, in this case

$$k_{2z} = n_2{}^2 k_{1z}, \tag{2.54a}$$

$$k_1{}^2 = k_0{}^2 = k_{1x}^2 + k_{1z}^2, \tag{2.54b}$$

$$k_2{}^2 = n_2{}^2 k_0{}^2 = k_{1x}^2 + k_{2z}^2, \tag{2.54c}$$

from which it will follow that

$$k_{1z} = \frac{k_0}{(1 + n_2{}^2)^{1/2}} = k'_{1z} - ik''_{1z}. \tag{2.54d}$$

If the conductivity of the lower medium is small, $\sigma_2 \ll \omega\varepsilon_2$, then the radical in (2.52b) can be expanded, retaining only the first two terms, and we find

$$n_2 = n_2' + in_2'', \tag{2.55a}$$

$$n_2' = \sqrt{\varepsilon_2}, \tag{2.55b}$$

$$n_2'' = \frac{\sigma_2}{2\omega\varepsilon_0\sqrt{\varepsilon_2}}. \tag{2.55c}$$

Since in this case $n_2'' \ll n_2'$, the expression for k_{1z} in (2.54d) can be approximated by

$$k_{1z}' = \frac{k_0}{(1 + n'^2)^{1/2}}, \tag{2.56a}$$

$$k_{1z}'' = \frac{k_0 n_2' n_2''}{(1 + n_2'^2)^{3/2}}, \tag{2.56b}$$

where we note in passing that k_{1z}'' as defined in (2.54d) is always positive. Similarly, we find

$$k_{1x} = k_{1x}' + ik_{1x}'', \tag{2.57a}$$

$$k_{2z} = k_{2z}' + ik_{2z}'', \tag{2.57b}$$

where

$$k_{1x}' = \frac{k_0 n_2'}{(1 + n_2'^2)^{1/2}}, \qquad k_{1x}'' = \frac{k_0 n_2''}{(1 + n_2'^2)^{3/2}}, \tag{2.57c}$$

$$k_{2z}' = \frac{k_0 n_2'^2}{(1 + n_2'^2)^{1/2}}, \qquad k_{2z}'' = \frac{k_0 n_2' n_2''(2 + n_2'^2)}{(1 + n_2'^2)^{3/2}}. \tag{2.57d}$$

We can now rewrite the fields in the upper and lower regions as follows.

$$E_{1x} = E_0 \exp[i(k_{1x}' x - k_{1z}' z)] \exp[-(k_{1x}'' x + k_{1z}'' z)], \qquad z > 0, \tag{2.58a}$$

$$E_{2x} = E_0 \exp[i(k_{1x}' x - k_{2z}' z)] \exp[-(k_{1x}'' x - k_{2z}'' z)], \qquad z < 0. \tag{2.58b}$$

Equations (2.58) show that the field attains its maximum value at the interface and decays exponentially on either side away from the interface. The field is also attenuated in the lateral direction, however, since

$$k_{1z}'' = n_2' \, k_{1x}'', \tag{2.59a}$$

$$k_{2z}'' = n_2' (2 + n_2'^2) k_{1x}'' \tag{2.59b}$$

and if $\varepsilon_2 \gg 1$, it will follow that the attenuation in the lateral direction will be much slower than that in the vertical direction. The wave of this type is called the *Zenneck surfacc wave*.

It will be instructive to inquire more deeply into the mechanics of the surface wave propagation. The planes of constant phase in the upper and lower medium, respectively, are given by

$$\frac{dx}{dz} = \frac{k_{1z}'}{k_{1x}'} = \frac{1}{n_2'} = \cot \theta_1', \qquad z > 0, \tag{2.60a}$$

and

$$\frac{dx}{dz} = \frac{k'_{2z}}{k'_{1x}} = n_2' = \tan \theta_1', \qquad z < 0, \tag{2.60b}$$

where θ_1' is the real part of the complex angle of incidence, $\theta_1 = \theta_1' + i\theta_1''$. Correspondingly, the planes of constant amplitude are

$$\frac{dx}{dz} = -\frac{k''_{1z}}{k''_{1x}} = -n_2' = -\tan \theta_1', \qquad z > 0, \tag{2.61a}$$

in the upper medium, and

$$\frac{dx}{dz} = \frac{k''_{2z}}{k''_{1x}} = n_2'(2 + n_2'^2), \qquad z < 0, \tag{2.61b}$$

in the lower medium.

It will follow that in the upper medium the planes of constant phase and constant amplitude are orthogonal to each other, whereas in the lower conducting medium they are not. This is in complete agreement with the conclusions of Section 2.3.

The direction of the power flow is determined by the real part of the complex Poynting vector (1.21). We find

$$\mathrm{Re}(S_{1x}) = \frac{k'_{1x}}{2\omega\varepsilon_0} |H_{1y}|^2, \tag{2.62a}$$

$$\mathrm{Re}(S_{1z}) = \frac{-k'_{1z}}{2\omega\varepsilon_0} |H_{1y}|^2, \tag{2.62b}$$

$$\mathrm{Re}(S_{2x}) = \frac{k'_{1x}}{2\omega\varepsilon_0 n_2'^2} |H_{2y}^2|^2, \tag{2.62c}$$

$$\mathrm{Re}(S_{2z}) = \frac{-k'_{1z}}{2\omega\varepsilon_0} |H_{2y}|^2, \tag{2.62d}$$

from which it follows that

$$\frac{\mathrm{Re}(S_{1x})}{\mathrm{Re}(S_{1z})} = -n_2' = -\tan \theta_1', \tag{2.63a}$$

$$\frac{\mathrm{Re}(S_{2x})}{\mathrm{Re}(S_{2z})} = -\frac{1}{n_2'} = -\cot \theta_1'. \tag{2.63b}$$

Thus, in the upper nonconducting medium the direction of the power flow coincides with the direction of the planes of constant amplitude and it is normal to the planes of constant phase. In the lower medium, the direction of the power flow is still orthogonal to the planes of constant phase, but it does not coincide with the direction of the constant amplitude planes since these two planes are not orthogonal to each other. The situation is depicted in Fig. 2.3 for the case when $n_2' > 1$.

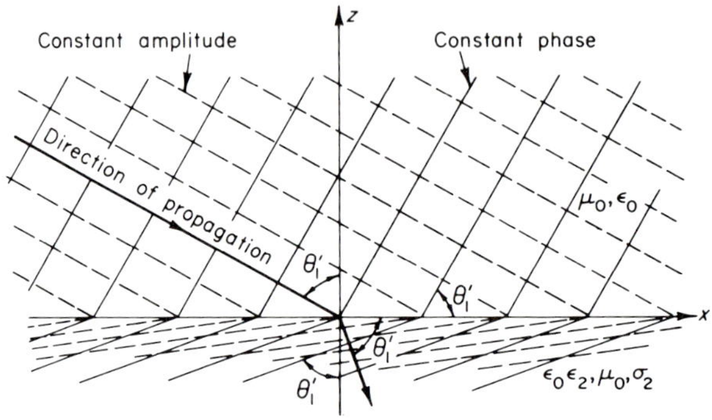

Fig. 2.3. Constant phase and constant amplitude planes for the Zenneck surface wave when $\varepsilon_2 \gg 1$ and $\sigma_2 \ll \omega\,\varepsilon_0\,\varepsilon_2$.

2.6 The Lateral Wave

Another interesting special case of plane wave behavior at an interface is one when no wave is transmitted into the lower medium. For the sake of concreteness, let us consider the case when the electric field vector is in the plane of incidence, the upper medium is air, the lower medium is nonconducting and its density is lower than that of air, $n_2 = \sqrt{\varepsilon_2} < 1$, and $\mu_2 = 1$. The fields in this case are

$$E_{1x} = E_0[\exp(-ik_{1z}z) + R_{\parallel}\exp(ik_{1z}z)]\exp(ik_{1x}x), \tag{2.64a}$$

$$E_{1z} = E_0\frac{k_{1x}}{k_{1z}}[\exp(-ik_{1z}z) - R_{\parallel}\exp(ik_{1z}z)]\exp(ik_{1x}x), \tag{2.64b}$$

$$H_{1y} = -E_0\frac{\omega\varepsilon_0}{k_{1z}}[\exp(-ik_{1z}z) - R_{\parallel}\exp(ik_{1z}z)]\exp(ik_{1x}x) \tag{2.64c}$$

in the upper region and

$$E_{2x} = E_0\,T_{\parallel}\exp[i(k_{1x}x - k_{2z}z)], \tag{2.65a}$$

$$E_{2z} = E_0\frac{k_{1x}}{k_{2z}}T_{\parallel}\exp[i(k_{1x}x - k_{2z}z)], \tag{2.65b}$$

$$H_{2y} = -E_0 \frac{\omega\varepsilon_0 \varepsilon_2}{k_{2z}} T_{\parallel} \exp[i(k_{1x}x - k_{2z}z)] \tag{2.65c}$$

in the lower region, where

$$R_{\parallel} = \frac{k_{2z} - \varepsilon_2 k_{1z}}{k_{2z} + \varepsilon_2 k_{1z}}, \tag{2.66a}$$

$$T_{\parallel} = 1 + R_{\parallel}. \tag{2.66b}$$

When $T_{\parallel} = 0$, $R_{\parallel} = -1$ and it will follow that $k_{2z} = 0$, which implies that $\theta_2 = \pi/2$ and $\sin\theta_1 = \sin\theta_c = \sqrt{\varepsilon_2}$ where θ_c is known as the *critical angle*. Let us focus our attention on the reflected fields in the upper region and the transmitted fields in the lower region. Using the transformation

$$x = d + g\sin\theta_c, \qquad z = g\cos\theta_c, \tag{2.67}$$

and noting that

$$k_{1x}x + k_{1z}z = k_0 g + k_2 d, \tag{2.68}$$

we obtain (with $Z_0 = (\mu_0/\varepsilon_0)^{1/2}$)

$$E_{1x}^r = -E_0 \exp[i(k_2 d + k_0 g)], \tag{2.69a}$$

$$E_{1z}^r = E_0 \tan\theta_c \exp[i(k_2 d + k_0 g)], \tag{2.69b}$$

$$H_{1y}^r = -\frac{E_0}{Z_0\cos\theta_c} \exp[i(k_2 d + k_0 g)] \tag{2.69c}$$

for the reflected fields in the upper region and

$$E_{2x} = 0, \tag{2.70a}$$

$$E_{2z} = 2E_0 \varepsilon_2^{-1} \tan\theta_c \exp(ik_2 x), \tag{2.70b}$$

$$H_{2y} = -\frac{2E_0}{Z_0\cos\theta_c} \exp(ik_2 x) \tag{2.70c}$$

for the transmitted fields in the lower region.

Note that the total phase change of the reflected fields in the upper region consists of two parts. The factor $k_2 d$ expresses a change in phase due to propagation of the wave in the lower medium a distance d, while the phase component $k_0 g$ expresses a phase change due to the wave's traversing a distance g in the upper medium. The geometrical interpretation of this phenomenon, called a *lateral wave*, is depicted in Fig. 2.4. It is seen that a

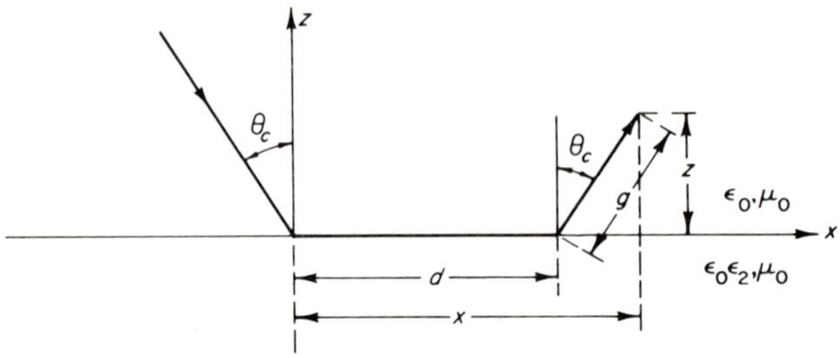

Fig. 2.4. Geometric representation of a lateral wave when $\varepsilon_2 < 1$.

wave of this sort is produced at the interface when the incidence angle of the incident plane wave is the critical angle, i.e., when $\theta_c = \arcsin\sqrt{\varepsilon_2}$ and in order for such an angle to be in the range zero to $\pi/2$ the condition $\varepsilon_2 < \varepsilon_1$ must be satisfied. Such a wave suffers a lateral displacement upon reflection, hence the name "lateral wave."

The direction of the power flow is determined by the real part of the complex Poynting vector (1.21). We find for the reflected power in the upper region

$$\operatorname{Re}(S^r_{1x}) = \frac{E_0{}^2 \sin\theta_c}{Z_0 \cos^2\theta_c}, \tag{2.71a}$$

$$\operatorname{Re}(S^r_{1z}) = \frac{E_0{}^2}{Z_0 \cos\theta_c} \tag{2.71b}$$

and in the lower region we have

$$\operatorname{Re}(S_{2x}) = \frac{4E_0{}^2 \sin\theta_c}{Z_0\,\varepsilon_2 \cos^2\theta_c}, \tag{2.72a}$$

$$\operatorname{Re}(S_{2z}) = 0. \tag{2.72b}$$

Hence in the upper region the energy is reflected at the angle θ_c, which is the same as the incident angle. In the lower region the energy is propagated parallel to the interface.

The direction of the energy flow in the upper and lower regions, in addition to phase interpretation of the reflected waves made earlier, substantiates the correctness of the geometrical interpretation of the lateral wave mechanism shown in Fig. 2.4. (Experimental evidence of this effect was first reported in 1938 by Schmidt, who performed experiments with pressure waves produced by a spark discharge at the interface of two liquid half spaces. The series of schlieren photographs obtained by him clearly exhibit the mechanism of lateral wave propagation.)

2.7 The Impedance Concept and Boundary Conditions

Let us return for a moment to the problem of reflection and refraction of Section 2.4. Taking the ratio of the tangential electric and magnetic fields on the upper side of the interface at $z = 0$, we obtain

$$\left(\frac{E_y}{H_x}\right)_1 = Z_1{}^{\perp}\left(\frac{1 + R_{\perp}}{1 - R_{\perp}}\right), \tag{2.73}$$

where

$$Z_1{}^{\perp} = \frac{Z_1}{\cos\theta_1}, \qquad Z_1 = \frac{\mu_1 Z_0}{n_1}, \tag{2.74}$$

and $Z_0 = (\mu_0/\varepsilon_0)^{1/2} = 377$ ohms is the intrinsic impedance of free space.

The corresponding ratio for the lower medium at $z = 0$ can be written

$$\left(\frac{E_y}{H_x}\right)_2 = Z_2{}^{\perp} \tag{2.75}$$

where

$$Z_2{}^{\perp} = \frac{Z_2}{\cos\theta_2}, \qquad Z_2 = \frac{\mu_2 Z_0}{n_2}. \tag{2.76}$$

Since at the interface $z = 0$ the tangential electric and magnetic field components must be continuous, (2.75) may be set equal to (2.73) and we can solve for the reflection coefficient to obtain

$$R_{\perp} = \frac{Z_2{}^{\perp} - Z_1{}^{\perp}}{Z_2{}^{\perp} + Z_1{}^{\perp}}, \tag{2.77}$$

which is an alternate and often more useful representation.

It can be easily shown that the corresponding representation for the parallel polarization has the same basic form, i.e.,

$$R_{\parallel} = \frac{Z_2{}^{\parallel} - Z_1{}^{\parallel}}{Z_2{}^{\parallel} + Z_1{}^{\parallel}}, \tag{2.78}$$

where

$$Z_j{}^{\parallel} = \cos\theta_j Z_j, \qquad Z_j = \frac{\mu_j Z_0}{n_j}. \tag{2.79}$$

The impedance concept formulated above is found useful in simplification of the boundary conditions at the interface $z = 0$ when the lower medium is much denser than the upper one. According to Snell's law, the angle of refraction in the lower medium can be expressed in terms of the angle of incidence as

$$\cos\theta_2 = (1 - n_{12}^2 \sin^3 \theta_1)^{1/2}. \tag{2.80}$$

If the upper medium is nonconducting, then $\sigma_1 = 0$ and, moreover, if the permeabilities of both media are the same, $\mu_2 = \mu_1$, it will follow that

$$n_{12}^2 = \frac{\varepsilon_1}{\varepsilon_2 + i\sigma_2/\omega\varepsilon_0}. \tag{2.81}$$

Thus, if either $\varepsilon_2 \gg \varepsilon_1$ or $\sigma_2/\omega\varepsilon_0 \gg \varepsilon_1$, then n_{12}^2 will be very small and consequently $\theta_2 \sim 0$. In this case, $Z_2^{\perp} \sim Z_2^{\parallel} \sim Z_2$ and the reflection coefficient becomes insensitive to the angle of incidence. Moreover, the continuity of the tangential electric and magnetic field components in the less dense region can be written in the form

$$\left(\frac{E_t}{H_t}\right)_1 = Z_2 \tag{2.82}$$

and Z_2 is a constant independent of the angle of incidence. Note that in this case we need not be concerned about the fields in the lower medium in order to solve the boundary value problem in the upper region.

The analysis above was carried out for a planar interface, but the concept has proved so useful that it is often applied to curved surfaces as well. Over a sufficiently small distance along an interface that includes no edges, the field can always be represented in terms of plane waves and the surface impedance concept should apply there. Indeed, it was shown by Senior (1960), and will be demonstrated in Chapter 10, that the concept of surface impedance can be applied to curved surfaces providing that $|n_2|k_0\rho \gg 1$ where ρ is the smallest radius of curvature of the surface and the upper medium is air.

2.8 System of Parallel Plane Layers

Consider a system of parallel plane layers as shown in Fig. 2.5. A plane wave with the electric field vector normal to the plane of incidence is incident on this system of plane layers. We wish to extend the analysis of Section 2.4 to the present problem and find the reflection coefficient in the upper semi-infinite region and the transmission coefficient in the lower semi-infinite region.

For concreteness we consider the case of perpendicular polarization and, consequently, we write for the upper semi-infinite region

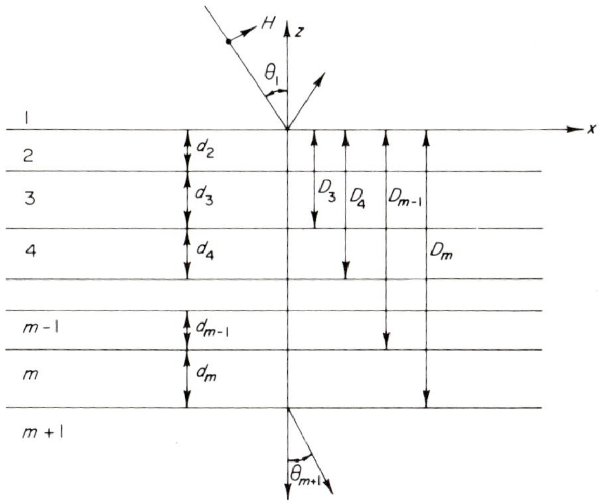

Fig. 2.5. A system of parallel plane layers.

$$E_{1y} = E_0[\exp(-ik_{1z}z) + R_\perp \exp(ik_{1z}z)] \exp(ik_{1x}x), \tag{2.83a}$$

$$H_{1x} = \frac{E_0}{Z_1^\perp} [\exp(-ik_{1z}z) - R_\perp \exp(ik_{1z}z)] \exp(ik_{1x}x), \tag{2.83b}$$

and for the lower semi-infinite region

$$E_{(m+1)y} = E_0 T_\perp \exp[ik_{1x}x - ik_{(m+1)z}(z + D_m)], \tag{2.84a}$$

$$H_{(m+1)x} = \frac{E_{(m+1)y}}{Z_{m+1}^\perp}. \tag{2.84b}$$

In any other intermediate finite layer there will be both the downward- and upward-traveling waves; therefore, we must write

$$E_{py} = E_0[A_p \exp(-ik_{pz}z) + B_p \exp(ik_{pz}z)] \exp(ik_{1x}x), \tag{2.85a}$$

$$H_{px} = \frac{E_0}{Z_p^\perp} [A_p \exp(-ik_{pz}z) - B_p \exp(ik_{pz}z)] \exp(ik_{1x}x). \tag{2.85b}$$

Let $\mathscr{Z}_p$ denote the input impedence at any of the interfaces, then at $z = 0$

$$\mathscr{Z}_1^\perp = \left.\frac{E_{1y}}{H_{1x}}\right|_{z=0} = \left(\frac{1 + R_\perp}{1 - R_\perp}\right) Z_1^\perp, \tag{2.86a}$$

$$= \left.\frac{E_{2y}}{H_{2x}}\right|_{z=0} = \left(\frac{A_2 + B_2}{A_2 - B_2}\right) Z_2^\perp. \tag{2.86b}$$

From (2.86a) it will follow immediately that the reflection coefficient can be expressed in terms of the impedances as follows

$$R_\perp = \frac{\mathscr{Z}_1{}^\perp - Z_1{}^\perp}{\mathscr{Z}_1{}^\perp + Z_1{}^\perp}. \tag{2.87}$$

Notice that the overall reflection coefficient $R_\perp$ has the same basic form as the corresponding reflection coefficient for a homogeneous half space found in (2.77) except that in the present case, the input impedance $\mathscr{Z}_1{}^\perp$ takes place of the impedance $Z_2{}^\perp$ of the lower medium. The input impedance $\mathscr{Z}_1{}^\perp$ is still an unknown of the problem, but it can be expressed in terms of the layers' properties and the angle of incidence by the following procedure.

Using (2.86b), we find

$$\mathscr{Z}_1{}^\perp = Z_2{}^\perp\left(\frac{1 + B_2/A_2}{1 - B_2/A_2}\right) \tag{2.88}$$

where the input impedance $\mathscr{Z}_1{}^\perp$ is now expressed in terms of the known impedance $Z_2{}^\perp$ and the unknown ratio B_2/A_2, which we shall determine next. At the interface $z = -d_2$, the input impedance to the third layer can be expressed as follows.

$$\mathscr{Z}_2{}^\perp = \left.\frac{E_{2y}}{H_{2x}}\right|_{z=-d_2} = Z_2{}^\perp\left(\frac{\exp(i2\psi_2) + B_2/A_2}{\exp(i2\psi_2) - B_2/A_2}\right) \tag{2.89}$$

where $\psi_2 = k_{2z}\, d_2$. Now we can solve (2.89) for the ratio B_2/A_2 and substitute back into (2.88) to obtain the input impedance transformation formula

$$\mathscr{Z}_1{}^\perp = Z_2{}^\perp\left(\frac{\mathscr{Z}_2{}^\perp - iZ_2{}^\perp \tan\psi_2}{Z_2 - i\mathscr{Z}_2{}^\perp \tan\psi_2}\right), \tag{2.90}$$

which expresses the input impedance $\mathscr{Z}_1{}^\perp$ in terms of the yet unknown input impedance $\mathscr{Z}_2{}^\perp$ and other known quantities.

The input impedance transformation relation (2.90) is a special case of the more general relation

$$\mathscr{Z}_{p-1}^\perp = Z_p{}^\perp\left(\frac{\mathscr{Z}_p{}^\perp - iZ_p{}^\perp \tan\psi_p}{Z_p{}^\perp - i\mathscr{Z}_p{}^\perp \tan\psi_p}\right), \tag{2.91a}$$

$$\psi_p = k_{pz}\, d_p \tag{2.91b}$$

where p is a number corresponding to the particular layer, $2 \le p \le m$. Since the input impedance at the last interface $z = -D_m$ is simply equal to the wave impedance in the last semi-infinite layer, $\mathscr{Z}_m{}^\perp = Z_{m+1}^\perp$, all input impedances can be found in terms of specified parameters and this renders the solution to the reflection coefficient (2.87) complete.

The transmission coefficient $T_\perp$ can be found in a similar manner. Starting at the last interface $z = -D_m$ and equating the tangential components of the electrical field, we obtain

$$T_\perp = \left(1 + \frac{B_m}{A_m}\exp(-i2k_{mz}D_m)\right)A_m \exp(ik_{mz}D_m). \tag{2.92}$$

The ratio B_m/A_m can be expressed in terms of the various impedances using (2.85) and can thus be considered known. The amplitude factor A_m can be expressed in terms of A_{m-1} by equating the tangential electric field components at $z = -D_{m-1}$ and eliminating B_{m-1}/A_{m-1} using (2.85). We obtain

$$A_m \exp(ik_{mz}D_m) = \frac{\mathscr{Z}^\perp_{m-1}(\mathscr{Z}_m{}^\perp + Z_m{}^\perp)A_{m-1}\exp(ik_{(m-1)z}D_{m-1})}{(\mathscr{Z}^\perp_{m-1} + Z^\perp_{m-1})(\mathscr{Z}_m{}^\perp \cos\psi_m - iZ_m{}^\perp \sin\psi_m)}. \tag{2.93}$$

Equation (2.93) can be considered a reduction formula for the amplitude factor A_p in any layer. It can be continued up to the second layer, whose amplitude factor A_2 can be expressed in terms of the amplitude of the incident wave. We find

$$T_\perp = \frac{2Z^\perp_{m+1}}{\mathscr{Z}_1{}^\perp + Z_1{}^\perp}\prod_{p=2}^{m}\frac{\mathscr{Z}^\perp_{p-1}}{\mathscr{Z}_p{}^\perp \cos\psi_p - iZ_p{}^\perp \sin\psi_p} \tag{2.94}$$

as the overall transmission coefficient.

Problems

2.1 Derive the expression for the reflection coefficient $R_\parallel$ as the ratio of the reflected and incident electric fields when **E** is in the plane of incidence.

2.2 (a) Draw a plot of $R_\perp$ in the complex $R_\perp$ plane when $\mu_2 = \mu_1$ and n_{21} is real for the two cases when $n_{21} > 1$ and $n_{21} < 1$. Denote the principal points, i.e., Brewster's angle and the critical angle.
(b) Repeat part (a) for $R_\parallel$.

2.3 Consider a plane interface of two media with a plane wave incident as shown in Fig. 2.6. If $\varepsilon_1 = \varepsilon_2$, $\sigma_1 = \sigma_2 = 0$, and $\mu_2 = \mu_1\mu_r$ where $\mu_r = \mu_r'$

Fig. 2.6.

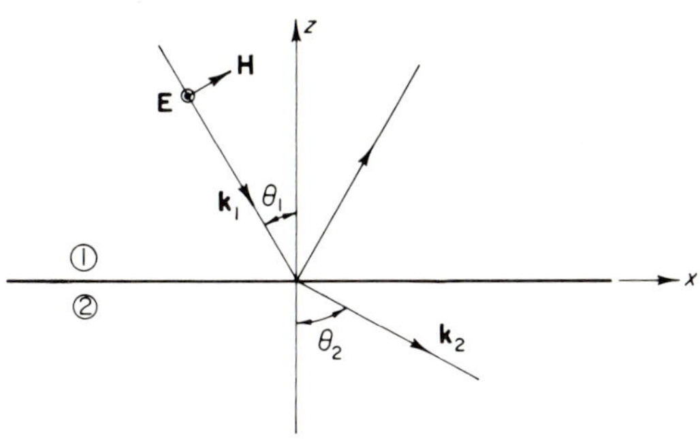

$+ i\mu_r''$, $\mu_r'' \ll \mu_r'$, and μ_1 is real, show that this interface will support a surface wave and this wave will be "tightly" bound to the interface when μ_r' is large and only "loosely" bound when μ_r' is small. Draw planes of constant phase and amplitude on each side of the interface and show the direction of propagation of the surface wave.

References

Born, M., and Wolf, E. (1964). "Principles of Optics," 2nd ed. Pergamon, Oxford.

Brekhovskikh, L. M. (1960). "Waves in Layered Media." Academic Press, New York.

Senior, T. B. A. (1960). Impedance boundary conditions for imperfectly conducting surfaces, *Appl. Sci. Res.* **B8**, 418–436.

Stratton., J. A. (1941). "Electromagnetic Theory." McGraw-Hill, New York.

von Schmidt, O. (1938). Über Knallwellenausbreitung in Flüssigkeiten und Festen Körpern, *Phys. Z.* **39**, 868–875.

Wait, J. R. (1962). "Electromagnetic Waves in Stratified Media." Macmillan, New York.

CHAPTER 3

PLANE WAVES IN HOMOGENEOUS ANISOTROPIC MEDIA

Anisotropic media in electromagnetic theory can be either electrically anisotropic or magnetically anisotropic. In the first case the medium is characterized by a tensor permittivity $\mathbf{D} = \varepsilon_0 \boldsymbol{\varepsilon}\mathbf{E}$, and in the second by a tensor permeability $\mathbf{B} = \mu_0 \boldsymbol{\mu}\mathbf{H}$. An example of a physical medium exhibiting electrical anisotropy is a gaseous plasma in a steady magnetic field, which will be referred to henceforth as a magnetoplasma. A medium of practical importance that exhibits magnetic anisotropy is a ferrite material in a steady magnetic field.

In what follows we begin with the general formulation of a plane wave solution for both of the two cases, showing that such a solution can be obtained without a detailed knowledge of the permittivity or permeability tensor. Later, by means of physical considerations we shall derive the components of the tensor in each case and provide motivation for solutions of some of the boundary value problems dealing with plane interfaces.

3.1 Electrically Anisotropic Media

Consider an electrically anisotropic medium that is characterized by the electric permittivity tensor of the form given by

$$\boldsymbol{\varepsilon} = \begin{pmatrix} \varepsilon_{11} & \varepsilon_{12} & \varepsilon_{13} \\ \varepsilon_{21} & \varepsilon_{22} & \varepsilon_{23} \\ \varepsilon_{31} & \varepsilon_{32} & \varepsilon_{33} \end{pmatrix}. \tag{3.1}$$

In a region free of charges and currents with harmonic time dependence $\exp(-i\omega t)$, the electric and the magnetic field vectors will satisfy the vector wave equations

$$(\nabla \times \nabla \times - k_0{}^2\mu\boldsymbol{\varepsilon})\mathbf{E} = 0, \tag{3.2a}$$

$$(\nabla \times \boldsymbol{\varepsilon}^{-1} \nabla \times - k_0{}^2\mu)\mathbf{H} = 0, \tag{3.2b}$$

and it will suffice to solve one of these equations since the other field can then be determined from Maxwell's equations.

Using the matrix representation of the vector differential operator $\nabla \times \nabla \times$ in (1.38), the differential equation (3.2a) becomes

$$\begin{pmatrix} -(\partial_y^2+\partial_z^2)-k_0^2\mu\varepsilon_{11} & \partial_x\,\partial_y-k_0^2\mu\varepsilon_{12} & \partial_x\,\partial_z-k_0^2\mu\varepsilon_{13} \\ \partial_x\,\partial_y-k_0^2\mu\varepsilon_{21} & -(\partial_x^2+\partial_z^2)-k_0^2\mu\varepsilon_{22} & \partial_y\,\partial_z-k_0^2\mu\varepsilon_{23} \\ \partial_x\,\partial_z-k_0^2\mu\varepsilon_{31} & \partial_y\,\partial_z-k_0^2\mu\varepsilon_{32} & -(\partial_x^2+\partial_y^2)-k_0^2\mu\varepsilon_{33} \end{pmatrix}\begin{pmatrix} E_x \\ E_y \\ E_z \end{pmatrix}=0. \tag{3.3}$$

Let us now seek a plane wave solution of the form

$$\mathbf{E}=\mathbf{E}_0\exp[i(\mathbf{k}\cdot\mathbf{r}-\omega t)], \tag{3.4}$$

where $\mathbf{E}_0$ is a constant vector and $\mathbf{k}$ is the propagation vector to be determined. Now, since $\partial/\partial x = ik_x$ and so on, the system of partial differential equations in (3.3) can be replaced by a system of three linear algebraic equations in three unknowns as follows.

$$\begin{pmatrix} k_y^2+k_z^2-k_0^2\mu\varepsilon_{11} & -(k_xk_y+k_0^2\mu\varepsilon_{12}) & -(k_xk_z+k_0^2\mu\varepsilon_{13}) \\ -(k_xk_y+k_0^2\mu\varepsilon_{21}) & k_x^2+k_z^2-k_0^2\mu\varepsilon_{22} & -(k_yk_z+k_0^2\mu\varepsilon_{23}) \\ -(k_xk_z+k_0^2\mu\varepsilon_{31}) & -(k_yk_z+k_0^2\mu\varepsilon_{32}) & k_x^2+k_y^2-k_0^2\mu\varepsilon_{33} \end{pmatrix}\begin{pmatrix} E_x \\ E_y \\ E_z \end{pmatrix}=0. \tag{3.5}$$

A unique nontrivial solution to the system of algebraic equations in (3.5) can be obtained if, and only if, the determinant of the square matrix on the left vanishes. This will be satisfied if

$$n^4-\frac{A}{D}n^2+\frac{|\boldsymbol{\varepsilon}|}{D}=0, \tag{3.6a}$$

$$k=k_0\,n\sqrt{\mu}, \tag{3.6b}$$

where n is the index of refraction and

$$\begin{aligned} A = {} & (\varepsilon_{11}\varepsilon_{22}-\varepsilon_{12}\varepsilon_{21})(1-\gamma^2)+(\varepsilon_{11}\varepsilon_{33}-\varepsilon_{13}\varepsilon_{31})(1-\beta^2) \\ & +(\varepsilon_{22}\varepsilon_{33}-\varepsilon_{23}\varepsilon_{32})(1-\alpha^2) \\ & -[\varepsilon_{21}\varepsilon_{32}+\varepsilon_{12}\varepsilon_{23}-\varepsilon_{22}(\varepsilon_{13}+\varepsilon_{31})]\alpha\gamma \\ & -[\varepsilon_{13}\varepsilon_{32}+\varepsilon_{23}\varepsilon_{31}-\varepsilon_{33}(\varepsilon_{12}+\varepsilon_{21})]\alpha\beta \\ & -[\varepsilon_{21}\varepsilon_{13}+\varepsilon_{31}\varepsilon_{12}-\varepsilon_{11}(\varepsilon_{23}+\varepsilon_{32})]\beta\gamma, \end{aligned} \tag{3.6c}$$

$$D=\varepsilon_{11}\alpha^2+\varepsilon_{22}\beta^2+\varepsilon_{33}\gamma^2+(\varepsilon_{12}+\varepsilon_{21})\alpha\beta+(\varepsilon_{13}+\varepsilon_{31})\alpha\gamma+(\varepsilon_{23}+\varepsilon_{32})\beta\gamma, \tag{3.6d}$$

$$\alpha=\sin\theta\cos\varphi, \qquad \beta=\sin\theta\sin\varphi, \qquad \gamma=\cos\theta, \tag{3.6e}$$

where θ and φ are the polar and the azimuthal angles describing the position of the propagation vector $\mathbf{k}$ with respect to a cartesian coordinate system x, y, z. It will readily follow that the electric field components are related by

$$E_y = -\frac{b_{23}}{b_{22}} E_z, \tag{3.7a}$$

$$E_x = \frac{a_{12} b_{23} - a_{13} b_{22}}{a_{11} b_{22}} E_z, \tag{3.7b}$$

where

$$b_{22} = a_{11} a_{22} - a_{12} a_{21}, \tag{3.7c}$$

$$b_{23} = a_{11} a_{23} - a_{13} a_{21}, \tag{3.7d}$$

and where the coefficients a_{ij} are the elements of the square matrix in (3.5).

Now, the index of refraction can be found by solving (3.6a) to give

$$n_{1,2} = \{(2D)^{-1}[A \mp (A^2 - 4|\varepsilon|D)^{1/2}]\}^{1/2}, \tag{3.8}$$

where we note two possible choices of sign and, hence, associate the subscripts 1 and 2. Media that are characterized by two different values of the refractive index are often referred to as *double refracting media.*

In view of what has been said, it is clear that the most general plane wave solution for an electrically anisotropic medium characterized by the electric permittivity tensor given in (3.1) can be written

$$\mathbf{E} = \mathbf{E}_1 \exp(i\mathbf{k}_1 \cdot \mathbf{r}) + \mathbf{E}_2 \exp(i\mathbf{k}_2 \cdot \mathbf{r}) \tag{3.9}$$

where $\mathbf{E}_{1,2}$ are constant vectors, and

$$\mathbf{k}_{1,2} = k_{1,2}(\mathbf{1}_x \sin\theta \cos\varphi + \mathbf{1}_y \sin\theta \sin\varphi + \mathbf{1}_z \cos\theta), \tag{3.10a}$$

$$k_{1,2} = n_{1,2} \sqrt{\mu} k_0 \tag{3.10b}$$

where $\mathbf{1}_x$, $\mathbf{1}_y$, and $\mathbf{1}_z$ are unit vectors in the positive x, y, and z directions, respectively. The magnetic field vector can be found from Maxwell's equations to be

$$\mathbf{H} = (\omega\mu_0 \mu)^{-1}[\mathbf{k}_1 \times E_1 \exp(i\mathbf{k}_1 \cdot \mathbf{r}) + \mathbf{k}_2 \times \mathbf{E}_2 \exp(i\mathbf{k}_2 \cdot \mathbf{r})]. \tag{3.11}$$

3.2 Magnetically Anisotropic Media

Consider a magnetically anisotropic medium that is characterized by the permeability tensor of the form

$$\boldsymbol{\mu} = \begin{pmatrix} \mu_{11} & \mu_{12} & \mu_{13} \\ \mu_{21} & \mu_{22} & \mu_{23} \\ \mu_{31} & \mu_{32} & \mu_{33} \end{pmatrix}. \tag{3.12}$$

In a region free of charges and currents with harmonic time dependence of the form $\exp(-i\omega t)$, the electric and magnetic field vectors will satisfy the vector wave equations

$$(\nabla \times \nabla \times - k_0{}^2 \varepsilon \boldsymbol{\mu})\mathbf{H} = 0, \tag{3.13a}$$

$$(\nabla \times \boldsymbol{\mu}^{-1} \nabla \times - k_0{}^2 \varepsilon)\mathbf{E} = 0. \tag{3.13b}$$

We note that (3.13a) and (3.13b) are identical in form to (3.2a) and (3.2b), respectively. Thus, if we define in this case

$$k_{1,2} = n_{1,2}\sqrt{\varepsilon}k_0, \tag{3.14}$$

then the refractive indices $n_{1,2}$ will be solutions to

$$n_{1,2} = \{(2D)^{-1}[A \mp (A^2 - 4|\boldsymbol{\mu}|D)^{1/2}]\}^{1/2}, \tag{3.15a}$$

where in this case

$$\begin{aligned} A = {} & (\mu_{11}\mu_{22} - \mu_{12}\mu_{21})(1 - \gamma^2) + (\mu_{11}\mu_{33} - \mu_{13}\mu_{31})(1 - \beta^2) \\ & + (\mu_{22}\mu_{33} - \mu_{23}\mu_{31})(1 - \alpha^2) - [\mu_{21}\mu_{32} + \mu_{12}\mu_{23} - \mu_{22}(\mu_{13} + \mu_{31})]\alpha\gamma \\ & - [\mu_{13}\mu_{32} + \mu_{23}\mu_{31} - \mu_{33}(\mu_{12} + \mu_{21})]\alpha\beta \\ & - [\mu_{21}\mu_{13} + \mu_{31}\mu_{12} - \mu_{11}(\mu_{23} + \mu_{32})]\beta\gamma, \end{aligned} \tag{3.15b}$$

$$\begin{aligned} D = {} & \mu_{11}\alpha^2 + \mu_{22}\beta^2 + \mu_{33}\gamma^2 + (\mu_{12} + \mu_{21})\alpha\beta + (\mu_{13} + \mu_{31})\alpha\gamma \\ & + (\mu_{23} + \mu_{32})\beta\gamma, \end{aligned} \tag{3.15c}$$

and α, β, and γ are given by (3.6e).

The plane wave solution to (3.13a) can be written by analogy with (3.9a) as follows.

$$\mathbf{H} = \mathbf{H}_1 \exp(i\mathbf{k}_1 \cdot \mathbf{r}) + \mathbf{H}_2 \exp(i\mathbf{k}_2 \cdot \mathbf{r}), \tag{3.16}$$

where $\mathbf{H}_{1,2}$ are constant vectors and $\mathbf{k}_{1,2}$ have the form of (3.10a) with $k_{1,2}$ given by (3.14).

The electric field components are readily found from Maxwell's equations as follows.

$$\mathbf{E} = -(\omega\varepsilon_0\varepsilon)^{-1}[\mathbf{k}_1 \times \mathbf{H}_1 \exp(i\mathbf{k}_1 \cdot \mathbf{r}) + \mathbf{k}_2 \times \mathbf{H}_2 \exp(i\mathbf{k}_2 \cdot \mathbf{r})]. \tag{3.17}$$

3.3 Plane Waves in Magnetoplasmas

The phenomena associated with electromagnetic wave propagation in a magnetoplasma are encountered in a number of important cases. These phenomena play an important role in radio wave propagation in the ionosphere and around the earth; in radio astronomy and space communication; and in propagation in various plasmas created under laboratory conditions.

The various parameters that determine the electromagnetic properties of plasma may vary by several orders of magnitude under diverse conditions. If N_e denotes the number of free electrons per cubic centimeter and T denotes the temperature in degrees Kelvin, then in the earth's ionosphere $10^3 < N_e < 3 \times 10^6$ and $300 < T < 3000$, whereas, for instance, in thermonuclear reaction devices $N_e \sim 10^{15}$ and $T \sim 10^7$.

In what follows we shall investigate the electromagnetic properties of plasma from the viewpoint of elementary theory first in the absence of collisions and a steady magnetic field; later, we introduce the effect of collisions, and finally we derive the permittivity tensor for a lossy plasma in a steady magnetic field.

3.3.1 Lossless Plasma, $H_{\mathrm{DC}} = 0$

Consider an electrically neutral gaseous medium consisting of an equal number of free electrons and positive ions, all of which are singly charged. In the absence of free charges or currents, the Maxwell equations with harmonic time dependence $\exp(-i\omega t)$ are

$$\nabla \times \mathbf{E} = i\omega\mu_0 \mathbf{H}, \tag{3.18a}$$

$$\nabla \times \mathbf{H} = -i\omega(\mathbf{P} + \varepsilon_0 \mathbf{E}), \tag{3.18b}$$

$$\nabla \cdot \mathbf{D} = -|e|(N_e - N_i) = 0, \tag{3.18c}$$

$$\nabla \cdot \mathbf{H} = 0, \tag{3.18d}$$

where N_e and N_i denote the number of electrons and ions per unit volume, respectively. Let $\mathbf{r}_k^{(e)}$ denote the displacement vector of the kth electron due to the action of the wave's electric field and $\mathbf{r}_k^{(i)}$ the corresponding displacement of a positive ion. Then the polarization $\mathbf{P}$ of the medium can be written

$$\mathbf{P} = -|e| \sum_{k=1}^{N} (\mathbf{r}_k^{(e)} - \mathbf{r}_k^{(i)}). \tag{3.19}$$

The force acting on each free electron is Newton's force, which for equilibrium must be set equal to Coulomb's force; hence

$$m\ddot{\mathbf{r}}_k^{(e)} = -|e|\mathbf{E} = -|e|\mathbf{E}_0 \exp(-i\omega t), \tag{3.20}$$

where $\mathbf{E}_0$ is the amplitude vector of the wave. Equation (3.20) can be integrated twice with respect to time and, assuming that the initial velocity and the initial displacement of the electron were zero, we obtain

$$\mathbf{r}_k^{(e)}(t) = \frac{|e|}{m\omega^2}\,\mathbf{E}. \tag{3.21}$$

The displacement of an ion is almost identical to that of an electron except for a change in sign of the first term on the right. Thus,

$$\mathbf{r}_k^{(i)}(t) = -\frac{|e|}{\omega^2 M}\,\mathbf{E}, \tag{3.22}$$

where M denotes the mass of an ion. Substituting (3.21) and (3.22) into (3.19) gives

$$\mathbf{P} = -\frac{|e|^2 N}{m\omega^2}\left(1 + \frac{m}{M}\right)\mathbf{E}. \tag{3.23}$$

Since the mass of an ion is almost always much greater than that of an electron, then $m/M \ll 1$ and can be neglected. Thus, we have

$$\mathbf{P} = -\frac{|e|^2 N}{m\omega^2}\,\mathbf{E} \tag{3.24}$$

as the effective polarization of the gaseous medium under the influence of an applied electric field $\mathbf{E}$. Since $\mathbf{D} = \mathbf{P} + \varepsilon_0\mathbf{E}$, we obtain

$$\mathbf{D} = \varepsilon_0\left(1 - \frac{|e|^2 N}{m\varepsilon_0\omega^2}\right)\mathbf{E} \tag{3.25}$$

and note that the quantity $|e|^2 N/m\varepsilon_0$ has the dimension of frequency squared. Thus, we denote

$$\omega_p{}^2 = \frac{|e|^2 N}{m\varepsilon_0} \tag{3.26}$$

and henceforth refer to ω_p as the *plasma frequency*. The electric permittivity in this case is

$$\varepsilon = 1 - \left(\frac{\omega_p}{\omega}\right)^2, \tag{3.27}$$

where

$$\left(\frac{\omega_p}{\omega}\right)^2 = 3.18 \times 10^9\,\frac{N}{\omega^2} = 8.06 \times 10^7\,\frac{N}{f^2}$$

and N is the number of free electrons per cubic centimeter.

3.3.2 Lossy Plasma, $H_{DC} = 0$

Within the framework of elementary theory, the effect of collisions can be taken into account by introducing a frictional force $g\dot{\mathbf{r}}$ in the equation of motion. Clearly, the expression $g\dot{\mathbf{r}}$ stands for the average change of momentum per second associated with the collisions. On the other hand, however, this change should be equal to $m\nu\dot{\mathbf{r}}$ where ν is the effective number of collisions per second. With each impact the electron imparts to a molecule or an ion on the average a momentum $m\dot{\mathbf{r}}$ where $\dot{\mathbf{r}}$ is the ordered velocity imparted to the electron by the impressed electric field.

If an electron of negligibly small effective radius moves among neutral particles in the gas, then it is reasonable to assume that a collision will occur when the center of the electron approaches within a distance a from the center of a neutral particle. The mean free path $l^{(n)}$ will then be given by

$$l^{(n)} = (\pi a^2 N_n)^{-1} = \frac{\bar{u}}{\nu^{(n)}}. \tag{3.28}$$

where N_n is the number of neutral particles per unit volume, $\bar{u}$ is the average velocity, and $\nu^{(n)}$ is the number of collisions per second. From elementary gas-kinetic theory we know that the relation between the average velocity and the temperature can be asserted via the expression

$$\frac{m\bar{u}^2}{2} = \frac{3}{2} kT, \tag{3.29}$$

where $k = 1.38042 \times 10^{-23}$ joule per degree Kelvin is the Boltzmann constant and T is the temperature in degrees Kelvin. From (3.28) and (3.29) it follows that

$$\begin{aligned} \nu^{(n)} &= \pi a^2 N_n \left(\frac{3kT}{m}\right)^{1/2} \\ &\sim 3.7 \times 10^{-9} N_n \left(\frac{T}{300}\right)^{1/2} \quad \text{per second.} \end{aligned} \tag{3.30}$$

Next consider the motion of an electron in a gas consisting of neutral molecules and positive ions. The neutral molecules influence its motion only when the electron finds itself within a distance a, denoting some molecular radius. A positive ion, however, will influence the motion of an electron at a much larger distance because of the electric field existing around it. Suppose that a deflection of the electron occurring at a distance r_{eff} from the center of an ion is counted as a collision, then we can write

$$\frac{m\bar{u}^2}{2} = \frac{3}{2}kT = \frac{e^2}{4\pi\varepsilon_0 r_{\text{eff}}}, \tag{3.31}$$

where we have equated the kinetic energy of the electron with the potential energy of an ion at a distance r_{eff} (Debye shielding radius). By analogy with the previous case, we can write

$$\begin{aligned} \nu^{(i)} &= \frac{e^4 N_i}{36\pi{\varepsilon_0}^2}\left(\frac{3}{m(kT)^3}\right)^{1/2} \\ &\sim 5.1 \times 10^{-4} N_i \left(\frac{300}{T}\right)^{3/2}. \end{aligned} \tag{3.32}$$

It is significant to note that $\nu^{(n)} \sim T^{1/2}$, whereas $\nu^{(i)} \sim T^{-3/2}$. This is so because the effective collision cross section of an ion (for collision with an electron) is given by πr_{eff}^2, and by virtue of (3.31) it is a function of electron velocity.

The foregoing order-of-magnitude calculations, as crude as they are, provide the correct form of the expression for collision frequency. An elaborate investigation due to Cowling (1945) results in the expression

$$\begin{aligned} \nu &= \nu^{(n)} + \nu^{(i)} \\ &= 1.8 \times 10^{-8} N_n \left(\frac{T}{300}\right)^{1/2} + 6.1 \times 10^{-3} N_i \left(\frac{300}{T}\right)^{3/2}. \end{aligned} \tag{3.33}$$

It will be noticed that as the temperature increases the effect of the ions becomes less important. The neutral molecules and the ions contribute equally to the effective collision frequency at a temperature of about 1.8×10^5 degrees Kelvin.

Since the positive ions, due to their immobility, do not contribute significantly to the result, we can neglect their contribution altogether and consider the electron motion only. In view of what has been said, their equation of motion, including collisions, can now be written in the form

$$m\ddot{\mathbf{r}}_k + m\nu\dot{\mathbf{r}}_k = -|e|\mathbf{E}_0 \exp(-i\omega t). \tag{3.34}$$

Assuming a solution of the form $\mathbf{r}_k = \mathbf{r}_k(0)\exp(-i\omega t)$, it readily follows that

$$\mathbf{r}_k = \frac{|e|}{m\omega(\omega + i\nu)}\mathbf{E}. \tag{3.35}$$

Comparing (3.35) with (3.21) reveals the addition of the term $(im\omega\nu)$ in the denominator as the only modification. Thus, by analogy with the previous lossless case, we write

$$\varepsilon = 1 - \frac{{\omega_p}^2}{\omega(\omega + i\nu)}, \tag{3.36a}$$

$$\varepsilon = 1 - \frac{{\omega_p}^2}{\omega^2 + \nu^2} + \frac{i\nu\omega_p}{\omega(\omega^2 + \nu^2)}. \tag{3.36b}$$

The following two limiting cases of (3.36) are worth noting. In the case when the wave frequency is much higher than the collision frequency, then

$$\varepsilon \sim 1 - \left(\frac{\omega_p}{\omega}\right)^2 + \frac{i\nu{\omega_p}^2}{\omega^3}, \qquad \omega \gg \nu. \tag{3.37a}$$

Conversely, if the collision frequency is higher than the wave frequency then

$$\varepsilon = 1 - \frac{{\omega_p}^2}{\nu^2} + \frac{i{\omega_p}^2}{\omega\nu}. \tag{3.37b}$$

3.3.3 Lossy Plasma, $H_{\mathrm{DC}} \neq 0$

Finally, let us consider the effect of a steady magnetic field H_{DC}. It will be useful to define the following quantities.

$$\omega_h = \frac{|e|\mu_0 H_{\mathrm{DC}}}{m} = 1.76 \times 10^7 H_{\mathrm{DC}} \quad \text{(gauss)}, \tag{3.38a}$$

$$\Omega_h = \frac{|e|\mu_0 H_{\mathrm{DC}}}{M} = 1.76 \times 10^7 \left(\frac{m}{M}\right) H_{\mathrm{DC}} \quad \text{(gauss)}, \tag{3.38b}$$

where ω_h is the electron cyclotron frequency and Ω_h is the ion cyclotron frequency. The frequencies ω_h and Ω_h are those at which free electrons and ions, respectively, rotate around the lines of force of the field H_{DC}. It is interesting to note that in the earth's ionosphere $\omega_h \sim 3 \times 10^6$ to 9×10^6 cycles per second, whereas for the O $\pm$ ions $\Omega_h \sim 100$ to 300 cycles per second.

The effect of ions is usually negligible under the condition $\omega \gg \Omega_h$, while the effect of the steady magnetic field on electron motion is negligible if $\omega \gg \omega_h$. The effect of the steady magnetic field is substantial, however, as far as the rotation of the plane of polarization is concerned—for instance, in radio astronomy, where a wave traverses enormous distances.

To derive the electric permittivity tensor, we begin with the equation of motion (3.34) but in this case we must add to the right side the term $-|e|(\dot{\mathbf{r}} \times \mathbf{B})$, which takes into account the change in the direction of the electron velocity. Thus, we have

$$\ddot{\mathbf{r}} + \nu\dot{\mathbf{r}} = \frac{-|e|}{m}[\mathbf{E}_0 \exp(-i\omega t) + \mu_0 \dot{\mathbf{r}} \times (\mathbf{H}_{\mathrm{DC}} + \mathbf{H}_0 \exp(-i\omega t))]. \tag{3.39}$$

Usually, even the smallest steady component of the magnetic field is substantially larger than the amplitude of the wave field, i.e., $H_{DC} \gg H_0$, so that the latter can be neglected. Moreover, if we assume a solution of the form $\mathbf{r}(t) = \mathbf{r}(0) \exp(-i\omega t)$, we obtain

$$\mathbf{r} = \frac{|e|}{\omega m} [(\omega + i\nu)\mathbf{I} - i\omega_h \mathbf{1}_{DC} \times]^{-1}\mathbf{E}, \tag{3.40}$$

where $\mathbf{I}$ is the "identity" matrix and $\mathbf{1}_{DC}$ is a unit vector in the direction of the steady magnetic field. If the effect of the positive ions on the polarization of the medium is neglected, then $\mathbf{P} = -|e|N\mathbf{r}$ and since $\mathbf{D} = \mathbf{P} + \varepsilon_0 \mathbf{E} = \varepsilon_0 \boldsymbol{\varepsilon}\mathbf{E}$, we obtain

$$\boldsymbol{\varepsilon} = \mathbf{I} - \frac{\omega_p{}^2}{\omega} [(\omega + i\nu)\mathbf{I} - i\omega_h \mathbf{1}_{DC} \times]^{-1}. \tag{3.41}$$

The expression in the square brackets on the right is

$$[\quad]^{-1} = \begin{pmatrix} \omega + i\nu & i\omega_h 1_z & -i\omega_h 1_y \\ -i\omega_h 1_z & \omega + i\nu & i\omega_h 1 \\ i\omega_h 1_y & -i\omega_h 1_x & \omega + i\nu \end{pmatrix}^{-1}, \tag{3.42}$$

where 1_x, 1_y, and 1_z are the components of the unit vector $\mathbf{1}_{DC}$ along the axes x, y, and z, respectively. We recall that the inverse of any square nonsingular matrix $\mathbf{A}$ is

$$\mathbf{A}^{-1} = (|\mathbf{A}|)^{-1}(\text{adj } \mathbf{A}), \tag{3.43}$$

where $(\text{adj } \mathbf{A}) = (\mathbf{A}_{ij})'$ where the prime denotes the transpose and $\mathbf{A}_{ij}$ is the cofactor of the element a_{ij} of the matrix $\mathbf{A}$. In cartesian coordinates

$$\begin{aligned} 1_x &= \sin\theta' \cos\varphi', \\ 1_y &= \sin\theta' \sin\varphi', \\ 1_z &= \cos\theta', \end{aligned} \tag{3.44}$$

and the components of the electric permittivity tensor become

$$\begin{aligned}
\varepsilon_{11} &= \varepsilon + (\zeta - \varepsilon) \sin^2\theta' \cos^2\varphi', \\
\varepsilon_{12} &= [(\zeta - \varepsilon)/2] \sin^2\theta' \sin 2\varphi' - i\eta \cos\theta', \\
\varepsilon_{13} &= [(\zeta - \varepsilon)/2] \sin 2\theta' \cos\varphi' + i\eta \sin\theta' \sin\varphi', \\
\varepsilon_{21} &= [(\zeta - \varepsilon)/2] \sin^2\theta' \sin 2\varphi' + i\eta \cos\theta', \\
\varepsilon_{22} &= \varepsilon + (\zeta - \varepsilon) \sin^2\theta' \sin^2\varphi', \\
\varepsilon_{23} &= [(\zeta - \varepsilon)/2] \sin 2\theta' \sin\varphi' - i\eta \sin\theta' \cos\varphi', \\
\varepsilon_{31} &= [(\zeta - \varepsilon)/2] \sin 2\theta' \cos\varphi' - i\eta \sin\theta' \sin\varphi', \\
\varepsilon_{32} &= [(\zeta - \varepsilon)/2] \sin 2\theta' \sin\varphi' + i\eta \sin\theta' \cos\varphi', \\
\varepsilon_{33} &= \zeta - (\zeta - \varepsilon) \sin^2\theta',
\end{aligned} \tag{3.45}$$

where

$$\zeta = 1 - \frac{\omega_p{}^2}{\omega^2 + \nu^2} + \frac{i\nu\omega_p{}^2}{\omega(\omega^2 + \nu^2)}, \tag{3.46a}$$

$$\varepsilon = 1 - \frac{\omega_p{}^2(\omega^2 - \omega_h{}^2 + \nu^2)}{\Delta} + \frac{i\nu\omega_p{}^2(\omega^2 + \omega_h{}^2 + \nu^2)}{\omega\Delta}, \tag{3.46b}$$

$$\eta = -\frac{\omega_h\omega_p{}^2(\omega^2 - \omega_h{}^2 - \nu^2)}{\omega\Delta} + \frac{i2\nu\omega_h\omega_p{}^2}{\Delta}, \tag{3.46c}$$

$$\Delta = [(\omega + \omega_h)^2 + \nu^2][(\omega - \omega_h)^2 + \nu^2]. \tag{3.46d}$$

In the case when the magnetostatic field is parallel to the z axis, $\theta' = 0$, the electric permittivity tensor becomes

$$\boldsymbol{\varepsilon} = \begin{pmatrix} \varepsilon & -i\eta & 0 \\ i\eta & \varepsilon & 0 \\ 0 & 0 & \zeta \end{pmatrix}. \tag{3.47}$$

We note in this case that $\varepsilon_{11} = \varepsilon_{22} = \varepsilon$, $\varepsilon_{21} = -\varepsilon_{12} = i\eta$, $\varepsilon_{33} = \zeta$, and all other elements are zero. Then from (3.6c) and (3.6d)

$$A = (\varepsilon_{11}^2 + \varepsilon_{12}^2)\sin^2\theta + \varepsilon_{11}\varepsilon_{33}(1 + \cos^2\theta),$$

$$D = \varepsilon_{11}\sin^2\theta + \varepsilon_{33}\cos^2\theta,$$

$$|\boldsymbol{\varepsilon}| = (\varepsilon_{11}^2 + \varepsilon_{12}^2)\varepsilon_{33};$$

hence, by (3.8) we can write

$$n_{1,2}^2 = \frac{\begin{gathered}[\varepsilon_{11}(\varepsilon_{11} - \varepsilon_{33}) + \varepsilon_{12}^2]\sin^2\theta + 2\varepsilon_{11}\varepsilon_{33} \\ \pm \{[\varepsilon_{11}(\varepsilon_{33} - \varepsilon_{11}) - \varepsilon_{12}^2]^2\sin^4\theta - 4\varepsilon_{12}^2\varepsilon_{33}^2\cos^2\theta\}^{1/2}\end{gathered}}{2(\varepsilon_{11}\sin^2\theta + \varepsilon_{33}\cos^2\theta)} \tag{3.48}$$

or alternately, in terms of the components of the permittivity matrix in (3.47),

$$n_{1,2}^2 = \frac{\begin{gathered}[\varepsilon(\varepsilon - \zeta) - \eta^2]\sin^2\theta + 2\varepsilon\zeta \\ \pm \{[\varepsilon(\zeta - \varepsilon) + \eta^2]^2\sin^4\theta + 4\eta^2\zeta^2\cos^2\theta\}^{1/2}\end{gathered}}{2(\varepsilon\sin^2\theta + \zeta\cos^2\theta)}. \tag{3.49}$$

Consider a few special cases of the expression for the index of refraction in (3.49). For the direction of propagation parallel to the magnetostatic field (z axis); $\theta = 0$ and we obtain

$$n_{1,2} = (\varepsilon \mp \eta)^{1/2}. \tag{3.50}$$

The field components can be determined from (3.5) by setting $k_x = k_y = 0$, $k_z = n_{1,2}k_0$, $\varepsilon_{11} = \varepsilon_{22} = \varepsilon$, $\varepsilon_{12} = -\varepsilon_{21} = -i\eta$, and $\varepsilon_{33} = \zeta$. Thus,

$$\begin{pmatrix} \varepsilon - n_{1,2}^2 & -i\eta & 0 \\ i\eta & \varepsilon - n_{1,2}^2 & 0 \\ 0 & 0 & \zeta \end{pmatrix} \begin{pmatrix} E_x \\ E_y \\ E_z \end{pmatrix} = 0, \tag{3.51}$$

from which it follows that $E_z = 0$ and the electric field is in the plane transverse to the direction of propagation. Moreover, using (3.50) we can readily show that

$$E_{x1} = iE_{y1}, \tag{3.52a}$$

$$E_{x2} = -iE_{y2}, \tag{3.52b}$$

where the subscripts 1 and 2 correspond to the subscripts in (3.50). It will be noted that (3.52a) designates a right-hand circularly polarized wave, whereas (3.52b) designates a left-hand circularly polarized wave. The complete solution for a wave traveling in the positive z direction can be written

$$\mathbf{E} = (\mathbf{1}_x - i\mathbf{1}_y)E_1 \exp(ik_0 n_1 z) + (\mathbf{1}_x + i\mathbf{1}_y)E_2 \exp(ik_0 n_2 z), \tag{3.53a}$$

$$\mathbf{H} = (Z_0)^{-1}(i\mathbf{1}_x + \mathbf{1}_y)n_1 E_1 \exp(ik_0 n_1 z) - (i\mathbf{1}_x - \mathbf{1}_y)n_2 E_2 \exp(ik_0 n_2 z). \tag{3.53b}$$

Next, consider the direction of propagation parallel to the x axis. In this case, $k_x = k_0 n_{1,2}$, $k_y = k_z = 0$ and we obtain

$$\begin{pmatrix} \varepsilon & -i\eta & 0 \\ i\eta & \varepsilon - n_{1,2}^2 & 0 \\ 0 & 0 & \zeta - n_{1,2}^2 \end{pmatrix} \begin{pmatrix} E_x \\ E_y \\ E_z \end{pmatrix} = 0; \tag{3.54}$$

it will follow that two groups of solutions are possible. The first one is

$$\begin{aligned} &n_1 = \sqrt{\zeta}, \qquad E_x = E_y = 0, \qquad E_z = E_1 \exp(ik_0 n_1 x), \\ &H_x = H_z = 0, \qquad H_y = \frac{-n_1}{Z_0} E_1 \exp(ik_0 n_1 x); \end{aligned} \tag{3.55a}$$

and the second is

$$\begin{aligned} &n_2 = \sqrt{\chi}, \qquad \chi = \frac{\varepsilon^2 - \eta^2}{\varepsilon}, \qquad E_z = 0, \\ &E_x = \frac{i\eta}{\varepsilon} E_y, \qquad E_y = E_2 \exp(ik_0 n_2 x), \\ &H_x = H_y = 0, \qquad H_z = \frac{n_2}{Z_0} E_2 \exp(ik_0 n_2 x). \end{aligned} \tag{3.55b}$$

Note that in the latter case the two modes can ex ist independently, whereas in the former case, (3.53), the two modes are co upled in the sense that they satisfy a single set of coupled second-order di fferential equations.

Next, let us investigate the propagation pa rallel to the y axis. In this case $k_y = k_0 n_{1,2}$, $k_x = k_z = 0$ and we obtain

$$\begin{pmatrix} \varepsilon - n_{1,2}^2 & -i\eta & 0 \\ i\eta & \varepsilon & 0 \\ 0 & 0 & \zeta - n_{1,2}^2 \end{pmatrix} \begin{pmatrix} E_x \\ E_y \\ E_z \end{pmatrix} = 0, \tag{3.56}$$

from which it follows that two solutions are possible. We get

$$\begin{aligned} n_1 &= \sqrt{\zeta}, \qquad E_x = E_y = 0, \qquad E_z = E_1 \exp(ik_0 n_1 y), \\ H_y &= H_z = 0, \qquad H_x = \frac{n_1}{Z_0} E_1 \exp(ik_0 n_1 y), \end{aligned} \tag{3.57a}$$

and

$$\begin{aligned} n_2 &= \sqrt{\chi}, \qquad E_z = 0, \qquad E_y = \frac{-i\eta}{\varepsilon} E_x, \qquad E_x = E_2 \exp(ik_0 n_2 y), \\ H_x &= H_y = 0, \qquad H_z = \frac{n_2}{Z_0} E_2 \exp(ik_0 n_2 y). \end{aligned} \tag{3.57b}$$

Finally, note that in the last two cases, (3.55b) and (3.57b), the electric field is no longer transverse to the direction of propagation, and in (3.55a) and (3.57a) the field is not affected by the presence of the magnetostatic field.

3.4 Plane Waves in Magnetized Ferrites

The factor that determines the character of electromagnetic wave propagation through any material medium is the medium's local electric or magnetic structure. If the medium is subjected to a steady field, one may expect changes in the character of wave propagation. Ferrite media are examples of those in which these changes are large enough to be experimentally significant.

The magnetic properties of ferrites are not much different from those of iron and other ferromagnetic materials. The factor that makes ferrites so outstanding and gives rise to their unique microwave behavior is their extremely low conductivity (typically from 10^{-6} to 10^{-4} mho per meter for ferrites, as compared to 10^7 mhos per meter for iron). Whereas in the case of iron, the electromagnetic wave "sees" an almost perfect reflector, in the case of ferrite, the wave can enter and pass through considerable amounts of the medium without excessive reflection or attenuation. Thus, the wave has an opportunity to interact with the electrons within the ferrite medium and these interactions are responsible for the various reciprocal and nonreciprocal effects produced.

The mathematical model of the motion of magnetization from which the microwave properties of ferrites can be adequately deduced is as follows. The spinning electrons in a ferrite medium, which can be looked upon as negatively charged spinning spheres, produce a magnetic moment **m** directed along the axis of the spin and opposite to the direction of the angular momentum **P**. If a steady magnetic field $\mathbf{H}_{\mathrm{DC}}$ is applied to the medium, the spinning electrons find themselves under the influence of the torque $\mathbf{T} = \mathbf{m} \times \mathbf{H}_{\mathrm{DC}}$, which causes them to precess about the lines of the steady magnetic field. If, in addition to the steady magnetic field $\mathbf{H}_{\mathrm{DC}}$ a time-varying field $\mathbf{H}\exp(-i\omega t)$ is present, the precession orbit will be in the form of a circle with a superposed sinusoidal deviation. The governing equation of motion can then be obtained by equating the effective torque with the time rate of change of the angular momentum **P**. Since the angular momentum can be related to the magnetic moment, $\mathbf{m} = \gamma \mathbf{P}$, where $\gamma = 1.76 \times 10^{10}$ ampere-meter per joule is the gyromagnetic ratio, we obtain

$$\frac{\partial \mathbf{M}_T}{\partial t} = \gamma(\mathbf{M}_T \times \mathbf{H}_T), \tag{3.58}$$

where $\mathbf{M}_T$ is the total magnetization per unit volume and $\mathbf{H}_T$ is the total magnetic field, each consisting of a steady and a time-varying component as follows.

$$\mathbf{M}_T = \mathbf{M}_{\mathrm{DC}} + \mathbf{M}e^{-i\omega t}; \tag{3.59a}$$

$$\mathbf{H}_T = \mathbf{H}_{\mathrm{DC}} + \mathbf{H}e^{-i\omega t}. \tag{3.59b}$$

Equation (3.58) does not account for losses that are known to exist in a ferrite medium. To this end, a damping force must be introduced that would tend to align the electron's axis of rotation with the direction of the magnetic field. This can be accomplished by writing the equation of motion as (Yager, 1950).

$$\frac{\partial \mathbf{M}_T}{\partial t} = \gamma(\mathbf{M}_T \times \mathbf{H}_T) - \frac{\gamma\alpha}{M_T}[\mathbf{M}_T \times (\mathbf{M}_T \times \mathbf{H}_T)]. \tag{3.60}$$

Equations (3.58) and (3.59) assume a ferrite sample fully saturated in the magnetostatic field, which is always the case in microwave applications. The right side of (3.60) has the nature of a torque. The force on $\mathbf{M}_T$ is always at right angles to $\mathbf{M}_T$, thus leaving the magnitude unchanged. The first term on the right in (3.60) is a vector normal to $\mathbf{M}_T$ and to $\mathbf{H}_T$; it is also responsible for precession. The second term on the right in (3.60) represents damping. It is a vector normal to $\mathbf{M}_T$ but situated in the plane of $\mathbf{M}_T$ and $\mathbf{H}_T$ in a sense such as to reduce the angle of the precession. The term $\mathbf{H}_T$ is the total field acting on the electronic spin and it consists of the magnetostatic field $\mathbf{H}_{\mathrm{DC}}$, applied externally, and the microwave carrier field $\mathbf{H}\exp(-i\omega t)$. The microwave field $\mathbf{H}$ in the ferrite medium is one of the unknowns of the problem of propagation and it will appear in the solutions to the Maxwell equations subject to the appropriate boundary conditions. The term $\mathbf{M}_T$ consists of the steady magnetization $\mathbf{M}_{\mathrm{DC}}$ (directed along $\mathbf{H}_{\mathrm{DC}}$ in a saturated sample) and a time-varying part $\mathbf{M}\exp(-i\omega t)$.

Equation (3.60) must now be solved for $\mathbf{M}$ in terms of $\mathbf{H}$. It is a nonlinear equation whose solution $\mathbf{M}$ will depend on $\mathbf{H}$ nonlinearly. In order to permit a solution, it is necessary to linearize the problem by assuming that the time-varying quantities are small in comparison with the steady quantities so that their products can be neglected and only linear terms considered.

In view of what has been said, Eq. (3.60) can be written

$$-\frac{i\omega}{\gamma}\mathbf{M} = \mathbf{M}_{\mathrm{DC}} \times \mathbf{H} + \mathbf{M} \times \mathbf{H}_{\mathrm{DC}} - \frac{\alpha}{M_{\mathrm{DC}}}[\mathbf{M}_{\mathrm{DC}} \times (\mathbf{M}_{\mathrm{DC}} \times \mathbf{H} + \mathbf{M} \times \mathbf{H}_{\mathrm{DC}})], \tag{3.61a}$$

or alternately

$$[\mathbf{I} + i\sigma(\mathbf{I} - \alpha\mathbf{1}_{\mathrm{DC}} \times)\mathbf{1}_{\mathrm{DC}} \times]\mathbf{M} = ip[(\mathbf{I} - \alpha\mathbf{1}_{\mathrm{DC}} \times)\mathbf{1}_{\mathrm{DC}} \times]\mathbf{H}, \tag{3.61b}$$

where

$$\sigma = \frac{\gamma H_{\mathrm{DC}}}{\omega}, \qquad p = \frac{\gamma M_{\mathrm{DC}}}{\omega}. \tag{3.61c}$$

When the magnetization $\mathbf{M}_{\mathrm{DC}}$ is parallel to the z axis, then $\mathbf{1}_{\mathrm{DC}} = \mathbf{1}_z$ and upon multiplying by the inverse of the matrix on the right in (3.61b), we obtain

$$\begin{pmatrix} M_x \\ M_y \\ M_z \end{pmatrix} = \frac{p}{\sigma^2 - (1 + i\alpha\sigma)^2}\begin{pmatrix} \sigma(1+\alpha^2) - i\alpha & i & 0 \\ -i & \sigma(1+\alpha^2) - i\alpha & 0 \\ 0 & 0 & 0 \end{pmatrix}\begin{pmatrix} H_x \\ H_y \\ H_z \end{pmatrix}. \tag{3.62}$$

Finally, we use the relation $\mathbf{B} = \mu_0(\mathbf{M} + \mathbf{H}) = \mu_0 \boldsymbol{\mu}\mathbf{H}$ and obtain, for the components of the permeability tensor,

$$\begin{aligned} \mu_{11} &= \mu_{22} = 1 + \frac{p[\sigma(1+\alpha^2) - i\alpha]}{\sigma^2 - (1 + i\alpha\sigma)^2} = \mu, \\ \mu_{12} &= -\mu_{21} = \frac{ip}{\sigma^2 - (1 + i\alpha\sigma)^2} = -i\rho, \\ \mu_{33} &= 1, \\ \mu_{13} &= \mu_{31} = \mu_{23} = \mu_{32} = 0. \end{aligned} \tag{3.63}$$

In a more general case, when the magnetostatic field is arbitrarily oriented, it can be shown (Tyras, 1959) that the permeability tensor has the following components

$$
\begin{aligned}
\mu_{11} &= \mu + (1-\mu)\sin^2\theta' \cos^2\varphi', \\
\mu_{12} &= [(1-\mu)/2]\sin^2\theta' \sin 2\varphi' - i\rho\cos\theta', \\
\mu_{13} &= [(1-\mu)/2]\sin 2\theta' \cos\varphi' + i\rho\sin\theta'\sin\varphi', \\
\mu_{21} &= [(1-\mu)/2]\sin^2\theta' \sin 2\varphi' + i\rho\cos\theta', \\
\mu_{22} &= \mu + (1-\mu)\sin^2\theta' \sin^2\varphi' \\
\mu_{23} &= [(1-\mu)/2]\sin 2\theta' \sin\varphi' - i\rho\sin\theta'\cos\varphi', \\
\mu_{31} &= [(1-\mu)/2]\sin 2\theta' \cos\varphi' - i\rho\sin\theta'\sin\varphi', \\
\mu_{32} &= [(1-\mu)/2]\sin 2\theta' \sin\varphi' + i\rho\sin\theta'\cos\varphi', \\
\mu_{33} &= 1 - (1-\mu)\sin^2\theta',
\end{aligned}
\tag{3.64}
$$

where θ' and φ' are the polar and azimuthal angles that the magnetostatic field makes with the z axis and x axis, respectively.

In the case of the magnetostatic field parallel to the z axis and the direction of propagation described by the polar angle θ and azimuthal angle φ, the index of refraction can be found by employing the results of (3.15a) and (3.15b). We obtain

$$
\begin{aligned}
A &= (\mu^2 - \rho^2)\sin^2\theta + \mu(1+\cos^2\theta), \\
D &= \mu\sin^2\theta + \cos^2\theta, \\
|\boldsymbol{\mu}| &= \mu^2 - \rho^2,
\end{aligned}
\tag{3.65}
$$

which upon substitution into (3.15a) gives

$$
n_{1,2}^2 = \frac{[\mu(\mu-1)-\rho^2]\sin^2\theta + 2\mu \mp \{[\mu(1-\mu)+\rho^2]^2\{\sin^4\theta + 4\rho^2\cos^2\theta\}^{1/2}}{2(\mu\sin^2\theta + \cos^2\theta)}, \tag{3.66}
$$

which is similar in form to the corresponding case of magnetoplasma in Eq. (3.49).

Let us now consider a few special cases of the expression for the index of refraction (3.66). For the direction of propagation parallel to the magnetostatic field (z axis), $\theta = 0$ and we obtain

$$
\begin{aligned}
n_{1,2} &= (\mu \mp \rho)^{1/2}, \qquad H_z = E_z = 0, \\
\mathbf{H} &= (\mathbf{1}_x - i\mathbf{1}_y)H_1 \exp(ik_0\sqrt{\varepsilon}n_1 z) + (\mathbf{1}_x + i\mathbf{1}_y)H_2 \exp(ik_0\sqrt{\varepsilon}n_2 z), \\
\mathbf{E} &= \frac{-Z_0}{\sqrt{\varepsilon}}[n_1(i\mathbf{1}_x + \mathbf{1}_y)H_1 \exp(ik_0\sqrt{\varepsilon}n_1 z) \\
&\quad + n_2(-i\mathbf{1}_x + \mathbf{1}_y)H_2 \exp(ik_0\sqrt{\varepsilon}n_2 z)].
\end{aligned}
\tag{3.67}
$$

For the propagation parallel to the x axis we obtain two independent solutions. The first is

$$n_1 = 1, \qquad H_x = H_y = 0, \qquad H_z = H_1 \exp(ik_0\sqrt{\varepsilon}x),$$
$$E_x = E_z = 0, \qquad E_y = \frac{Z_0}{\sqrt{\varepsilon}} H_1 \exp(ik_0\sqrt{\varepsilon}x), \tag{3.68}$$

and the second is

$$n_2 = (\chi_m)^{1/2}, \qquad \chi_m = \frac{\mu^2 - \rho^2}{\mu}, \qquad H_z = 0,$$
$$H_x = \frac{i\rho}{\mu} H_y, \qquad H_y = H_2 \exp(ik_0\sqrt{\varepsilon}\, n_2 x), \tag{3.69}$$
$$E_x = E_y = 0, \qquad E_z = \frac{-Z_0 n_2}{\sqrt{\varepsilon}} H_2 \exp(ik_0\sqrt{\varepsilon}\, n_2 x).$$

In the case of propagation parallel to the y axis we again obtain two independent solutions. The first is

$$n_1 = 1, \qquad H_x = H_y = 0, \qquad H_z = H_1 \exp(ik_0\sqrt{\varepsilon}\, y),$$
$$E_y = E_z = 0, \qquad E_x = -\frac{Z_0}{\sqrt{\varepsilon}} H_1 \exp(ik_0\sqrt{\varepsilon}\, y), \tag{3.70}$$

while the second one is

$$n_2 = \sqrt{\chi_m}, \qquad H_z = 0, \quad H_y = \frac{-i\rho}{\mu} H_x, \qquad H_x = H_2 \exp(ik_0\sqrt{\varepsilon}\, n_2 y), \tag{3.71}$$
$$E_x = E_y = 0, \qquad E_z = \frac{-Z_0 n_2}{\sqrt{\varepsilon}} H_2 \exp(ik_0\sqrt{\varepsilon}\, n_2 y).$$

Problems

3.1 Consider the case of a plane wave in magnetoplasma when the direction of propagation is parallel to the magnetostatic field. Show that the propagation constant of this wave is $k_0(n_1 + n_2)/2$ and the plane of polarization of the wave rotates at a rate of $k_0(n_1 - n_2)/2$ radians per meter.

Hint: Assume that at $z = 0$, $E_y = 0$, and then find the angle that the electric field vector makes with the x axis after the wave has traversed a distance $z = z_0$.

3.2 Consider a gaseous medium consisting only of equal numbers of free electrons and positive ions. Assume that the ions carry a single charge and that the steady magnetic field is parallel to the z axis. On the basis of the elementary theory of plasma, show that when the motion of the ions as well as electrons is taken into account, the resulting components of the permittivity tensor are

$$\varepsilon_{11} \pm i\varepsilon_{12} = 1 - \frac{\omega_p^2}{(\omega \mp \omega_h)(\omega \pm \Omega_h) + i\nu\omega},$$

$$\varepsilon_{33} = 1 - \frac{\omega_p^2}{\omega(\omega + i\nu)}, \qquad \varepsilon_{21} = -\varepsilon_{12}, \qquad \varepsilon_{13} = \varepsilon_{32} = \varepsilon_{31} = \varepsilon_{33} = 0,$$

where

$$\omega_h = |e|\mu_0 H_{DC}/m, \quad \Omega_h = |e|\mu_0 H_{DC}/M, \quad \omega_p^2 = |e|^2 N/m\varepsilon_0,$$

and ν is the average number of collisions that an electron undergoes with a positive ion.

Hints: (a) Note that the frictional force exerted by the electrons on the ions is $m\nu(\dot{\mathbf{r}} - \dot{\mathbf{R}})$ and it must be equal and opposite to the force of friction exerted by the ions on the electrons. (b) The terms of the order m/M and $(m/M)^{1/2}$ can be neglected as compared with unity in the factors multiplying the variables.

3.3 A plane wave linearly polarized along the x axis is incident normally on an infinite plasma slab of thickness d as shown in Fig. 3.1. Assuming that the permittivity tensor of the plasma is of the form

$$\boldsymbol{\varepsilon} = \begin{pmatrix} \varepsilon & i\eta & 0 \\ -i\eta & \varepsilon & 0 \\ 0 & 0 & \zeta \end{pmatrix}$$

when the magnetostatic field is in the z direction, find the overall reflection and transmission coefficients when the magnetostatic field is directed along the (i) z axis; (ii) y axis; (iii) x axis. Discuss the polarization of the reflected and transmitted waves for each of these three orientations of the magnetostatic field.

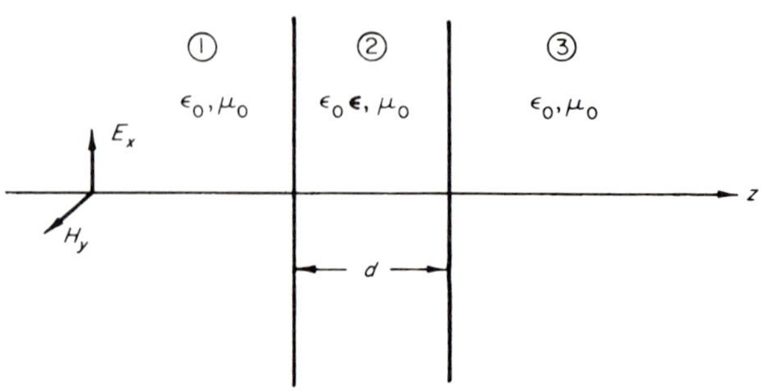

Fig. 3.1

References

Allis, W. P., Buchsbaum, S. J., and Bers, A. (1963). "Waves in Anisotropic Plasmas." M.I.T. Press, Cambridge, Massachusetts.

Cowling T. G. (1945). The electric conductivity of an ionized gas in a magnetic field, with applications to the solar atmosphere and the ionosphere, *Proc. Roy. Soc.* **A183,** 453.

Ginzburg, V. L. (1961). "Propagation of Electromagnetic Waves in Plasma." Gordon & Breach, New York.

Holt, E. H., and Haskel, R. E. (1965). "Foundations of Plasma Dynamics." Macmillan, New York.

Ratcliffe, J. A. (1959). "The Magneto-Ionic Theory and Its Applications to the Ionosphere." Cambridge Univ. Press, London and New York.

Suhl, H., and Walker, L. R. (1959). Topics in guided-wave propagation through gyromagnetic media, *Bell Syst. Tech. J.* **33**, 579–659.

Tyras, G. (1959). The permeability matrix for a ferrite medium magnetized at an arbitrary direction and its eigenvalues, *IRE Trans. Microwave Theory and Techniques* **MTT-7,** 176–177.

Yager, W. A., Galt, J. K., Merritt, F. R., and Wood, E. A. (1950). *Phys. Rev.* **80**, 744.

CHAPTER 4

PLANE WAVES IN INHOMOGENEOUS STRATIFIED MEDIA

In the previous chapters we limited ourselves to the study of plane waves in (a) unbounded homogeneous media and (b) homogeneous media bounded by parallel planes. We shall devote this chapter to the study of plane waves in media that are not homogeneous along one axis of the cartesian coordinate system. Such media are called plane-stratified inhomogeneous media.

There are many media in nature that can be classified as plane-stratified inhomogeneous media, including the earth's atmosphere, the sea, and the core of the earth, which can be looked on as plane-stratified inhomogeneous media since their dielectric properties vary appreciably in the vertical direction while the variations in the horizontal direction are almost negligible.

One approach to the problem of plane-stratified inhomogeneous media was presented in Section 2.8, where we considered an arbitrary number of plane layers as shown in Fig. 2.5. With this approach, an arbitrary plane-stratified medium can be approximated with a desired degree of accuracy by making the individual layers sufficiently thin. The results of this approach are not very elegant inasmuch as one must resort to a digital computer for numerical answers.

In this chapter we shall follow a direct approach to the problem of plane wave propagation in plane-stratified inhomogeneous media. This approach will consist of attempting to solve the pertinent differential equation directly or by approximate methods. It will be shown that a direct solution is possible in a limited number of stratification profiles, and that a general approximate solution can be obtained when the parameters of the medium vary slowly.

4.1 The Characteristic Differential Equations

Since most of the media of interest exhibit inhomogeneities in their electric rather than magnetic properties, we shall concentrate on the former and assume in the remainder of this chapter that $\mu = 1$.

In an isotropic electrically inhomogeneous medium free of charges and currents, the electric and the magnetic field vectors, according to (1.30) and (1.31), must satisfy the differential equations

$$\nabla^2\mathbf{E} + \nabla\left(\mathbf{E}\cdot\frac{\nabla\varepsilon}{\varepsilon}\right) + k_0{}^2\varepsilon\mathbf{E} = 0, \tag{4.1a}$$

$$\nabla^2\mathbf{H} + \frac{\nabla\varepsilon}{\varepsilon}\times\nabla\times\mathbf{H} + k_0{}^2\varepsilon\mathbf{H} = 0, \tag{4.1b}$$

where the harmonic time dependence $\exp(-i\omega t)$ has been assumed and suppressed throughout. Note that in the case when the electric properties of the medium vary slowly, the gradient $\nabla\varepsilon$ approaches zero and we obtain the usual wave equations.

If the electric permittivity properties of the medium vary only along one cartesian coordinate, say z, and the wave motion takes place in the x-z plane so that $\partial/\partial y = 0$, then the problem can be formulated in terms of a single component of the electric or magnetic field.

Thus, in the case of the electric field perpendicular to the plane of incidence, we obtain

$$(\nabla_{xz}^2 + k^2(z))E_y = 0, \tag{4.2a}$$

$$H_x = \frac{i}{\omega\mu_0}\frac{\partial E_y}{\partial z}, \tag{4.2b}$$

$$H_z = -\frac{i}{\omega\mu_0}\frac{\partial E_y}{\partial x}, \tag{4.2c}$$

$$k^2(z) = k_0{}^2\varepsilon(z) = k_0{}^2n^2(z). \tag{4.2d}$$

In the case that the electric field is in the plane of incidence, the problem can be formulated in terms of a single component of the magnetic field as follows.

$$\left(\nabla_{xz}^2 - \frac{\varepsilon'}{\varepsilon}\frac{\partial}{\partial z} + k^2(z)\right)H_y = 0, \tag{4.3a}$$

$$E_x = -\frac{i}{\omega\varepsilon_0\,\varepsilon(z)}\frac{\partial H_y}{\partial z}, \tag{4.3b}$$

$$E_z = \frac{i}{\omega\varepsilon_0\,\varepsilon(z)}\frac{\partial H_y}{\partial x}. \tag{4.3c}$$

Alternately, the differential equation (4.3a) can be somewhat simplified if we put

$$H_y(x, z) = k(z)V(x, z), \tag{4.4}$$

which upon substitution into (4.3) gives

$$(\nabla_{xz}^2 + K^2)V = 0, \tag{4.5a}$$

$$E_x = -\frac{i}{\omega\varepsilon_0\,\varepsilon}\frac{\partial}{\partial z}(kV), \tag{4.5b}$$

$$E_z = \frac{ik}{\omega\varepsilon_0\,\varepsilon}\frac{\partial V}{\partial x}, \tag{4.5c}$$

where

$$\begin{aligned} K^2 &= k^2 + \frac{1}{k}\frac{d^2k}{dz^2} - \frac{2}{k^2}\left(\frac{dk}{dz}\right)^2 \\ &= k^2\left(1 - \frac{1}{k}\frac{d^2}{dz^2}\left(\frac{1}{k}\right)\right). \end{aligned} \tag{4.5d}$$

Note that when $k(z)$ is sufficiently large in the region of interest but varies slowly with z, then the various derivatives in (4.5d) can be neglected and the differential equation (4.5a) becomes identical with (4.2a).

If a plane wave originates in a homogeneous medium, say free space, and its direction of propagation makes an angle θ_0 with the z axis, then it will follow that its x dependence must be of the form $\exp(\pm ik_0 \sin\theta_0)$ everywhere in the homogeneous as well as in the adjoining inhomogeneous medium, since Snell's law

$$k_0 \sin\theta_0 = k(z)\sin\theta(z) \tag{4.6}$$

must be satisfied everywhere. Thus, we must seek solutions of the form

$$E_y = \psi_\perp(z)\exp(\pm ik_0 \sin\theta_0\, x) \tag{4.7}$$

and

$$H_y = k(z)\cdot\psi_\parallel(z)\exp(\pm ik_0 \sin\theta_0\, x). \tag{4.8}$$

By virtue of (4.2a) and (4.5a) it will follow that $\psi_\perp(z)$ and $\psi_\parallel(z)$ must satisfy the differential equations

$$\left[\frac{d^2}{dz^2} + k_0{}^2(n^2(z) - \sin^2\theta_0)\right]\psi_\perp = 0 \tag{4.9}$$

and

$$\left[\frac{d^2}{dz^2} + k_0{}^2(N^2(z) - \sin^2\theta_0)\right]\psi_\parallel = 0 \tag{4.10}$$

where

$$N^2(z) = n^2 + \left[\frac{1}{n^2}\frac{d^2n}{dz^2} - \frac{2}{n^2}\left(\frac{dn}{dz}\right)^2\right]\frac{1}{k_0{}^2}$$

$$= n^2\left[1 - \frac{1}{k_0{}^2 n}\frac{d^2}{dz^2}\left(\frac{1}{n}\right)\right]. \tag{4.11}$$

4.2 General Solution

We note that the central problem in the two cases considered is the solution to the differential equation

$$\frac{d^2\psi}{dz^2} + g^2(z)\psi = 0, \tag{4.12}$$

which we shall now investigate in detail.

Let us consider the most general second-order differential equation

$$\xi(1-\xi)\frac{d^2F}{d\xi^2} - [(\alpha+\beta+1)\xi - \gamma]\frac{dF}{d\xi} - \alpha\beta F = 0, \tag{4.13}$$

which is known as the hypergeometric or Gauss equation. A solution that is convergent for $|\xi| < 1$ is

$$F = 1 + \frac{\alpha\beta}{\gamma}\xi + \frac{\alpha\beta(\alpha+1)(\beta+1)}{1\cdot 2\gamma(\gamma+1)}\xi^2 + \frac{\alpha\beta(\alpha+1)(\beta+1)(\alpha+2)(\beta+2)}{1\cdot 2\cdot 3\gamma(\gamma+1)(\gamma+2)}\xi^3 + \cdots. \tag{4.14}$$

It can be shown that under the transformation

$$F = r(z)\psi(z), \qquad \xi = P(z), \tag{4.15}$$

where $P(z)$ is an arbitrary function of z and

$$r(z) = r_0\,\xi^{-\gamma/2}\cdot\left(\frac{d\xi}{dz}\right)^{1/2}\cdot(1-\xi)^{(\gamma-\alpha-\beta)/2}. \tag{4.16}$$

Equation (4.13) can be transformed into (4.12) where

$$g^2(z) = \left(\frac{1}{2}\frac{d^2}{dz^2}\left(\ln\frac{dP}{dz}\right) - \left\{\frac{1}{4}\left[\frac{d}{dz}\left(\ln\frac{dP}{dz}\right)\right]^2\right\}\right) - \left(\frac{d}{dz}\ln P\right)^2\left[K_1 + K_2\left(\frac{P}{1-P}\right) + K_3\frac{P}{(1-P)^2}\right] \tag{4.17}$$

and

$$\begin{aligned} 4K_1 &= \gamma(\gamma - 2), \\ 4K_2 &= 1 - (\alpha - \beta)^2 + \gamma(\gamma - 2), \\ 4K_3 &= (\alpha + \beta - \gamma)^2 - 1. \end{aligned} \tag{4.18}$$

Seemingly, our problem has been solved but this is rather deceiving. First, the hypergeometric functions are not tabulated to a sufficient extent. Second, what is usually known in the problem is the function $g(z)$ and in order to use the solution (4.14) we must express ξ in terms of $g(z)$ via Eqs. (4.15), (4.17), and (4.18), which may prove to be a formidable task. In practice the procedure to follow must be this. Start by assuming some kind of a functional dependence for $P(z)$; working back, we obtain $g(z)$, which may or may not coincide with the actual $g(z)$ profile for which we wanted to solve the problem.

The hypergeometric equation (4.13) can be transformed into the confluent hypergeometric equation by making the substitution $\xi = z/\beta$ and then letting β approach zero We obtain in this case

$$z\frac{d^2F}{dz^2} + (\gamma - z)\frac{dF}{dz} - \alpha F = 0, \tag{4.19a}$$

which under the transformation $F = \psi(z)z^{-r/2}e^{z/2}$ becomes

$$\left[\frac{d^2}{dz^2} + \left(-\frac{1}{4} + \frac{l}{z} + \frac{\frac{1}{4} - m^2}{z^2}\right)\right]\psi = 0 \tag{4.19b}$$

with $l = \alpha - \gamma/2$ and $m = (\gamma - 1)/2$. It will be noted that (4.19b) is of the same form as (4.12).

When $2m$ is not an integer, the two solutions of (4.19b) that are regular near zero and valid for all finite values of z are given by the series

$$M_{l,\pm m} = z^{1/2 \pm m}e^{-z/2}\left[1 + \frac{\frac{1}{2} \pm m - l}{1!(1 \pm 2m)}z + \frac{(\frac{1}{2} \pm m - l)(\frac{3}{2} \pm m - l)}{2!(1 \pm 2m)(2 \pm 2m)}z^2 + \cdots\right]. \tag{4.20}$$

These solutions are not, however, the most convenient to use because of the disappearance of one of them when $2m$ is an interger. A more convenient solution is the so-called second Whittaker function, defined by

$$W_{l,m}(z) = \frac{\pi}{\sin 2m\pi}\left[\frac{-M_{l,m}(z)}{\Gamma(\frac{1}{2} - m + l)\Gamma(1 + 2m)} + \frac{M_{l,-m}(z)}{\Gamma(\frac{1}{2} + m - l)\Gamma(1 - 2m)}\right] \tag{4.21}$$

for all values real or complex of l, m, and z. The function $W_{l,m}(z)$ can alternately be defined in terms of the contour integral

$$W_{l,m}(z) = \frac{1}{2\pi i}\Gamma(l + \frac{1}{2} - m)e^{-z/2}z^{l}\int_{C} t^{-l-1/2+m}\left(1 - \frac{t}{z}\right)^{l-1/2+m} e^{t}\, dt \tag{4.22}$$

where C is the contour shown in Fig. 4.1. The integral (4.22) can be transformed into a real integral whenever $\mathrm{Re}(m + \frac{1}{2} - l) \geq 0$ and $l - \frac{1}{2} - m$ is not

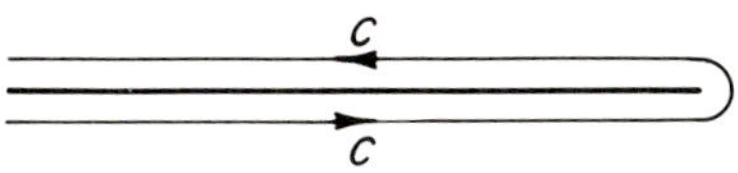

Fig. 4.1. The contour of integration for the second Whittaker function.

a negative integer to give (see Problem 4.9)

$$W_{l,m}(z) = \frac{e^{-z/2}z^{l}}{\Gamma(\frac{1}{2} - l + m)}\int_{0}^{\infty} x^{-l-1/2+m}e^{-x}\left(1 + \frac{x}{z}\right)^{l-1/2+m} dx. \tag{4.23}$$

The formulation above is very useful since it is quite general inasmuch as many special functions of mathematical physics are special cases of the Whittaker function $W_{l,m}(z)$ We note in particular

$$W_{0,m}(z) = \left(\frac{z}{\pi}\right)^{1/2} K_m\left(\frac{z}{2}\right), \tag{4.24a}$$

$$W_{0,m}(2iz) = \left(\frac{\pi z}{2}\right)^{1/2} \exp\left[\frac{-i\pi(2m+1)}{4}\right] H_m^{(2)}(z), \tag{4.24b}$$

$$W_{0,m}(-2iz) = \left(\frac{\pi z}{2}\right)^{1/2} \exp\left[\frac{i\pi(2m+1)}{4}\right] H_m^{(1)}(z), \tag{4.24c}$$

$$W_{0,1/2}(\mp 2z) = \exp(\pm z), \tag{4.24d}$$

$$W_{-1/4,\pm 1/4}(z^2) = 2\sqrt{z}\exp\left(\frac{z^2}{2}\right)\mathrm{erfc}(z), \tag{4.24e}$$

$$W_{\nu/2+1/4,\pm 1/4}(z^2) = 2^{-\nu/2}\left(\frac{z^2}{2}\right)^{1/4} D_\nu(z), \tag{4.24f}$$

$$W_{-1/2,0}(z) = -\exp\left(\frac{z}{2}\right)\sqrt{z}Ei(-z). \tag{4.24g}$$

The various functions listed in (4 24) are (a) modified Bessel function of the second kind; (b) Hankel function of the second kind; (c) Hankel function of the first kind; (d) exponential function; (e) complementary error function; (f) parabolic cylinder or Weber function; (g) exponential integral function. For completeness we also write the asymptotic expansion of $W_{l,m}(z)$:

$$W_{l,m}(z) \sim z^l e^{-z/2}$$
$$\times \left\{1 + \frac{m^2 - (l-\frac{1}{2})^2}{1!z} + \frac{[m^2 - (l-\frac{1}{2})^2][m^2 - (l-\frac{3}{2})^2]}{2!z^2} + \cdots\right\}, \quad (4.25a)$$

which terminates if $(m + l - \frac{1}{2})$ is an integer ≥ 0 or if $(m - l + \frac{1}{2})$ is an integer ≤ 0. Also,

$$W_{-l,m}(-z) \sim (-z)^{-l} e^{z/2}$$
$$\times \left\{1 - \frac{m^2 - (l+\frac{1}{2})^2}{1!z} + \frac{[m^2 - (l+\frac{1}{2})^2][m^2 - (l+\frac{3}{2})^2]}{2!z^2} + \cdots\right\}, \quad (4.25b)$$

which terminates if $(m + l + \frac{1}{2})$ is an integer ≤ 0 or $(m - l - \frac{1}{2})$ is an integer ≥ 0.

4.3 Specific Inhomogeneity Profiles

There exists a limited number of inhomogeneity profiles for which the governing differential equations can be solved exactly. In this section we shall consider some of these, particularly as they apply to the problem of an inhomogeneous transition region connecting two homogeneous half spaces.

4.3.1 Linear Increasing Transition Profile

Consider a transition profile of the electric permittivity that connects two homogeneous half spaces as shown in Fig. 4.2. A plane wave polarized normally to the plane of incidence and originating in the lower region is incident on the structure at an angle θ_0. The only component of the electric

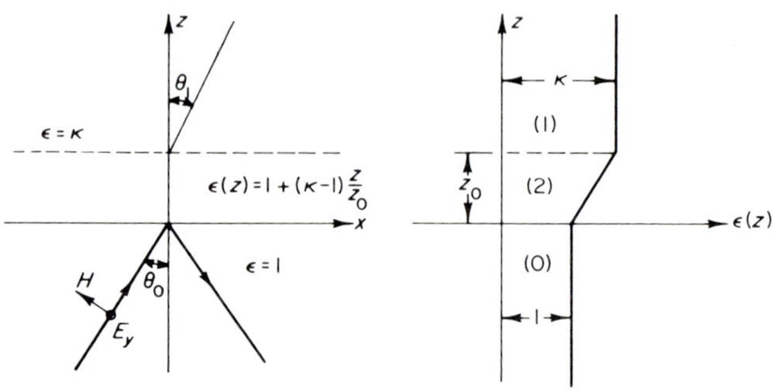

Fig. 4.2. Linear transition profile.

field is E_y and if we put

$$E_y = \psi_\perp(z) \exp(ik_0 \sin \theta_0 x), \tag{4.26}$$

then it will follow that for the lower homogeneous region we can write

$$\psi_0 = \exp(ik_0 \cos \theta_0 z) + R_\perp \exp(-ik_0 \cos \theta_0 z), \tag{4.27a}$$

$$H_{x0} = \frac{i}{\omega\mu_0} \cdot \frac{\partial \psi_0}{\partial z} \exp(ik_0 \sin \theta_0 x), \tag{4.27b}$$

while for the upper homogeneous region we choose

$$\psi_1 = T_\perp \exp[ik_1 \cos \theta_1(z - z_0)], \tag{4.28a}$$

$$H_{x1} = \frac{i}{\omega\mu_0} \frac{\partial \psi_1}{\partial z}. \tag{4.28b}$$

In the transition region $0 \leq z \leq z_0$ according to (4.9), the function $\psi_2(z)$ must satisfy

$$\left[\frac{d^2}{dz^2} + k_0{}^2(\cos^2 \theta_0 + az)\right]\psi_2 = 0, \tag{4.29a}$$

where

$$a = \frac{\kappa - 1}{z_0}. \tag{4.29b}$$

Under the transformation

$$z = \frac{1}{a}\left[\left(\frac{a}{k_0}\right)^{2/3} \eta - \cos^2 \theta_0\right] \tag{4.30a}$$

it will follow that ψ_2 will satisfy

$$\left(\frac{d^2}{d\eta^2} + \eta\right)\psi_2 = 0. \tag{4.30b}$$

Equation (4.30b) is in the form of a Bessel differential equation whose solution can be written in terms of a Bessel function of order $\frac{1}{3}$ or alternately in terms of Airy functions. For our purpose, it will be perhaps more convenient to adhere to the Bessel function and write

$$\psi_2 = w^{1/3}[AH_{1/3}^{(1)}(w) + BH_{1/3}^{(2)}(w)], \tag{4.31a}$$

where

$$w = \frac{2}{3}\eta^{3/2}$$

$$= \frac{2}{3}\left(\frac{k_0}{a}\right)(az + \cos^2 \theta_0)^{3/2} \tag{4.31b}$$

and the tangential magnetic field can be found from

$$H_{x2} = \frac{i}{\omega\mu_0}\frac{\partial\psi_2}{\partial z}. \tag{4.31c}$$

Applying the continuity of the tangential electric field at $z = 0$, z_0, we obtain

$$1 + R_\perp = w_0^{1/3}[AH_{1/3}^{(1)}(w_0) + BH_{1/3}^{(2)}(w_0)], \tag{4.32a}$$

$$T_\perp = w_1^{1/3}[AH_{1/3}^{(1)}(w_1) + BH_{1/3}^{(2)}(w_1)] \tag{4.32b}$$

where

$$w_0 = \frac{2}{3}\left(\frac{k_0}{a}\right)\cos^3\theta_0, \tag{4.32c}$$

$$w_1 = \frac{2}{3}\left(\frac{k_0}{a}\right)(az_0 + \cos^2\theta_0)^{3/2}. \tag{4.32d}$$

To apply the continuity of the tangential magnetic field, we must differentiate ψ_2 in (4.31a) with respect to z. Using the relation $d/dz(z^\nu\mathscr{C}_\nu(z)) = z^\nu\mathscr{C}_{\nu-1}(z)$ where $\mathscr{C}_\nu(z)$ is any Bessel function, we obtain

$$1 - R_\perp = -iw_0^{1/3}[AH_{-2/3}^{(1)}(w_0) + BH_{-2/3}^{(2)}(w_0)], \tag{4.33a}$$

$$T_\perp = -iw_1^{1/3}[AH_{-2/3}^{(1)}(w_1) + BH_{-2/3}^{(2)}(w_1)]. \tag{4.33b}$$

Equations (4.32) and (4.33) can be solved simultaneously for the reflection and transmission coefficients to give

$$R_\perp = \frac{F_+^{(1)}(w_0)F_+^{(2)}(w_1) - F_+^{(2)}(w_0)F_+^{(1)}(w_1)}{F_-^{(1)}(w_0)F_+^{(2)}(w_1) - F_-^{(2)}(w_0)F_+^{(1)}(w_1)} \tag{4.34a}$$

and

$$T_\perp = \frac{8}{\pi w_1}\left(\frac{w_1}{w_0}\right)^{1/3}[F_-^{(1)}(w_0)F_+^{(2)}(w_1) - F_-^{(2)}(w_0)F_+^{(1)}(w_1)]^{-1}, \tag{4.34b}$$

where

$$F_\pm^{(1,2)}(w) = H_{1/3}^{(1,2)}(w) \pm iH_{-2/3}^{(1,2)}(w). \tag{4.34c}$$

When z_0 is very small, so that the transition profile approaches a sharply bounded interface, then w_0 and w_1 are small and the form of the reflection and transmission coefficients can be simplified as follows.

$$R_\perp \sim \frac{J_{-2/3}(w_1)J_{-1/3}(w_0) - J_{-2/3}(w_0)J_{-1/3}(w_1)}{J_{-2/3}(w_1)J_{-1/3}(w_0) + J_{-2/3}(w_0)J_{-1/3}(w_1)}, \tag{4.35a}$$

$$T_\perp \sim \frac{2\sqrt{3}}{\pi w_1}\left(\frac{w_1}{w_0}\right)^{1/3}[J_{-2/3}(w_1)J_{-1/3}(w_0) + J_{-2/3}(w_0)J_{-1/3}(w_1)]^{-1}, \tag{4.35b}$$

and in the limit as $z_0 \to 0$, (4.35a) and (4.35b) take on the form of the reflection and transmission coefficients pertinent to a sharply bounded interface as given by (2.48a) and (2.48b), respectively.

Another important special case of the linear profile is that of an inhomogeneous half space. If we let $\kappa \to \infty$ and $z_0 \to \infty$ in such a way that the slope a remains constant, then $w_1 \sim \frac{2}{3} k_0 \sqrt{\kappa} z_0 \to \infty$. Assuming that κ is slightly lossy and $\mathrm{Im}(w_1) > 0$, the Hankel functions can be replaced by their asymptotic forms

$$H_\nu^{(1)}(w) \sim \left(\frac{2}{\pi w}\right)^{1/2} \exp\left[i\left(w - \frac{\nu\pi}{2} - \frac{\pi}{4}\right)\right]\left[1 + \frac{i(4\nu^2 - 1)}{8w} + \cdots\right],$$

$$-\pi < \arg w < 2\pi, \qquad (4.36a)$$

$$H_\nu^{(2)}(w) \sim \left(\frac{2}{\pi w}\right)^{1/2} \exp\left[-i\left(w - \frac{\nu\pi}{2} - \frac{\pi}{4}\right)\right]\left[1 - \frac{i(4\nu^2 - 1)}{8w} + \cdots\right],$$

$$-2\pi < \arg w < \pi, \qquad (4.36b)$$

and it will follow that $F_+^{(1)}(w_1) \sim (2/\pi w_1)^{1/2} \exp(iw_1) \to 0$, whereas $F_+^{(2)}(w_1) \sim (2/\pi w_1)^{1/2} \exp(-iw_1) \to \infty$. Hence, by (4.34a),

$$R_\perp \sim \frac{H_{1/3}^{(1)}(w_0) + iH_{-2/3}^{(1)}(w_0)}{H_{1/3}^{(1)}(w_0) - iH_{-2/3}^{(1)}(w_0)}. \qquad (4.37)$$

Two important subcases are of interest. If $w_0 \gg 1$; which will be realized when the angle of incidence is not too close to $\pi/2$, a slowly increasing electric permittivity (small a), or to a high frequency, or both; we can use the asymptotic forms of the Hankel functions (4.36) to show that

$$R_\perp \sim -ia/8k_0 \cos^3\theta_0, \qquad (4.38)$$

from which it follows that the reflection coefficient is proportional to the slope of the increasing permittivity profile, the wavelength, and the increasing angle of incidence.

In the case $w_0 \ll 1$, which can be realized when the angle of incidence is close to $\pi/2$, a rapidly increasing profile (large a) or low frequency, or both, we can use the small argument approximation for Hankel's functions. First, we note

$$H_\nu^{(1)}(w) = \frac{i}{\sin \nu\pi}\left[e^{-i\nu\pi} J_\nu(w) - J_{-\nu}(w)\right] \qquad (4.39a)$$

and

$$J_\nu(w) = \left(\frac{w}{2}\right)^\nu \sum_0^\infty \frac{(-1)^m}{m!\Gamma(\nu + m + 1)}\left(\frac{w}{2}\right)^{2m}, \qquad (4.39b)$$

which can be used to find the desired expansions of Hankel's functions. Substituting into (4.37) and retaining only the first two largest terms, we obtain

$$R_\perp \sim -\left[1 + (\sqrt{3} - i)\frac{\Gamma(\frac{1}{3})}{\Gamma(\frac{2}{3})}\left(\frac{k_0}{3a}\right)^{1/3} \cos\theta_0\right]. \tag{4.40}$$

Note that as the angle of incidence approaches $\pi/2$, the reflection coefficient approaches -1.

4.3.2 Linear Decreasing Transition Profile

Referring back to Fig. 4.2, when $\kappa < 1$, the slope of the inhomogeneous permittivity will be negative. The reflection and transmission coefficients in this case are formally identical to (4.34), but the arguments of the various Henkel functions are negative. Transition to positive arguments can be accomplished by using analytic continuation for Hankel's functions. First, let us define

$$a = -a', \qquad a' = \frac{1-\kappa}{z_0}, \tag{4.41a}$$

$$w_0 = -w_0', \qquad w_0' = \frac{2}{3}\left(\frac{k_0}{a'}\right)\cos^3\theta_0, \tag{4.41b}$$

$$w_1 = -w_1', \qquad w_1' = \frac{2}{3}\left(\frac{k_0}{a'}\right)(\kappa - \sin^2\theta_0)^{3/2}, \tag{4.41c}$$

where the primed quantities are now positive. Noting that

$$H_\nu^{(1)}(xe^{i\pi}) = -e^{i\nu\pi}H_\nu^{(2)}(x), \tag{4.42a}$$

$$H_\nu^{(2)}(xe^{i\pi}) = 2\cos\nu\pi H_\nu^{(2)}(x) + e^{i\nu\pi}H_\nu^{(1)}(x), \tag{4.42b}$$

we find

$$F_\pm^{(1)}(-w) = -\exp\left(-\frac{i\pi}{3}\right)F_\mp^{(2)}(w), \tag{4.43a}$$

$$F_\pm^{(2)}(-w) = F_\mp^{(2)}(w) + \exp\left(\frac{i\pi}{3}\right)F_\mp^{(1)}(w), \tag{4.43b}$$

where the various functions F are given in (4.34c). Using the foregoing results, we obtain for the reflection and transmission coefficients in this case

$$R_\perp = \frac{F_-^{(1)}(w_0')F_-^{(2)}(w_1') - F_-^{(2)}(w_0')F_-^{(1)}(w_1')}{F_+^{(1)}(w_0')F_-^{(2)}(w_1') - F_+^{(2)}(w_0')F_-^{(1)}(w_1')}, \tag{4.44a}$$

$$T_\perp = \frac{-8}{\pi w_1'}\left(\frac{w_1'}{w_0'}\right)^{1/3}[F_+^{(1)}(w_0')F_-^{(2)}(w_1') - F_+^{(2)}(w_0')F_-^{(1)}(w_1')]^{-1}. \tag{4.44b}$$

An important special subcase is one when $\kappa \to -\infty$, so that we are dealing with an inhomogeneous half space with a linearly decreasing electric permittivitiy. In this case, w_1' in (4.41c) must be written

$$w_1' = w_1'' \exp\left(\frac{i3\pi}{2}\right), \tag{4.45a}$$

$$w_1'' = \frac{2}{3}\left(\frac{k_0}{a'}\right)(\sin^2\theta - \kappa)^{3/2}, \tag{4.45b}$$

so that w_1'' is real. Next, we wish to expand asymptotically the various Hankel functions contained in $F(w_1')$. To this end we note, using (4.42),

$$H_\nu^{(1)}(w_1') = -e^{-i\nu\pi}H_\nu^{(2)}(iw_1''), \tag{4.46a}$$

$$H_\nu^{(2)}(w_1') = 2\cos\nu\pi H_\nu^{(2)}(iw_1'') + e^{i\nu\pi}H_\nu^{(1)}(iw_1''). \tag{4.46b}$$

The last term on the right in (4.46b) can be dropped since for $w_1'' \to \infty$ it is much smaller than the first. Thus, we find

$$F_-^{(1)}(w_1') = -\exp\left(-\frac{i\pi}{3}\right)F_-^{(2)}(w_1'), \tag{4.47a}$$

$$F_-^{(2)}(w_1') = H_{1/3}^{(2)}(iw_1'') + iH_{-2/3}^{(2)}(iw_1''), \tag{4.47b}$$

which upon substitution into (4.44a) gives

$$R_\perp = \frac{H_{1/3}^{(2)}(w_0') - iH_{-2/3}^{(2)}(w_0') + \exp(i\pi/3)[H_{1/3}^{(1)}(w_0') - iH_{-2/3}^{(1)}(w_0')]}{H_{1/3}^{(2)}(w_0') + iH_{-2/3}^{(2)}(w_0') + \exp(i\pi/3)[H_{1/3}^{(1)}(w_0') + iH_{-2/3}^{(1)}(w_0')]}. \tag{4.48}$$

Alternately, (4.48) can be expressed in terms of the Bessel functions:

$$R_\perp = \frac{[J_{1/3}(w_0') + J_{-1/3}(w_0')] + i[J_{2/3}(w_0') - J_{-2/3}(w_0')]}{[J_{1/3}(w_0') + J_{-1/3}(w_0')] - i[J_{2/3}(w_0') - J_{-2/3}(w_0')]}. \tag{4.49}$$

It is clear from (4.48), but even more so from (4.49), that in the case when w_0' is real (κ real), the magnitude of $R_\perp$ is unity and the reflection is total. The phase of $R_\perp$ in this case is found to be

$$R_\perp \sim -\exp\left[-i\left(2w_0' - \frac{\pi}{6}\right)\right], \qquad w_0' \gg 1, \tag{4.50a}$$

$$R_\perp \sim \exp\left[-i2\arctan\left(\frac{\Gamma(\frac{2}{3})}{\Gamma(\frac{1}{3})}\left(\frac{2}{w_0'}\right)^{1/3}\right)\right]; \qquad w_0' \ll 1. \tag{4.50b}$$

4.3.3 Exponential Half Space

Consider an exponential profile of the electric permittivity adjoining a homogeneous half space as shown in Fig. 4.3. In the upper half space the

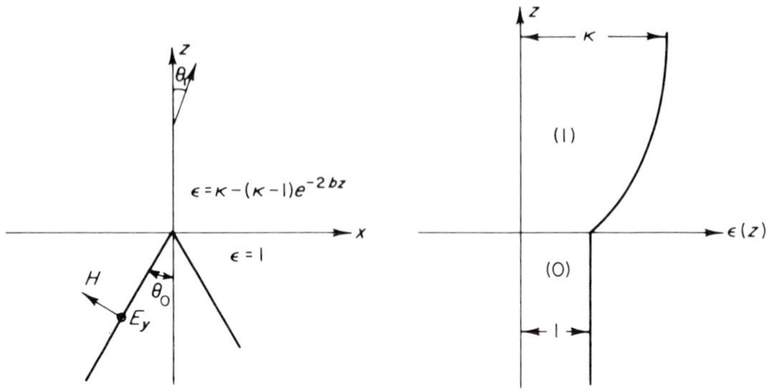

Fig. 4.3. Exponential half space.

electric permittivity is given by

$$\varepsilon(z) = \kappa - (\kappa - 1)e^{-2bz}, \qquad z > 0 \tag{4.51}$$

and the resulting wave equation becomes

$$\left\{\frac{d^2}{dz^2} - b^2[(ae^{-bz})^2 + (iv)^2]\right\}\psi_1(z) = 0, \tag{4.52}$$

where in this case

$$a = \frac{k_0(\kappa - 1)^{1/2}}{b}, \qquad v = \frac{k_1 \cos\theta_1}{b},$$
$$k_1 = k_0(\kappa)^{1/2}, \qquad \cos\theta_1 = (\kappa - \sin^2\theta_0)^{1/2}. \tag{4.53}$$

Equation (4.52) can be transformed to the standard form of the Bessel differential equation by putting $x = a\exp(-bz)$. The resulting solutions are in the form of modified Bessel functions of imaginary order, $I_{\pm iv}(a\exp(-bz))$. In choosing the proper solution, the determining factor is the behavior for large values of z, in which case the argument of the Bessel function approaches zero. Noting that $I_v(x) \to (x/2)^v/\Gamma(1 + v)$ as $x \to 0$, we must choose $I_{-iv}(\ )$ since

$$I_{-iv}(a\exp(-bz)) \xrightarrow[z\to\infty]{} \frac{(a/2)^{-iv}}{\Gamma(1 - iv)}\exp(ik_1\cos\theta_1 z), \tag{4.54}$$

which has the form of a wave travelling in the positive z direction. Thus, we write

$$\psi_1(z) = \Gamma(1 - i\nu) \exp\left(i\nu \ln \frac{a}{2}\right) I_{-i\nu}(a \exp(-bz)) T_\perp , \tag{4.55}$$

where the normalizing factor has been inserted in (4.40) so that $\psi_1(z)$ takes on the standard form for large z. The corresponding tangential magnetic field will be found from $H_{x1} = (i/\omega\mu_0)\, \partial\psi_1/\partial z$ and the fields in the lower homogeneous region are given by (4.27).

Applying the continuity of the tangential fields at $z = 0$, we obtain the following two equations.

$$1 + R_\perp = \Gamma(1 - i\nu) \exp\left(i\nu \ln \frac{a}{2}\right) I_{-i\nu}(a) T_\perp , \tag{4.56a}$$

$$1 - R_\perp = \frac{i(\kappa - 1)^{1/2}}{\cos\theta_0} \Gamma(1 - i\nu) \exp\left(i\nu \ln \frac{a}{2}\right) I'_{-i\nu}(a) T_\perp , \tag{4.56b}$$

which can be solved simultaneously to give

$$R_\perp = \frac{\cos\theta_0 I_{-i\nu}(a) - i(\kappa - 1)^{1/2} I'_{-i\nu}(a)}{\cos\theta_0 I_{-i\nu}(a) + i(\kappa - 1)^{1/2} I'_{-i\nu}(a)}, \tag{4.57a}$$

$$T_\perp = \frac{\exp(-i\nu \ln a/2)}{\Gamma(1 - i\nu)} \frac{2\cos\theta_0}{\cos\theta_0 I_{-i\nu}(a) + i(\kappa - 1)^{1/2} I'_{-i\nu}(a)} . \tag{4.57b}$$

If b is allowed to become very large, the configuration in Fig. 4.3 becomes equivalent to a sharply bounded interface. In this case $a \to 0$ and, using the small argument approximation for the Bessel functions, we find that (4.57) becomes identical with (2.48).

4.3.4 Other Profiles

Several additional profiles of the electric permittivity for which the wave equation can be solved exactly have been reported in the literature (Burman and Gould, 1963, 1964; Gould and Burman, 1964). They are

$$\begin{aligned}
&\varepsilon(z) = (a + bz^2,) \\
&\psi_\parallel(z) = [k_0(\varepsilon(z))^{1/2}]^{-1/2} W_{\pm p,\, 3/4}\left(\pm \frac{ik_0}{b} \varepsilon(z)\right), \\
&\psi_\perp(z) = [k_0(\varepsilon(z))^{1/2}]^{-1/2} W_{\pm p,\, 1/4}\left(\pm \frac{ik_0}{b} \varepsilon(z)\right), \\
&p = \frac{ik_0}{4} \sin^2\theta_0 ;
\end{aligned} \tag{4.58}$$

$$
\begin{aligned}
\varepsilon(z) &= [(a + bz)^2]^{-1}, \\
\psi_{\|,\perp} &= (\varepsilon(z))^{1/4}\mathscr{C}_\nu\left[\frac{ik_0 \sin\theta_0}{b(\varepsilon(z))^{1/2}}\right], \\
\nu^2 &= \frac{1}{4} - \left(\frac{k_0}{b}\right)^2;
\end{aligned}
\tag{4.59}
$$

$$
\begin{aligned}
\varepsilon(z) &= (a + bz)^{-1}, \\
\psi_{\|} &= W_{\pm p,0}\left[\pm 2k_0 \sin\theta_0\left(\frac{a}{b} + z\right)\right], \\
\psi_{\perp} &= W_{\pm p,1/2}\left[\pm 2k_0 \sin\theta_0\left(\frac{a}{b} + z\right)\right], \\
p &= \frac{k_0}{2b\sin\theta_0};
\end{aligned}
\tag{4.60}
$$

$$
\begin{aligned}
\varepsilon(z) &= c^2 \sec^2(a + bz), \\
\psi_{\|} &= \begin{matrix} P_\nu^{\ \mu} \\ Q_\nu^{\ \mu} \end{matrix}\{\tanh[i(a + bz)]\}, \\
\nu &= \tfrac{1}{2}[-1 \pm (1 - 4k_0^{\ 2}c^2/a^2)^{1/2}], \\
\mu &= \pm(1 - k_0^{\ 2}\sin^2\theta_0/b^2)^{1/2}, \\
\psi_{\perp} &= \begin{matrix} P_\nu^{\ \mu} \\ Q_\nu^{\ \mu} \end{matrix}\{\tanh[i(a + bz)]\}, \\
\nu &= \tfrac{1}{2}[-1 \pm (1 \quad 4k_0^{\ 2}c^2/a^2)^{1/2}]. \\
\mu &= \pm ik_0 \sin\theta_0/b;
\end{aligned}
\tag{4.61}
$$

$$
\begin{aligned}
\varepsilon(z) &= a^2 e^{2cz}, \\
\psi_{\|} &= \mathscr{C}_\nu\left(\frac{k_0 a}{c}\, e^{cz}\right), \\
\nu &= \left(1 + \frac{k_0^{\ 2}\sin^2\theta_0}{c^2}\right)^{1/2}
\end{aligned}
\tag{4.62}
$$

$$\varepsilon(z) = [(ae^{cz} + be^{-cz})^2]^{-1},$$

$$\psi_{\parallel} = \frac{P_m{}^n}{Q_m{}^n}\left(\frac{ae^{cz} - be^{-cz}}{ae^{cz} + be^{-cz}}\right),$$

$$n(n+1) = \frac{k_0{}^2}{4abc^2},$$

$$m = \left(1 + \frac{k_0{}^2 \sin^2\theta_0}{c^2}\right)^{1/2}, \tag{4.63}$$

$$\psi_{\perp} = \frac{P_m{}^n}{Q_m{}^n}\left(\frac{ae^{cz} - be^{-cz}}{ae^{cz} + be^{-cz}}\right),$$

$$n(n+1) = \frac{k_0{}^2}{4abc^2},$$

$$m = \frac{k_0 \sin\theta_0}{c}.$$

4.4 The WKB Approximation

The electric permittivity profile often cannot be described in terms of a known function for which the wave equation can be solved exactly. In such cases the only remaining possibility is an approximate solution, which can be realized by application of the WKB method, named after Wentzel, Kramers, and Brillouin, who about 1926 independently constructed an approximate solution to the pertinent differential equation.

4.4.1 Formal Development

As we have shown before, the central problem of finding the fields in a stratified medium is the solution to the second-order ordinary differential equation

$$\psi'' + g^2(z)\psi = 0. \tag{4.64}$$

Motivated by the fact that if $g(z)$ is a constant, the known solutions are of the form $A\exp(\pm igz)$, we attempt to find a solution to (4.64) in the form

$$\psi(z) = A(z)\exp(i\Phi(z)), \tag{4.65}$$

where $A(z)$ is the amplitude factor and $\Phi(z)$ is the phase. Substitution of (4.65) into (4.64) gives

$$A'' + 2i\Phi'A' + i\Phi''A + [g^2(z) - (\Phi')^2]A = 0. \tag{4.66}$$

Let L denote a distance over which the function $A(z)$ and $\Phi'(z)$ vary significantly. If $L \gg \lambda(z)$ where $\lambda(z)$ is the local wavelength, then it can be assumed that the medium is *slowly varying*. Let us now estimate the *order of smallness* of the various terms in (4.66). It will follow that

$$A' \sim \frac{A}{L}, \qquad A'' \sim \frac{A}{L^2}, \qquad \Phi'' \sim \frac{\Phi'}{L}, \tag{4.67}$$

and we argue that the first term in (4.66) will be much smaller than any other term and, consequently, can be dropped. The remaining three terms have two different orders of smallness, so they must be grouped accordingly and set to zero separately. Thus, we obtain

$$g^2(z) - (\Phi')^2 = 0, \tag{4.68a}$$

$$2\Phi'A' + \Phi''A = 0. \tag{4.68b}$$

From (4.68a) it will follow that $\Phi(z) = \pm\int_0^z g(z)\,dz$, whereas from (4.68b) $A = C[g(z)]^{-1/2}$ where C is a constant. Thus, by (4.65) the complete solution is

$$\psi(z) = (g(z))^{-1/2}\left[C_1 \exp\left(i\int_0^z g(z)\,dz\right) + C_2 \exp\left(-i\int_0^z g(z)\,dz\right)\right]. \tag{4.69}$$

Equation (4.69) is called the WKB (Wentzel–Kramers–Brillouin) solution or the geometrical optics approximation. The first term in (4.69) can be identified as an outgoing wave, whereas the second term represents an oncoming wave. The physical meaning of (4.69) is apparent if $g(z)$ is real. The exponent then represents a total phase change of a wave that traveled a distance z. A choice of reference level other than 0 does not alter the form of (4.69) since it produces a constant that can be lumped with C_1 and C_2. The meaning of the term $(g(z))^{-1/2}$ in (4.69) can be understood from the consideration of the power flow. Since the power flow is proportional to $P \sim \psi^*\,\partial\psi/\partial z$ and

$$\frac{d\psi}{dz} = \frac{C}{\sqrt{g}}\left[ig - \frac{1}{2g}\frac{dg}{dz}\right]\exp\left(\pm i\int_0^z g\,dz\right)$$

$$\sim iC\sqrt{g}\exp\left(\pm i\int_0^z g\,dz\right) \tag{4.70}$$

since $g(z)$ is slowly varying and $g' \sim 0$. Then it follows that the power densities in the outgoing and oncoming waves are constant, that is,

$$\begin{aligned} P\,(\text{outgoing}) &\sim |C_1|^2, \\ P\,(\text{oncoming}) &\sim |C_2|^2. \end{aligned} \tag{4.71}$$

In the case of perpendicular polarization $g^2(z) = k^2(z) - k_0{}^2 \sin^2\theta$ and since the Snell's law must hold everywhere in the medium, then $k(z)\sin\theta(z) = k_0 \sin\theta_0$ and consequently $g(z) = k(z)\cos\theta(z)$. Moreover, multiplying by the factor $\exp(ik_0 \sin\theta_0 x)$ the electric field becomes

$$E_y = \frac{\exp(ik_0 \sin\theta_0 x)}{[k(z)\cos\theta(z)]^{1/2}} [C_1 \exp(i\Phi(z)) + C_2 \exp(-i\Phi(z))], \quad (4.72a)$$

where

$$\Phi(z) = \int_0^z k(z)\cos\theta(z)\, dz. \quad (4.72b)$$

The associated magnetic field components H_x and H_z can be found by using (4.2) and (4.72a). Noting that since $k(z)$ is varying slowly, then $\partial/\partial z \sim \pm ik(z)\cos\theta(z)$, and we obtain

$$H_x = \frac{-[k(z)\cos\theta(z)]^{1/2}}{\omega\mu_0} [C_1 \exp(i\Phi(z)) - C_2 \exp(-i\Phi(z))] \times \exp(ik_0 \sin\theta_0 x), \quad (4.72c)$$

$$H_z = \frac{k_0 \sin\theta_0}{\omega\mu_0} E_y. \quad (4.72d)$$

In the case of the electric field vector in the plane of incidence we note that according to (4.5d), $K(z) \sim k(z)$ for a slowly varying medium and consequently $g(z) \sim k(z)\cos\theta(z)$, as before. Noting (4.8), we write

$$H_y = \left(\frac{k(z)}{\cos\theta(z)}\right)^{1/2} [D_1 \exp(i\Phi(z)) + D_2 \exp(-i\Phi(z))] \exp(ik_0 \sin\theta_0 x) \quad (4.73a)$$

with $\Phi(z)$ given by (4.72b). Moreover, using (4.5), we find that

$$E_x = \omega\mu_0 \left(\frac{\cos\theta(z)}{k(z)}\right)^{1/2} [D_1 \exp(i\Phi(z)) - D_2 \exp(-i\Phi(z))] \times \exp(ik_0 \sin\theta_0 x), \quad (4.73b)$$

$$E_z = -\frac{\omega\mu_0 \sin\theta(z)}{k(z)} H_y. \quad (4.73c)$$

4.4.2 Restrictions on Validity

The formulas we just derived give us no clue regarding their accuracy. It is possible to show (Bremmer, 1951) that the foregoing formulas are only the first approximation of the exact solution to the problem of propagation in inhomogeneous media and at the same time obtain an estimate of the error involved in retaining only the first term. In what follows we shall consider the case in which the electric field is parallel to the plane of incidence and the only component of the magetic field is H_y. From Maxwell's equations it follows that

$$E_x = -\frac{i\omega\mu_0}{k^2(z)}\frac{\partial H_y}{\partial z}, \qquad \frac{\partial E_x}{\partial z} - \frac{\partial E_z}{\partial x} = i\omega\mu_0 H_y,$$
$$E_z = \frac{i\omega\mu_0}{k^2(z)}\frac{\partial H_y}{\partial x}, \qquad \frac{\partial E_x}{\partial x} + \frac{\partial E_z}{\partial z} = E_z \frac{d}{dz}\left(\ln \frac{1}{k^2(z)}\right). \tag{4.74}$$

We shall now seek a solution to the foregoing differential equations in the form

$$H_y = \left(\frac{k(z)}{\cos\theta(z)}\right)^{1/2} [A_1(z)\exp(i\Phi(z)) + A_2(z)\exp(-i\Phi(z))],$$
$$\times \exp(ik_0 \sin\theta_0 x), \tag{4.75a}$$

$$E_x = \omega\mu_0\left(\frac{\cos\theta(z)}{(z)}\right)^{1/2} [A_1(z)\exp(i\Phi(z)) - A_2(z)\exp(-i\Phi(z))]$$
$$\times \exp(ik_0 \sin\theta_0 x), \tag{4.75b}$$

$$E_z = -\frac{\omega\mu_0 \sin\theta(z)}{[k(z)\cos\theta(z)]^{1/2}} [A_1(z)\exp(i\Phi(z)) + A_2(z)\exp(-i\Phi(z))]$$
$$\times \exp(ik_0 \sin\theta_0 x), \tag{4.75c}$$

where $\Phi(z)$ has the same meaning as before. We attempt to find the two functions $A_1(z)$ and $A_2(z)$ in such a way that the field components H_y, E_x, and E_z in (4.75) satisfy the differential equations (4.74). We obtain

$$A_1' e^{i\Phi} + A_2' e^{-i\Phi} = (A_1 e^{i\Phi} + A_2 e^{-i\Phi})\left[\frac{d}{dz}\ln\left(\frac{\cos\theta}{k}\right)^{1/2}\right], \tag{4.76a}$$

$$A_1' e^{i\Phi} - A_2' e^{-i\Phi} = -(A_1 e^{i\Phi} - A_2 e^{-i\Phi})\left[\frac{d}{dz}\ln\left(\frac{\cos\theta}{k}\right)^{1/2}\right]. \tag{4.76b}$$

These equations can be solved simultaneously to give

$$\frac{dA_1}{dz} = \frac{d}{dz}\left[\ln\left(\frac{\cos\theta}{k}\right)^{1/2}\right] A_2 e^{-i2\Phi}, \tag{4.77a}$$

$$\frac{dA_2}{dz} = \frac{d}{dz}\left[\ln\left(\frac{\cos\theta}{k}\right)^{1/2}\right] A_1 e^{i2\Phi}. \tag{4.77b}$$

For a slowly varying medium the right side of (4.77a) and (4.77b) will be small and the equations can be solved by the method of successive approximations. For the zeroth-order approximation we set the right side of (4.77a) and (4.77b) equal to zero and obtain $A_1^{(0)} = D_1$ and $A_2^{(0)} = D_2$ where D_1 and D_2 are constants. Substituting these results into (4.75), we obtain (4.73). It is clear, therefore, that the term containing $A_1^{(0)}$ describes a wave propagating in the positive z direction ("direct" wave), whereas the term containing $A_2^{(0)}$ describes a wave propagating in the negative z direction ("return" wave).

Higher-order approximations can be generated as follows.

$$\frac{dA_1}{dz} = \delta V_1(z) A_2(z), \tag{4.78a}$$

$$\frac{dA_2}{dz} = \delta V_2(z) A_1(z), \tag{4.78b}$$

where δ is a small parameter and $V_{1,2}(z) = \delta^{-1}\, d/dz[\ln(\cos\theta/k)^{1/2}]e^{\mp i2\Phi}$. Now we seek a solution in the form of convergent series as follows.

$$A_1 = A_1^{(0)} + \delta A_1^{(1)} + \delta^2 A_1^{(2)} + \cdots, \tag{4.79a}$$

$$A_2 = A_2^{(0)} + \delta A_2^{(1)} + \delta^2 A_2^{(2)} + \cdots. \tag{4.79b}$$

Putting the foregoing expansions into the differential equations (4.78) and comparing terms of like powers of δ, we secure

$$\frac{dA_1^{(m)}}{dz} = V_1 A_2^{(m-1)}, \tag{4.80a}$$

$$\frac{dA_2^{(m)}}{dz} = V_2 A_1^{(m-1)}, \tag{4.80b}$$

and note that for $m = 1$, $dA_1^{(1)}/dz = D_2 V_1(z)$ and $dA_2^{(1)}/dz = D_1 V_2(z)$.

These results can be interpreted as follows. As a result of the medium's inhomogeneity and continuous reflections, the zeroth-order direct wave produces a first-order return wave traveling in the opposite direction and the zeroth-order return wave produces a first-order direct wave. To obtain all of the contributions produced by the direct wave, we must integrate from $-\infty$ to some point z in the medium. Similarly, to obtain all of the contributions produced by the return wave, we must integrate from $+\infty$ to some point z in the medium. Thus,

$$A_1^{(1)}(z) = D_2 \int_{-\infty}^{z} V_1(\zeta)\, d\zeta, \tag{4.81a}$$

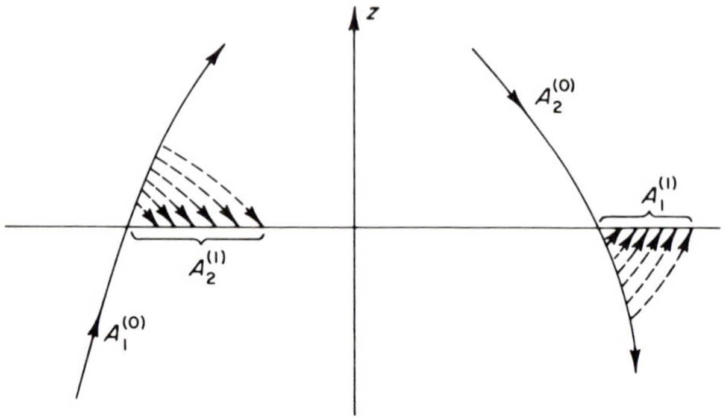

Fig. 4.4. The generation of secondary waves by primary waves.

$$A_2^{(1)}(z) = D_1 \int_{\infty}^{z} V_2(\zeta)\, d\zeta, \tag{4.81b}$$

and the situation is depicted graphically in Fig. 4.4. In a similar way we can argue that the higher-order waves of one kind are produced by the lower-order waves of the other kind, in agreement with the differential equations in (4.80). Thus, for the second-order terms we obtain

$$A_1^{(2)} = D_1 \int_{-\infty}^{z} d\zeta_1\, V_1(\zeta_1) \int_{\infty}^{\zeta_1} V_2(\zeta_2)\, d\zeta_2, \tag{4.82a}$$

$$A_2^{(2)} = D_2 \int_{\infty}^{z} d\zeta_1\, V_2(\zeta_1) \int_{-\infty}^{\zeta_1} V_1(\zeta_2)\, d\zeta_2. \tag{4.82b}$$

Higher-order terms can be generated by the continuation of this method. Figure 4.5 is a pictorial representation of this process.

The foregoing analysis makes it possible to assess the limits of validity of the WKB method of solution. Obviously, the WKB approximation will

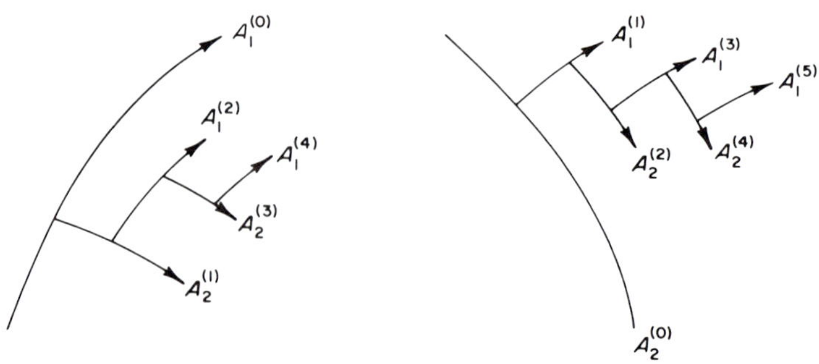

Fig. 4.5. Successive generation of higher-order waves in an inhomogeneous medium.

be valid if the amplitudes of the "secondary waves" $\delta A_1^{(1)}$ and $\delta A_2^{(1)}$ can be neglected in comparison with the amplitudes of the "primary waves" $A_1^{(0)}$ and $A_2^{(0)}$. Thus, we must have

$$\frac{\delta A_1^{(1)}}{A_1^{(0)}} = \frac{1}{2}\int_{-\infty}^{z} \left[\frac{d}{dz}\left(\ln \frac{\cos\theta}{k}\right)\right] \exp(-i2\Phi(z))\, dz \ll 1. \tag{4.83}$$

The maximum value of the integral in (4.83) may be estimated as follows. Since the exponential is a periodic function of z, we can obtain the maximum value of the integral by multiplying the factor $d/dz(\ln \cos\theta/k)$ by an interval over which the integrand does not change sign, i.e., the half period or $\Phi(z) = \pi/2$. Thus,

$$\frac{\delta A_1^{(1)}}{A_1^{(0)}} < \frac{1}{2}\frac{d}{dz}\left(\ln \frac{\cos\theta}{k}\right)L, \tag{4.84}$$

where L is the half period, which can be estimated as follows. For a slowly varying medium

$$\begin{aligned} \frac{\pi}{2} &= \int_0^z k(z)\cos\theta(z)\, dz \\ &\sim k\cos\theta \cdot L \end{aligned} \tag{4.85}$$

or $L \sim \pi/2k\cos\theta$. Using this result in (4.84) and taking account of (4.83), we state the criterion for the validity of the WKB approximation:

$$\frac{\lambda_0}{4\cos^2\theta}\frac{d}{dz}\left(\frac{\cos\theta}{n}\right) \ll 1, \tag{4.86}$$

where λ_0 is the free space wavelength.

The foregoing criterion may not be valid in the case of a periodically stratified medium, where the reflection from various layers may add in phase and produce a reflected wave that is of a magnitude comparable to that of the incident wave. This is the special case of the well-known Bragg reflection.

4.4.3 Expansion about a Turning Point

It is clear from (4.69) that the conventional WKB method fails in a region in which $g(z)$ goes through a zero, that is, if for some z, $k(z)$ assumes a value of $\sin\theta_0$. We must, in this case, develop an appropriate asymptotic expansion about the critical (turning) point, say $z = z_0$, $g(z_0) = 0$, and connect it on either side of $z = z_0$ with the previously found solutions.

Suppose that in the vicinity of the turning point the function $g(z)$ can be approximated by

$$g^2(z) \sim k_0{}^2 a(z_0 - z) = g_0{}^2(z); \tag{4.87}$$

then (4.12) could be written

$$\psi''(z) + k_0{}^2 a(z_0 - z)\psi(z) = 0. \tag{4.88}$$

Using the transformation $\eta = (k_0{}^2 a)^{1/3}(z_0 - z)$, we can recast (4.88) into the form of (4.30b). Hence, the most general solution of (4.88) is

$$\psi(z) = \left[\frac{\pi(z_0 - z)}{3}\right]^{1/2} \times \left\{AJ_{1/3}\left[\frac{2k_0\sqrt{a}}{3}(z_0 - z)^{3/2}\right] + BJ_{-1/3}\left[\frac{2k_0\sqrt{a}}{3}(z_0 - z)^{3/2}\right]\right\}. \tag{4.89}$$

where A and B are arbitrary constants and the factor $(\pi/3)^{1/2}$ has been included for future convenience. Since $J_\nu(x) \sim (x/2)^\nu/\Gamma(1 + \nu)$ as $x \to 0$, it will follow that the right side of (4.89) has no singularity at $z = z_0$ and the choice of branch is arbitrary. Let us fix $\arg(z_0 - z) = 0$ for $z < z_0$ and $\arg(z_0 - z) = \pi$ for $z > z_0$ and consequently denote

$$w_< = w, \qquad z < z_0, \tag{4.90a}$$

$$w_> = (iw)e^{i\pi}, \qquad z > z_0, \tag{4.90b}$$

$$w = \frac{2k_0\sqrt{a}}{3}|z_0 - z|^{3/2}. \tag{4.90c}$$

We note the form of the asymptotic expansion of the Bessel function $J_{\pm 1/3}(z)$ valid when $|\arg z| < \pi$:

$$J_{1/3}(w) = \left(\frac{2}{\pi w}\right)^{1/2} \cos\left(w - \frac{5\pi}{12}\right), \tag{4.91a}$$

$$w \to \infty, \ |\arg w| < \pi,$$

$$J_{-1/3}(w) = \left(\frac{2}{\pi w}\right)^{1/2} \cos\left(w - \frac{\pi}{12}\right), \tag{4.91b}$$

which is evidently applicable to (4.89) when $z < z_0$ and k_0 is large. We obtain in this case

$$\psi_<(z) = (g_0(z))^{-1/2}\left[A \cos\left(w - \frac{5\pi}{12}\right) + B \cos\left(w - \frac{\pi}{12}\right)\right]. \tag{4.92}$$

In the case $z > z_0$, the formulas (4.91) are not applicable, and we must use analytic continuation:

$$J_{1/3}(iwe^{i\pi}) = \exp\left(\frac{i\pi}{3}\right)J_{1/3}(iw), \tag{4.93a}$$

$$J_{-1/3}(iwe^{i\pi}) = \exp\left(-\frac{i\pi}{3}\right)J_{-1/3}(iw). \tag{4.93b}$$

Using (4.93) together with (4.91) and (4.89), we obtain

$$\psi_>(z) = \frac{1}{2}(|g_0(z)|)^{-1/2}$$
$$\times \left\{(B - A)\exp(w) + \left[B\exp\left(-\frac{i\pi}{6}\right) + A\exp\left(\frac{i\pi}{6}\right)\right]\exp(-w)\right\}. \tag{4.94}$$

The remaining task is the matching of the asymptotic forms (4.92) and (4.94) with the solution obtained by the WKB method:

$$\psi(z) = (g(z))^{-1/2}\left[C_1\exp\left(i\int_0^z g(z)\,dz\right) + C_2\exp\left(-i\int_0^z g(z)\,dz\right)\right], \qquad [4.69]$$

where $g(z)$ in (4.69) stands for $k_0(n^2(z) - \sin^2\theta_0)^{1/2}$.

In the case considered here the zero at $z = z_0$ involves an infinite value of (4.69) since $n(z) < \sin\theta_0$ as $z \to \infty$ was assumed. For this reason we must set $C_2 = 0$ for $z > z_0$, since otherwise the second term in (4.69) would increase exponentially in violation of physical conditions. Inasmuch as the WKB approximation is not capable of providing appropriate representation in the vicinity of $z = z_0 > 0$, this does not preclude the possibility that $C_2 \neq 0$ for $z < z_0$. In fact, it can be seen immediately from (4.12) that both fundamental solutions must be oscillatory when $g(z)$ is real.

What we encountered here is an example of *Stokes's phenomenon*, whose characteristic is a discontinuous change in the coefficients of a linear combination of two asymptotic series. The present situation requires the determination of the connecting equations for C_1 and C_2 for $z < z_0$ in such a way that only C_1 shall exist for $z > z_0$. It is clear from (4.94) that the latter condition will be ensured if we set $B = A$. Thus, we obtain

$$\psi_<(z) = A\left(\frac{3}{g_0(z)}\right)^{1/2}\cos\left(w - \frac{\pi}{4}\right), \qquad z < z_0, \tag{4.95a}$$

$$\psi_>(z) = \frac{A}{2}\left(\frac{3}{|g_0(z)|}\right)^{1/2}\exp(-w), \qquad z > z_0. \tag{4.95b}$$

Noting that

$$w = \int_z^{z_0} g_0(z)\,dz, \qquad z < z_0, \tag{4.96a}$$

$$w = \int_{z_0}^{z} |g_0(z)|\,dz, \qquad z > z_0, \tag{4.96b}$$

and that for $z \sim z_0$, $g_0(z) = g(z)$, we can write

$$\psi_<(z) = A\left(\frac{3}{g(z)}\right)^{1/2} \cos\left(\int_z^{z_0} g(z)\,dz - \frac{\pi}{4}\right), \qquad z < z_0, \tag{4.97a}$$

$$\psi_>(z) = \frac{A}{2}\left(\frac{3}{|g(z)|}\right)^{1/2} \exp\left(-\int_{z_0}^{z} |g(z)|\,dz\right), \qquad z > z_0. \tag{4.97b}$$

It is evident from (4.97) that the rising and descending waves contained in the cosine term of (4.97a) combine into the single exponentially damped wave given by (4.97b). Equation (4.97a) can be used to find the reflection coefficient from the decreasing index of the refraction profile. Noting that

$$\int_z^{z_0} g(z)\,dz = \int_0^{z_0} g(z)\,dz - \int_0^{z} g(z)\,dz, \tag{4.98}$$

we can write (4.97a) as

$$\begin{aligned}\psi_<(z) = \frac{A}{2}\left(\frac{3}{g(z)}\right)^{1/2} &\exp\left(-i\int_0^{z_0} g(z)\,dz + \frac{i\pi}{4}\right)\\ &\times\left\{\exp\left(i\int_0^{z} g(z)\,dz\right) + \left[-i\exp\left(i2\int_0^{z_0} g(z)\,dz\right)\right]\exp\left(-i\int_0^{z} g(z)\,dz\right)\right\}.\end{aligned} \tag{4.99}$$

Compare (4.99) with (4.69)! It is clear that the quantity in square brackets in (4.99) must be identified with the term C_2/C_1 in (4.69), which is the reflection coefficient at $z = 0$. Hence, we have

$$R = -i\exp\left[i2k_0\int_0^{z_0} (n^2(z) - \sin^2\theta_0)^{1/2}\,dz\right]. \tag{4.100}$$

Inasmuch as $|R| = 1$ in (4.100), it follows that there is total reflection at $z = z_0$. The presence of the factor $-i$ indicates a phase shift of $-\pi/2$ of the reflected field, while the factor 2 is due to the identical contributions from the rising and descending rays.

4.5 Geometrical Optics Approximation

Another approach to the problem of propagation in inhomogeneous media is that of geometrical optics. When the frequency of the electromagnetic wave is sufficiently high, then the finiteness of the wavelength may be neglected to the first order of approximation. In this case the propagation characteristics can be formulated in the language of geometry. The energy may be thought of as transported along certain curves, called "rays," and the field can be expected to have essentially the same character everywhere as a plane. This extends the validity of the laws of plane wave refraction and reflection to surfaces of finite curvatures.

4.5.1 The Eikonal Equation

We start with the time-harmonic homogeneous Maxwell equations in the form

$$\nabla \times \mathbf{E} = ik_0(Z_0\mathbf{H}), \tag{4.101a}$$

$$\nabla \times (Z_0\mathbf{H}) = -ik_0\,\varepsilon(\mathbf{r})\mathbf{E}, \tag{4.101b}$$

$$\nabla \cdot \varepsilon(\mathbf{r})\mathbf{E} = 0, \tag{4.101c}$$

$$\nabla \cdot \mathbf{H} = 0, \tag{4.101d}$$

where $Z_0 = (\mu_0/\varepsilon_0)^{1/2}$ is the intrinsic impedance of free space and $k_0 = 2\pi/\lambda_0$. Now we seek solutions of the form

$$\mathbf{E} = \mathbf{e}(\mathbf{r})\exp(ik_0\,\varphi(\mathbf{r})), \tag{4.102a}$$

$$\mathbf{H} = (Z_0)^{-1}\mathbf{h}(\mathbf{r})\exp(ik_0\,\varphi(\mathbf{r})), \tag{4.102b}$$

in the limit as $k \to \infty$ or $\lambda_0 \to 0$. We limit ourselves to lossless media and, hence, the function $\varphi(\mathbf{r})$ will be considered as a real function of position and $\mathbf{e}(\mathbf{r})$ and $\mathbf{h}(\mathbf{r})$ as vector functions of position that may be complex to account for other than a linearly polarized signal. Noting that $\nabla \exp(ik_0\,\varphi) = ik_0 \exp(ik_0\,\varphi)\,\nabla\varphi$ and using the vector identity

$$\begin{aligned} \nabla \times (a\mathbf{A}) &= \nabla a \times \mathbf{A} + a\nabla \times \mathbf{A}, \\ \nabla \cdot (a\mathbf{A}) &= \mathbf{A} \cdot \nabla a + a\nabla \cdot \mathbf{A}, \end{aligned} \tag{4.103}$$

we obtain from (4.101)

$$\nabla\varphi \times \mathbf{h} + \varepsilon\mathbf{e} = \frac{i}{k_0}\nabla \times \mathbf{h}, \tag{4.104a}$$

$$\nabla\varphi \times \mathbf{e} - \mathbf{h} = \frac{i}{k_0}\nabla \times \mathbf{e}, \tag{4.104b}$$

$$\mathbf{e} \cdot \nabla\varphi = \frac{i}{k_0}(\mathbf{e} \cdot \nabla \ln \varepsilon + \nabla \cdot \mathbf{e}), \tag{4.104c}$$

$$\mathbf{h} \cdot \nabla\varphi = \frac{i}{k_0}\nabla \cdot \mathbf{h}, \tag{4.104d}$$

which in the limit as $k_0 \to \infty$ give

$$\nabla\varphi \times \mathbf{h} = -\varepsilon\mathbf{e}, \tag{4.105a}$$

$$\nabla\varphi \times \mathbf{e} = \mathbf{h}, \tag{4.105b}$$

$$\mathbf{e} \cdot \nabla\varphi = 0, \tag{4.105c}$$

$$\mathbf{h} \cdot \nabla\varphi = 0. \tag{4.105d}$$

Note that the last two equations are not independent since they can be obtained from the preceding two by multiplication by $\nabla\varphi$. Substituting from (4.105b) into (4.105a) and noting the vector identity

$$\mathbf{A} \times (\mathbf{A} \times \mathbf{C}) = (\mathbf{A} \cdot \mathbf{C})\mathbf{A} - \mathbf{A}^2\mathbf{C}, \tag{4.106}$$

we obtain

$$(\nabla\varphi)^2 = n^2, \tag{4.107}$$

which is the *eikonal equation*, the basic equation of geometrical optics.

The geometrical optics approximation and, hence, the eikonal equation are justified provided that the magnitudes of the changes (derivatives) in **e** and **h** are small compared with magnitudes of **e** and **h** themselves over a distance of the order of wavelength λ_0. This condition is evidently violated at the boundaries of shadows, since across such boundaries the fields **e** and **h** will be expected to vary rapidly. In the vicinity of points where **e** and **h** attain very sharp maxima, e.g., caustic, the geometrical optics approximation is again not adequate.

A natural extension of the validity of the geometrical optics method to these critical regions would be solutions in terms of asymptotic series as follows.

$$\mathbf{E} = \exp(ik_0\varphi) \sum_0^\infty \frac{\mathbf{e}_m(\mathbf{r})}{(ik_0)^m}, \tag{4.108a}$$

$$\mathbf{H} = \frac{1}{Z_0} \exp(ik_0\varphi) \sum_0^\infty \frac{\mathbf{h}_m(\mathbf{r})}{(ik_0)^m}. \tag{4.108b}$$

In vacuum, $n = 1$ and the terms of the series can be shown to satisfy the differential equations

$$(\nabla\varphi)^2 = 1, \tag{4.109a}$$

$$2(\nabla\varphi \cdot \nabla)\mathbf{e}_m + (\nabla^2\varphi)\mathbf{e}_m + \nabla^2\mathbf{e}_{m-1} = 0. \tag{4.109b}$$

Alternately, it was proposed by Keller that the contribution of the higher-order terms could be studied by means of a model that is an extension of ordinary geometrical optics. Keller introduced in his theory the concept of a diffracted ray that obeys a generalized Fermat's principle. An appropriate field is associated with such a ray, and this field is assumed to satisfy the same propagation laws as the geometrical optics field.

4.5.2 Energy Propagation and Rays

Let w_e and w_m denote the electric and magnetic energy densities per unit volume, respectively. Then, for an electromagnetic wave, the time-average energy densities will be given by

$$\langle w_e \rangle = \frac{\varepsilon_0\,\varepsilon}{4}\,|\mathbf{E}|^2, \tag{4.110a}$$

$$\langle w_m \rangle = \frac{\mu_0}{4}\,|\mathbf{H}|^2, \tag{4.110b}$$

the factor $\frac{1}{4}$ arising from the fact that the time averaging is done over one period. The time-averaged Poynting vector is

$$\langle \mathbf{S} \rangle = \tfrac{1}{2}(\mathbf{E} \times \mathbf{H}^*). \tag{4.110c}$$

In terms of the geometrical optics field, using (4.102) and (4.105), we obtain equivalently for the energy densities

$$\langle w_e \rangle = \langle w_m \rangle = \frac{\varepsilon_0}{4}\,[\boldsymbol{\nabla}\varphi \cdot (\mathbf{e} \times \mathbf{h}^*)], \tag{4.111}$$

where the vector relation $\mathbf{a} \cdot (\mathbf{b} \times \mathbf{c}) = \mathbf{b} \cdot (\mathbf{c} \times \mathbf{a})$ was used. Equation (4.111) points out that *within the accuracy of the geometrical optics approximation the time-averaged electric and magnetic densities are equal.* Similarly, we find for the time-averaged Poynting vector

$$\langle \mathbf{S} \rangle = \frac{2c}{n^2}\,\langle w_e \rangle \boldsymbol{\nabla}\varphi, \tag{4.112}$$

where $c = (\mu_0\,\varepsilon_0)^{-1/2}$ is the velocity of light in vacuum. Alternately, if we denote by $\langle w \rangle = \langle w_e \rangle + \langle w_m \rangle = 2\langle w_e \rangle$ the total time-averaged energy density, then

$$\langle \mathbf{S} \rangle = \frac{c}{n^2}\,\langle w \rangle \boldsymbol{\nabla}\varphi. \tag{4.113}$$

Let $\mathbf{s}$ denote a unit vector; then from the eikonal equation it will follow that

$$\boldsymbol{\nabla}\varphi = n\mathbf{s}, \tag{4.114}$$

which upon substitution into (4.113) gives

$$\langle \mathbf{S} \rangle = v \langle w \rangle \mathbf{s}, \tag{4.115}$$

where $v = c/n$ is the velocity of light in the medium of the refractive index $n(\mathbf{r})$. It follows from (4.115) that the direction of the time-averaged Poynting vector is normal to the geometrical wave front $\varphi = \text{const.}$, while its magnitude is the product of the time-averaged energy density and the velocity of propagation in the medium.

Let $\mathbf{r}(\sigma)$ denote a position vector from some arbitrary origin 0 to a point on the ray and σ the length of arc on the ray as shown in Fig. 4.6. It will

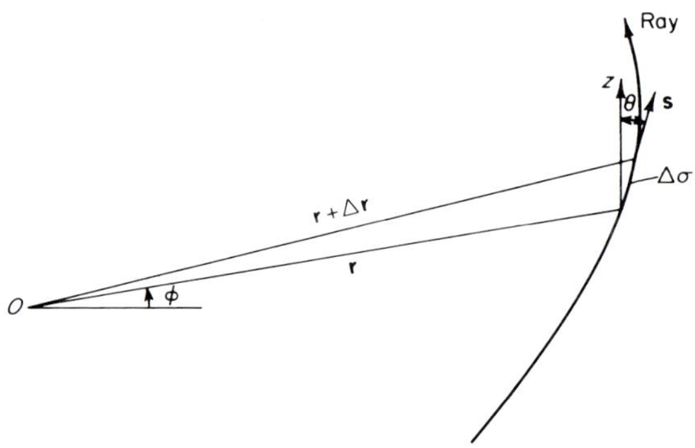

Fig. 4.6. Geometry of a ray and the position vector $\mathbf{r}(\sigma)$.

follow from the geometry shown that $(\Delta\sigma)\mathbf{s} = (\mathbf{r} + \Delta\mathbf{r}) - \mathbf{r} = \Delta\mathbf{r}$ and in the limit as $\Delta\sigma \to 0$, we obtain

$$\mathbf{s} = \lim_{\Delta\sigma \to 0} \frac{\Delta\mathbf{r}}{\Delta\sigma} = \frac{d\mathbf{r}}{d\sigma}, \tag{4.116}$$

which upon substitution into (4.114) gives

$$\nabla\varphi = n \frac{d\mathbf{r}}{d\sigma}, \tag{4.117}$$

or alternately

$$\frac{d\varphi}{d\sigma} = \frac{d\mathbf{r}}{d\sigma} \cdot \nabla\varphi = n. \tag{4.118}$$

Note that when the medium is stratified in the z direction, then $d\varphi = n(z) \cos \theta(z)\, dz$ and the phase change along the ray is identical with the corresponding expression found using the WKB method (note: $\Phi = k_0 \varphi$).

In general, integrating (4.118) between the points P_1 and P_2 on the ray, we obtain

$$\int_{P_1}^{P_2} n\, d\sigma = \varphi(P_2) - \varphi(P_1), \tag{4.119a}$$

$$\int_{P_1}^{P_2} n\, d\sigma = c[t(P_2) - t(P_1)], \tag{4.119b}$$

since $n\, d\sigma = (c/v)\, ds = c\, dt$. The expression on the right in (4.119a) denotes the optical length and the one on the right in (4.119b) denotes the total time by the energy (light) along the ray.

From the definition of the Poynting vector, it can be shown (problem) that within the geometrical optics approximation

$$\nabla \cdot \langle \mathbf{S} \rangle = 0, \tag{4.120}$$

which characterizes a solenoidal field, i.e., the flow lines are continuous and close upon themselves. Consequently, we can think of a tube of rays as shown in Fig. 4.7. We can integrate (4.120) over the volume of the tube and use Gauss's theorem to obtain

$$0 = \oiint \langle \mathbf{S} \rangle \cdot d\mathbf{A}. \tag{4.121}$$

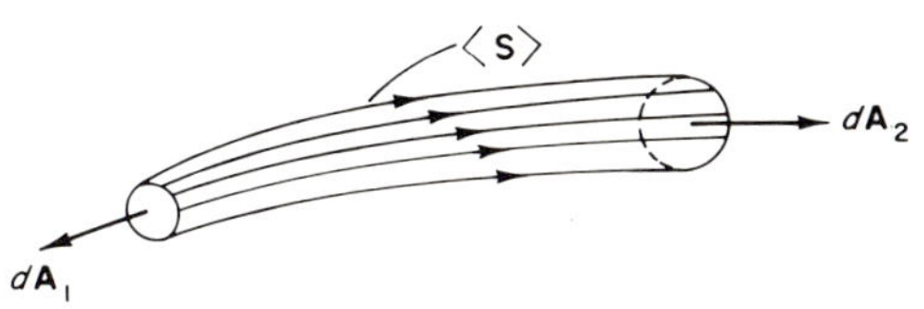

Fig. 4.7. Tube of rays.

Since on the cylindrical surface the two vectors are orthogonal there is no contribution therefrom, the only contribution arising from the ends. Thus, we obtain

$$|\langle \mathbf{S}_1 \rangle|\, dA_1 = |\langle \mathbf{S}_2 \rangle|\, dA_2, \tag{4.122}$$

which expresses the fact that the total amount of energy remains constant along a tube of rays.

If the medium is homogeneous, the rays are straight lines and if, in addition, all rays have a common origin, then it will follow from (4.122) that $|\langle \mathbf{S} \rangle| = \text{const.}/R^2$, where R is the distance measured from the common point of intersection of all rays to the surface considered.

4.5.3 Fermat's Principle

Euclidean geometry is based on the theorem of Pythagoras according to which the line element $d\sigma$ is related to its projections dx, dy, dz in the cartesian system as follows.

$$(d\sigma)^2 = (dx)^2 + (dy)^2 + (dz)^2. \tag{4.123}$$

The shortest line between two points is given by the stationary value of the variational integral

$$\sigma = \int_{P_1}^{P_2} d\sigma = \int_{z_1}^{z_2} (\dot{x}^2 + \dot{y}^2 + 1)^{1/2}\, dz \to \text{stationary}, \tag{4.124}$$

where z is considered to be the independent variable and $\dot{x} = dx/dz$ and $\dot{y} = dy/dz$. This fact motivated Fermat to postulate that the shortest optical path should be given by the integral

$$F = \int_{P_1}^{P_2} n(\mathbf{r})\, d\sigma = \int_{P_1}^{P_2} f(x, y, z; \dot{x}, \dot{y})\, dz \to \text{stationary}, \tag{4.125}$$

where $f(x, y, z; \dot{x}, \dot{y}) = n(x, y, z)(\dot{x} + \dot{y}^2 + 1)^{1/2}$. In reference to the foregoing variational problem we recall the Euler–Lagrange equations

$$\frac{\partial f}{\partial x_k} - \frac{d}{dz}\left(\frac{\partial f}{\partial \dot{x}_k}\right) = 0, \tag{4.126}$$

from which it follows immediately that

$$\frac{\partial n}{\partial x} = \frac{d}{d\sigma}\left(n \frac{dx}{d\sigma}\right), \tag{4.127a}$$

$$\frac{\partial n}{\partial y} = \frac{d}{d\sigma}\left(n \frac{dy}{d\sigma}\right), \tag{4.127b}$$

and it can be shown (problem) that the third equation has an identical form:

$$\frac{\partial n}{\partial z} = \frac{d}{d\sigma}\left(n \frac{dz}{d\sigma}\right), \tag{4.127c}$$

Collecting the results above, we obtain

$$\nabla n = \frac{d}{d\sigma}\left(n \frac{d\mathbf{r}}{d\sigma}\right), \tag{4.128}$$

which is the same as the differential equation for the rays found earlier in (4.118) when the latter is operated on by $\nabla\cdot$. Inasmuch as

$$F = \int_{P_1}^{P_2} n\, d\sigma = c \int_{t_1}^{t_2} dt \to \text{stationary}, \tag{4.129}$$

Fermat's principle is sometimes referred to as the "principle of least time."

4.5.4 Reflection and Refraction

In a homogeneous medium, n is a constant and the Fermat principle simply states

$$\delta \int_{P_1}^{P_2} dl = 0; \tag{4.130}$$

moreover, we know that in this case the rays are straight lines.

Consider the problem of reflection at a boundary separating two arbitrary homogeneous and isotropic media characterized by the refractive indices n_1 and n_2 as shown in Fig. 4.8. The possible rays between the points P_1 and P_2 are characterized in a way prescribed by Fermat's principle, which specifies that the rays shall be curves along which the optical path length is stationary with respect to an infinitesimal variation in the path. The plural "rays" is used to include the possibility that between P_1 and P_2 there may be a direct and a reflected ray.

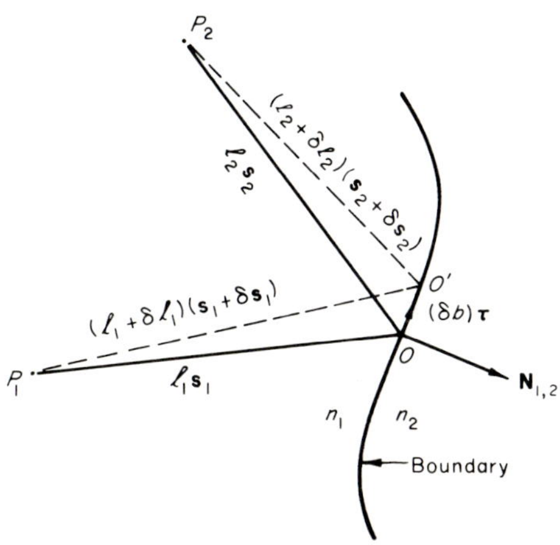

Fig. 4.8. Reflection at a boundary of two homogeneous isotropic media.

The difference (variation) between the paths P_1OP_2 and $P_1O'P_2$ to the first order in δ is

$$l_1(\delta\mathbf{s}_1) + (\delta l_1)\mathbf{s}_1 = (\delta b)\boldsymbol{\tau}, \tag{4.131a}$$

$$l_2(\delta\mathbf{s}_2) + (\delta l_2)\mathbf{s}_2 = -(\delta b)\boldsymbol{\tau}. \tag{4.131b}$$

Multiplying (4.131a) by $\mathbf{s}_1\cdot$ and (4.131b) by $\mathbf{s}_2\cdot$ and noting that $\mathbf{s}_{1,2}\cdot(\delta\mathbf{s}_{1,2}) = 0$ since $(\mathbf{s}_{1,2} + \delta\mathbf{s}_{1,2})$ are unit vectors, we obtain, upon adding the two equations,

$$\delta(l_1 + l_2) = (\mathbf{s}_1 - \mathbf{s}_2)\cdot\boldsymbol{\tau}(\delta b). \tag{4.132}$$

The left side of (4.132) is the total variation in the path δl, which, according to (4.130), must be zero. Moreover, since δb is arbitrary, we obtain

$$(\mathbf{s}_1 - \mathbf{s}_2) \cdot \boldsymbol{\tau} = 0 \tag{4.133}$$

for every $\boldsymbol{\tau}$. It follows that the plane defined by the unit vectors $\mathbf{s}_1$ and $\mathbf{s}_2$ is normal to the tangent plane, from which we conclude that the incident ray, reflected ray, and surface normal lie in the same plane. From (4.133) it also follows that $\cos(\mathbf{s}_1, \boldsymbol{\tau}) = \cos(\mathbf{s}_2, \boldsymbol{\tau})$, or alternately, that the reflected and incident angles are equal:

$$(\mathbf{s}_1, \boldsymbol{\tau}) = (\mathbf{s}_2, \boldsymbol{\tau}). \tag{4.134}$$

To derive the law of refraction, we consider the geometry of Fig. 4.9. To the first order the difference between the paths P_1OP_2 and $P_1O'P_2$ in this case is

$$n_1[l_1(\delta \mathbf{s}) + (\delta l_1)\mathbf{s}_1] = n_1(\delta b)\boldsymbol{\tau}, \tag{4.135a}$$

$$n_2[l_2(\delta \mathbf{s}_2) + (\delta l_2)\mathbf{s}_2] = n_2(\delta b)\boldsymbol{\tau}. \tag{4.135b}$$

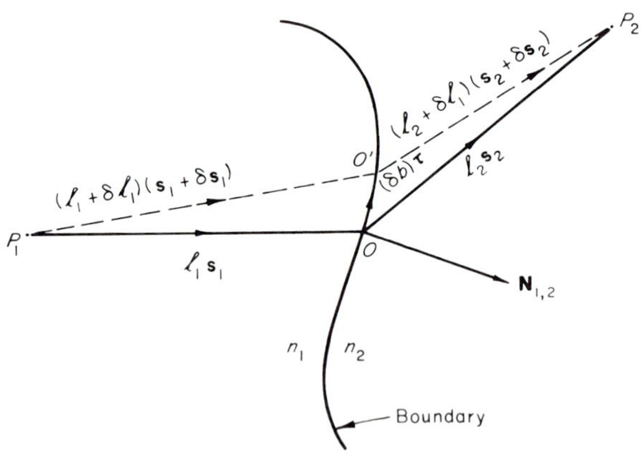

Fig. 4.9. Refraction at a boundary of two homogeneous isotropic media.

Multiplying (4.135a) by $s_1 \cdot$ and (4.135b) by $\mathbf{s}_2 \cdot$ and adding, we obtain

$$n_1(\delta l_1) + n_2(\delta l_2) = (n_1\mathbf{s}_1 - n_2\mathbf{s}_2) \cdot \boldsymbol{\tau}(\delta b). \tag{4.136}$$

Again, the right side of (4.136) is the total variation in the optical path between points P_1 and P_2; hence, according to (4.130), it must vanish. Moreover, since δb is arbitrary, we have

$$(n_1\mathbf{s}_1 - n_2\mathbf{s}_2) \cdot \boldsymbol{\tau} = 0 \tag{4.137}$$

for every $\boldsymbol{\tau}$. It follows from (4.137) that the incident ray, refracted ray, and surface normal lie in a plane and, moreover,

$$n_1 \cos(\mathbf{s}_1, \boldsymbol{\tau}) = n_2 \cos(\mathbf{s}_2, \boldsymbol{\tau}),$$

or alternately

$$n_1 \sin(\mathbf{N}_{1,2}, \mathbf{s}_1) = n_2 \sin(\mathbf{N}_{1,2}, \mathbf{s}_2), \tag{4.138}$$

which expresses the well-known Snell law of refraction.

Problems

4.1 Prove (4.37) and (4.48) by starting with half spaces of linearly increasing and decreasing refractive index, respectively, as shown in Fig. 4.10.

Fig. 4.10.

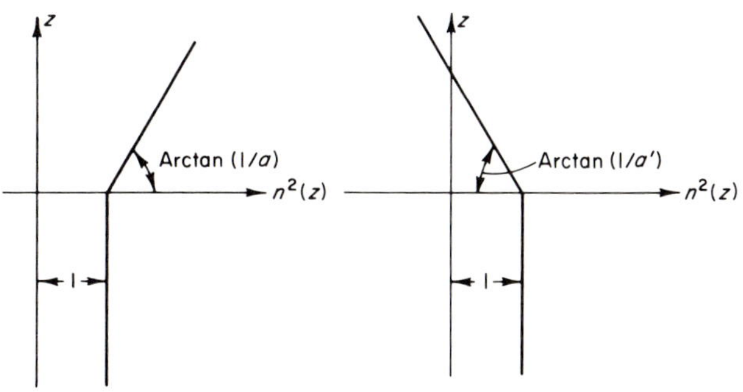

4.2 Using the transformation $x = a\exp(-bz)$, show that (4.52) is a variant of the Bessel differential equation.

4.3 Find the reflection and transmission coefficients for the exponential half space as shown in Fig. 4.11.

Fig. 4.11.

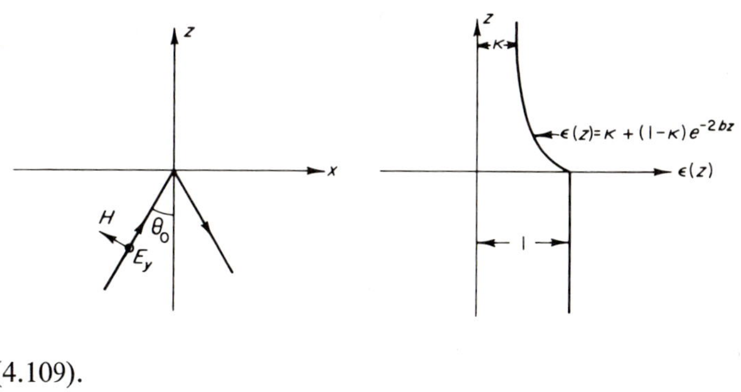

4.4 Derive (4.109).

4.5 Prove (4.120).

4.6 Starting with the differential equation for the rays (4.128), show that:
(a) in a homogeneous medium the rays are straight lines;
(b) in a spherically stratified medium $n(\mathbf{r}) = n(r)$, the ray equation is given by

$$\varphi = A \int^{r} \frac{dr}{r(n^2r^2 - A^2)^{1/2}};$$

(c) in a cylindrically stratified medium $n(\mathbf{r}) = n(\rho)$, $\partial n/\partial\varphi = 0$, the ray equation is given by

$$\varphi = A \int^{\rho} \frac{d\rho}{\rho(n^2\rho^2 - A^2)^{1/2}}.$$

4.7 (a) Derive (4.128) starting with (4.118).
(b) Prove (4.127c).

4.8 Show that the radius of curvature of a ray is given by

$$\rho = (\mathbf{v} \cdot \nabla \ln(n))^{-1}$$

where $\mathbf{v}$ is a unit vector normal to the ray and pointing into the region of increasing $n(\mathbf{r})$.

4.9 Starting with (4.22), prove (4.23). *Hint:* Note that above the branch cut we can set $t = xe^{i\pi}$ and below the branch cut $t = xe^{-i\pi}$. Moreover, in the vicinity of the branch point, set $t = \varepsilon e^{i\theta}$, $-\pi/2 < \theta < \pi/2$, and show that this contribution to the integral vanishes when $\varepsilon \to 0$ as long as $\mathrm{Re}(m + \frac{1}{2} - l) \geq 0$. Also recall the relation $\Gamma(\nu)\Gamma(1 - \nu) = \pi/\sin \pi\nu$. Why is the restriction, $l - \frac{1}{2} + m$ not a negative integer, necessary?

References

Born, M., and Wolf, E. (1964). "Principles of Optics," 2nd ed. Pergamon, Oxford.

Brekhovskikh, L. M. (1960). "Waves in Layered Media." Academic Press, New York.

Bremmer, H. (1951). The WKB approximation as the first term of a geometric optics series, *Comm. Pure Appl. Math.* **4**, 105–115.

Burman, R., and Gould, R. N. (1963). On the propagation of vertically polarized electromagnetic waves in a horizontally stratified medium, *J. Atm. Terr. Phys.* **25**, 543–549.

Burman, R., and Gould, R. N. (1964). On certain exact wave functions for electromagnetic fields in planar stratified media, *J. Atm. Terr. Phys.* **26**, 1127–1129.

Carrier, G. F., Krook, M., and Pearson, C. E. (1966). "Functions of a Complex Variable." McGraw-Hill, New York.

Gould, R. N., and Burman, R. (1964). Some electromagnetic wave functions for propagation in stratified media, *J. Atm. Terr. Phys.* **26**, 335–340.

Jones, D. S. (1964). "The Theory of Electromagnetism." Macmillan, New York.

Keller, J. B. (1953). The geometrical theory of diffraction, *Proc. Symp. Microwave Optics* (*McGill Univ.*); also, The American Mathematical Society (1958). "Calculus of Variations and Its Applications." McGraw-Hill, New York.

Kline, M., and Kay, I. W. (1965). "Electromagnetic Theory and Geometrical Optics." Wiley(Interscience), New York.

Slater, L. J. (1960). "Confluent Hypergeometric Functions." Cambridge Univ. Press, London and New York.

Wait, J. R. (1962). "Electromagnetic Waves in Stratified Media." Macmillan, New York.

Watson, G. N. (1962). "A Treatise on the Theory of Bessel Functions." Cambridge Univ. Press, London and New York.

Whittaker, E. T., and Watson, G. N. (1958). "A Course of Modern Analysis," 4th ed. Cambridge Univ. Press, London and New York.

CHAPTER 5

SPECTRAL REPRESENTATION OF ELEMENTARY SOURCES

Thus far, we have discussed the propagation, reflection, and refraction of plane electromagnetic waves in various isotropic and anisotropic, homogeneous and inhomogeneous media without inquiring how these waves may be produced. We shall devote this chapter to the subject of radiation from elementary isolated electromagnetic sources, and in due course it will become apparent that such sources may be represented as the superposition of plane waves studied in the preceding chapters.

5.1 Integration of the Wave Equation

In Chapter 1 we defined the Hertzian vectors $\mathbf{\Pi}^{(e)}$ and $\mathbf{\Pi}^{(m)}$ of the electric and magnetic type, respectively. The sources of these vector potentials are the electric and magnetic dipole distributions, which can be expressed in terms of electric and magnetic currents as follows.

$$\mathbf{P}_0 = \frac{i}{\omega}\,\mathbf{J}^{(e)}, \tag{5.1a}$$

$$\mathbf{M}_0 = \frac{i}{\omega\mu_0}\,\mathbf{J}^{(m)}; \tag{5.1b}$$

and the Hertzian vectors satisfy

$$(\nabla^2 + k^2)\mathbf{\Pi}^{(e)} = -\frac{i\mu}{\omega\varepsilon_0 n^2}\,\mathbf{J}^{(e)}, \tag{5.2a}$$

$$\mathbf{H} = -\frac{i\omega\varepsilon_0 n^2}{\mu}\,\nabla \times \mathbf{\Pi}^{(e)}, \tag{5.2b}$$

$$\mathbf{E} = k^2\mathbf{\Pi}^{(e)} + \nabla(\nabla \cdot \mathbf{\Pi}^{(e)}), \tag{5.2c}$$

and

$$(\nabla^2 + k^2)\mathbf{\Pi}^{(m)} = -\frac{i}{\omega\mu_0\,\mu}\,\mathbf{J}^{(m)}, \tag{5.3a}$$

$$\mathbf{E} = i\omega\mu_0\,\mu\nabla \times \mathbf{\Pi}^{(m)}, \tag{5.3b}$$

$$\mathbf{H} = k^2\mathbf{\Pi}^{(m)} + \nabla(\nabla \cdot \mathbf{\Pi}^{(m)}). \tag{5.3c}$$

From (5.2a) and (5.3a) it is clear that each cartesian component of $\mathbf{\Pi}^{(e)}$ and $\mathbf{\Pi}^{(m)}$ satisfies

$$(\nabla^2 + k^2)\Pi = -\frac{i}{Q}\,J, \tag{5.4}$$

where $Q = \omega\varepsilon_0\,n^2/\mu$ in the electric case and $Q = \omega\mu_0\,\mu$ in the magnetic case. Suppose that a scalar function $G(\mathbf{r}, \mathbf{r}')$ can be found that satisfies the same boundary conditions as π and the differential equation

$$(\nabla^2 + k^2)G(\mathbf{r}, \mathbf{r}') = -\delta(\mathbf{r} - \mathbf{r}'), \tag{5.5}$$

where $\mathbf{r}'$ denotes the source coordinates and $\mathbf{r}$ denotes the observation coordinates. Multiplying (5.4) by G and (5.5) by Π and subtracting one from the other, we obtain

$$\Pi\,\nabla^2 G - G\,\nabla^2\Pi = -\Pi\,\delta(\mathbf{r} - \mathbf{r}') + \frac{i}{Q}\,JG. \tag{5.6}$$

We integrate the entire expression over the volume $V(\mathbf{r})$, noting that by Green's theorem the left-hand side can be written

$$\int_V (\Pi\,\nabla^2 G - G\,\nabla^2\Pi)\,dv = \int_S \left(\Pi\,\frac{\partial G}{\partial n} - G\,\frac{\partial \Pi}{\partial n}\right)\,ds, \tag{5.7}$$

where $\partial/\partial n$ is the derivative in the direction of the positive normal and S is the surface enclosing the volume V. Suppose that $\Pi(S)$ and $G(S)$ satisfied one of the homogeneous boundary conditions given by

$$\Pi(S) = 0, \tag{5.8a}$$

$$\frac{\partial \Pi(S)}{\partial n} = 0, \tag{5.8b}$$

$$a\Pi(S) + b\,\frac{\partial \Pi(S)}{\partial n} = 0, \tag{5.8c}$$

where a and b are constants. Since G satisfies the same boundary conditions as $\mathbf{\Pi}$, it follows that the surface integral in (5.7) vanishes. Thus, from (5.6) we have

$$\Pi(\mathbf{r}') = \frac{i}{Q} \int_V J(\mathbf{r}) G(\mathbf{r}, \mathbf{r}')\, dv \tag{5.9}$$

and since $G(\mathbf{r}, \mathbf{r}') = G(\mathbf{r}', \mathbf{r})$, we obtain

$$\Pi(\mathbf{r}) = \frac{i}{Q} \int_{V'} J(\mathbf{r}') G(\mathbf{r}, \mathbf{r}')\, dv'. \tag{5.10}$$

We can generalize the result above by writing

$$\Pi_\nu(\mathbf{r}) = \frac{i}{Q} \int_{V'} J_\nu(\mathbf{r}') G_\nu(\mathbf{r}, \mathbf{r}')\, dv' \tag{5.11}$$

with the understanding that the result above is valid for a cartesian component of the vector $\mathbf{\Pi}$.

Evidently, the problem is solved once we know the function $G(\mathbf{r}, \mathbf{r}')$, which, incidentally, is known as *Green's function.* We now concentrate on finding such a function that is a solution to (5.5) and, moreover, satisfies the prescribed boundary conditions (5.8) imposed on a cartesian component of the vector $\mathbf{\Pi}$.

5.1.1 The Green Function for Spherical Waves in Free Space

Since the region of interest is unbounded we can use a triple Fourier transform defined by

$$\tilde{\tilde{\tilde{F}}}(\boldsymbol{\alpha}) = (2\pi)^{-3/2} \iiint_{-\infty}^{\infty} F(\mathbf{r}) \exp(-i\boldsymbol{\alpha} \cdot \mathbf{r})\, d\mathbf{r} \tag{5.12a}$$

and its inverse

$$F(\mathbf{r}) = (2\pi)^{-3/2} \iiint_{-\infty}^{\infty} \tilde{\tilde{\tilde{F}}}(\boldsymbol{\alpha}) \exp(i\boldsymbol{\alpha} \cdot \mathbf{r})\, d\boldsymbol{\alpha}, \tag{5.12b}$$

where

$$\boldsymbol{\alpha} = \mathbf{a}_1 \alpha_1 + \mathbf{a}_2 \alpha_2 + \mathbf{a}_3 \alpha_3, \tag{5.12c}$$

$$d\mathbf{r} = dx\, dy\, dz, \tag{5.12d}$$

$$d\alpha = d\alpha_1\, d\alpha_2\, d\alpha_3. \tag{5.12e}$$

We can transform (5.5) in the transform space by multiplying by $(2\pi)^{-3/2}$ $\exp(-i\boldsymbol{\alpha} \cdot \mathbf{r})\, d\mathbf{r}$ and integrating from $-\infty$ to $+\infty$. The right side becomes

$-(2\pi)^{-3/2}\exp(-i\boldsymbol{\alpha}\cdot\mathbf{r}')$ while the left side must be integrated by parts. It is easy to show that in the case

$$\left[\exp(-i\alpha_\nu x_\nu)\left(\frac{\partial G}{\partial x_\nu}+i\alpha_\nu G\right)\right]_{-\infty}^{\infty}=0 \tag{5.13}$$

the transformed equation (5.5) can be written

$$\approx{G}(\boldsymbol{\alpha},\mathbf{r}')=(2\pi)^{-3/2}\frac{\exp(-i\boldsymbol{\alpha}\cdot\mathbf{r}')}{\alpha^2-k^2}. \tag{5.14}$$

We can now invert (5.14) by means of the inversion formula (5.12b) to obtain

$$G(\mathbf{r},\mathbf{r}')=(2\pi)^{-3}\iiint_{-\infty}^{\infty}\frac{\exp[i\boldsymbol{\alpha}\cdot(\mathbf{r}-\mathbf{r}')]}{\alpha^2-k^2}\,d\boldsymbol{\alpha}. \tag{5.15}$$

To evaluate (5.15) we first put $\mathbf{R}=\mathbf{r}-\mathbf{r}'$ where the new vector $\mathbf{R}$ joins the "source point" with the "field point" as shown in Fig. 5.1. Now we can write

$$G(\mathbf{r},\mathbf{r}')=(2\pi)^{-3}\iiint_{-\infty}^{\infty}\frac{\exp(i\alpha R\cos\theta)}{\alpha^2-k^2}\,d\boldsymbol{\alpha}, \tag{5.16}$$

where θ is the angle between the vectors $\boldsymbol{\alpha}$ and $\mathbf{R}$. Inasmuch as the relative orientation of the transform space to the configuration space is arbitrary,

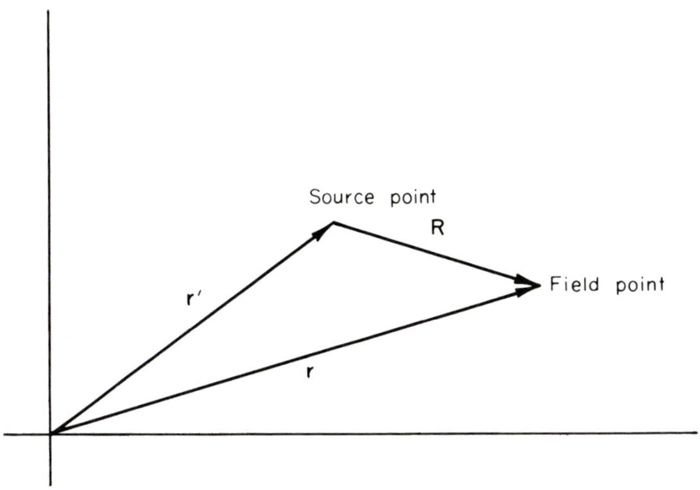

Fig. 5.1. The geometry of the "source point" and "field point" and the associated vectors.

we can rotate one system or the other in such a way as to superpose the vector **R** on one of the axes of the configuration space, say α_3, as shown in Fig. 5.2.

The components of the vector $\boldsymbol{\alpha}$ can then be expressed as

$$\begin{aligned} \alpha_1 &= \alpha \sin\theta \cos\varphi, \\ \alpha_2 &= \alpha \sin\theta \sin\varphi, \\ \alpha_3 &= \alpha \cos\theta, \end{aligned} \tag{5.17}$$

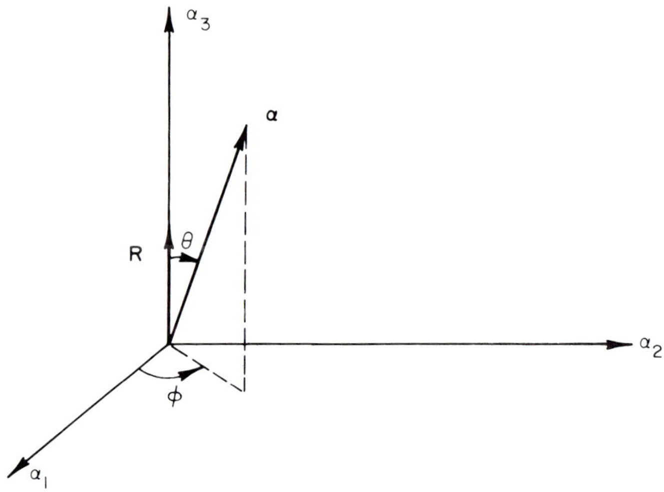

Fig. 5.2. The geometry of the vector **R** and the transform space.

while the differential volume element $d\boldsymbol{\alpha} = \alpha^2 \sin\theta \, d\alpha \, d\theta \, d\varphi$. The integral of (5.16) now becomes

$$\begin{aligned} G(\mathbf{r}, \mathbf{r}') &= (2\pi)^{-3} \int_{\alpha=0}^{\infty} \int_{\theta=0}^{\pi} \int_{\varphi=0}^{2\pi} \frac{e^{i\alpha R \cos\theta}}{\alpha^2 - k^2} \alpha^2 \, d\alpha \sin\theta \, d\theta \, d\varphi \\ &= (4\pi^2 R)^{-1} \int_{-\infty}^{\infty} \frac{\sin(\alpha R)}{\alpha^2 - k^2} \alpha \, d\alpha \\ &= -\frac{i}{8\pi^2 R} \left[\int_{-\infty}^{\infty} \frac{e^{i\beta} \beta \, d\beta}{(\beta + \gamma)(\beta - \gamma)} - \int_{-\infty}^{\infty} \frac{e^{-i\beta} \beta \, d\beta}{(\beta + \gamma)(\beta - \gamma)} \right], \end{aligned} \tag{5.18}$$

where $\beta = \alpha R$ and $\gamma = kR$.

The last two integrals in (5.18) can be evaluated by allowing β to be complex and closing the contour in the upper half plane for the first integral and in the lower half plane for the second integral, as shown in Fig. 5.3. Then the integrand has two poles in the complex β plane at the points $\beta = \pm\gamma$. If the conductivity of the medium is nonzero, then $\gamma = \gamma' + i\gamma''$ and the poles are found in the first and third quadrant, as shown in Fig. 5.3. When the conductivity vanishes, $\gamma'' = 0$, then the poles lie on the axis of the reals and the contour C_1 must be indented above $\beta = -\gamma$ and below $\beta = \gamma$.

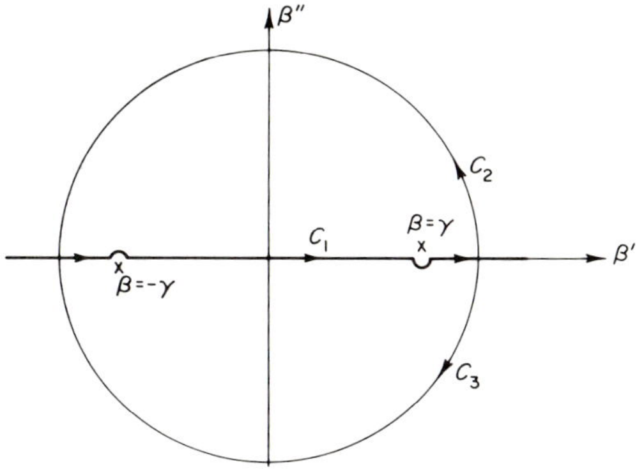

Fig. 5.3. The complex β plane.

It is easy to show that in the limit as the radius of the semicircles becomes infinite, the contribution therefrom to the integrals becomes zero. The contour integrals along $C_1 + C_2$ and $C_1 + C_3$ assume, respectively, the form of the first and second integrals in (5.18). Since each one of the integrals is equal to the residue at the corresponding pole, we obtain

$$G(\mathbf{r}, \mathbf{r}') = \frac{e^{ikR}}{4\pi R}. \tag{5.19}$$

The representation (5.19) is a correct solution to (5.5) providing that the condition (5.13) is fulfilled. Now we find

$$\exp(-i\alpha_1 x)\left(\frac{\partial G}{\partial x} + i\alpha_1 G\right)\Bigg|_{-\infty}^{\infty}$$
$$= \exp(-i\alpha_1 x)\left[\frac{x}{R}\left(ik - \frac{1}{R}\right) + i\alpha_1\right]\frac{\exp(ikR)}{R}\Bigg|_{x=-\infty}^{x=\infty} = 0 \tag{5.20}$$

for all real α_1. Since the same can readily be proven for the two other coordinates as well, it follows that (5.19) is the correct solution to (5.5) and satisfies the homogeneous boundary conditions (5.8). We note that the rather complex condition (5.13) can be satisfied by demanding simply that $G \to 0$ as $R \to \infty$.

The representation (5.19) is not a useful one for the solution of certain boundary value problems involving more than one region. We shall now derive several other representations that will prove useful from time to time. Let us make a change of variable in the integral of (5.15) as follows.

$$\tau = [k^2 - (\alpha_1{}^2 + \alpha_2{}^2)]^{1/2}; \tag{5.21a}$$

then we can write

$$G(\mathbf{r}, \mathbf{r}') = (2\pi)^{-3} \iint_{-\infty}^{\infty} d\alpha_1\, d\alpha_2 \exp\{i[\alpha_1(x - x') + \alpha_2(y - y')]\} \times \int_{-\infty}^{\infty} \frac{\exp[i\alpha_3(z - z')]}{(\alpha_3 + \tau)(\alpha_3 - \tau)}\, d\alpha_3. \tag{5.21b}$$

The last integral on the right in (5.21b) can be readily evaluated by assuming α_3 to be complex and integrating about a contour similar to that in Fig. 5.3, where the contribution along C_2 vanishes for $z > z'$ and the contribution along C_3 vanishes for $z < z'$. Thus, we obtain

$$\int_{-\infty}^{\infty} \frac{\exp[i\alpha_3(z - z')]}{\alpha_3{}^2 - \tau^2}\, d\alpha_3 = \pi i \frac{\exp(i|z - z'|\tau)}{\tau}, \tag{5.22a}$$

which when substituted into (5.21b) gives

$$G(\mathbf{r}, \mathbf{r}') = \frac{i}{8\pi^2} \iint_{-\infty}^{\infty} \frac{\exp\{i[\alpha_1(x - x') + \alpha_2(y - y') + \tau|z - z'|]\}}{\tau}\, d\alpha_1\, d\alpha_2. \tag{5.22b}$$

Now if we make the substitutions

$$\begin{aligned} \alpha_1 &= \lambda \cos\beta, \qquad & x - x' &= \rho\cos\varphi, \\ \alpha_2 &= \lambda \sin\beta, \qquad & y - y' &= \rho\sin\varphi, \end{aligned} \tag{5.23a}$$

we can put (5.22b) in the form

$$G(\mathbf{r}, \mathbf{r}') = \frac{i}{8\pi^2} \int_0^{\infty} d\lambda \frac{\lambda \exp[i(k^2 - \lambda^2)^{1/2}|z - z'|]}{(k^2 - \lambda^2)^{1/2}} \times \int_{-\pi}^{\pi} \exp[i\lambda\rho\cos(\beta - \varphi)]\, d\beta. \tag{5.23b}$$

The last integral on the right can be identified as an integral representation of the Bessel function. Noting the formula

$$J_n(z) = \frac{1}{2\pi} \int_{-\pi}^{\pi} \exp\left[iz\cos(\beta - \varphi) + in\left(\beta - \varphi - \frac{\pi}{2}\right)\right] d\beta, \tag{5.24}$$

where φ is an arbitrary angle, we obtain

$$G(\mathbf{r}, \mathbf{r}') = \frac{i}{4\pi} \int_0^{\infty} \frac{J_0(\lambda\rho) \exp[i(k^2 - \lambda^2)^{1/2}|z - z'|]}{(k^2 - \lambda^2)^{1/2}}\, \lambda\, d\lambda. \tag{5.25}$$

The integral in (5.25) can be recast into one running from $-\infty$ to $+\infty$ by expressing the Bessel function in terms of Hankel functions $J_\nu(z) = \frac{1}{2}\{H_\nu^{(1)}(z) + H_\nu^{(2)}(z)\}$ and using analytic continuation to express the Hankel function of the second kind in terms of one of the first kind, i.e., $H_\nu^{(2)}(z\exp(-i\pi)) = -\exp(i\nu\pi)H_\nu^{(1)}(z)$. We obtain in this case

$$G(\mathbf{r}, \mathbf{r}') = \frac{i}{8\pi}\int_{-\infty}^{\infty} \frac{H_0^{(1)}(\lambda\rho)\exp[i(k^2-\lambda^2)^{1/2}|z-z'|]}{(k^2-\lambda^2)^{1/2}}\,\lambda\,d\lambda, \qquad (5.26)$$

which is known as the *Sommerfeld–Ott formulation*. The path of integration runs above the negative real axis and along the positive real axis as shown in Fig. 5.4.

To arrive at the representation (5.26) of the solution, we have introduced a double-valued function τ in the integrand via the transformation (5.21a). Since the fields, and hence the Green function G, must be single valued, we

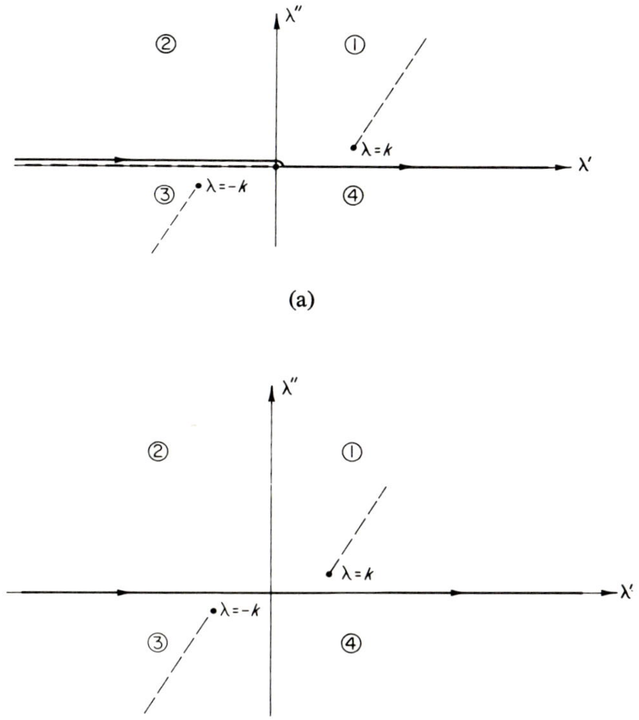

Fig. 5.4. (a) Path of integration for the Sommerfeld–Ott representation. (b) Modified path of integration for the Sommerfeld–Ott representation.

must inquire into the possibility of determining the proper branch of the function $(k^2 - \lambda^2)^{1/2}$ on which radiation condition (5.13) is satisfied.

As remarked earlier, the radiation condition (5.13) will be satisfied if $G \to 0$ as $R \to \infty$. In terms of the representation (5.25) the last condition will be satisfied if the integrand vanishes for each one of the configuration space variables x, y, or z while the other two are held constant. Now since

$$H_0^{(1)}(\lambda\rho) \xrightarrow[|\lambda\rho| \to \infty]{} \left(\frac{2}{\pi\lambda\rho}\right)^{1/2} \exp\left[i\left(\lambda\rho - \frac{\pi}{4}\right)\right]. \tag{5.27}$$

the convergence of the integrand in (5.26) is assured as $x \to \pm\infty$ or as $y \to \pm\infty$ inasmuch as $\mathrm{Im}(\lambda) \geq 0$ everywhere on the path of integration. Evidently, in order to ensure the convergence as $z \to \pm\infty$, we must require

$$\mathrm{Im}(k^2 - \lambda^2)^{1/2} > 0 \tag{5.28a}$$

and

$$(k^2 - \lambda^2)^{1/2} \xrightarrow[\lambda \to 0]{} k = k' + ik'', \qquad k' > 0, \quad k'' \geq 0. \tag{5.28b}$$

Let us now examine the effect of the conditions in (5.28) on the path of integration in (5.26). Let us put

$$(k^2 - \lambda^2)^{1/2} = \tau' + i\tau'' \tag{5.29a}$$

and find

$$\tau'' = \frac{k'k'' - \lambda'\lambda''}{\tau'} \tag{5.29b}$$

$$\tau' = \left\{\left[\left(\frac{k'^2 - k''^2 - \lambda'^2 + \lambda''^2}{2}\right)^2 + (k'k'' - \lambda'\lambda'')^2\right]^{1/2} + \frac{k'^2 - k''^2 - \lambda'^2 + \lambda''^2}{2}\right\}^{1/2} > 0. \tag{5.29c}$$

Note that in (5.29c) we must choose the plus sign in front of the brace to ensure that (5.28b) is satisfied. Thus, τ' is always positive. On the path of integration $\tau'' > 0$ and it will follow from (5.29b) that if $k'' \neq 0$, we will satisfy this requirement if we set $\lambda'' = 0$ on the path. If the medium is lossless, $k'' = 0$, we must have $\lambda'' < 0$ when $\lambda' > 0$ and $\lambda'' > 0$ when $\lambda' < 0$. Note that in the last case, the requirement $\lambda'' < 0$, $\lambda' > 0$ conflicts with the convergence condition for the Hankel function $H_0^{(1)}(\lambda\rho)$ in (5.27). In order to avoid the mathematical complexities arising therefrom, we shall always assume that k has a nonzero imaginary part, and a "lossless case" will be looked upon as a limiting form of the pertinent expression as $k'' \to 0$.

Since the Hankel function $H_0^{(1)}(\lambda\rho)$ approaches zero even if $\lambda'' = 0$, the path of integration in (5.26) can be modified by superposing it on the entire real axis, as shown in Fig. 5.4b.

A still different representation of the function $G(\mathbf{r}, \mathbf{r}')$ can be obtained by transforming (5.22b) to spherical coordinates, using

$$\alpha_1 = \lambda \cos \varphi, \qquad \alpha_2 = \lambda \sin \varphi \tag{5.30a}$$

to obtain

$$G(\mathbf{r}, \mathbf{r}') = \frac{i}{8\pi^2} \int_0^\infty \int_{-\pi}^{\pi} \times \frac{\exp\{i[(x - x')\lambda \cos \varphi + (y - y')\lambda \sin \varphi + (k^2 - \lambda^2)^{1/2}|z - z'|]\}}{(k^2 - \lambda^2)^{1/2}} \lambda \, d\lambda \, d\varphi; \tag{5.30b}$$

now, by substituting

$$\lambda = k \sin \beta, \tag{5.30c}$$

we obtain the final result

$$G(\mathbf{r}, \mathbf{r}') = \frac{ik}{8\pi^2} \int_{\Gamma_0} \int_{\varphi=-\pi}^{\varphi=\pi} \times \exp\{ik[(x - x') \sin \beta \cos \varphi + (y - y') \sin \beta \sin \varphi + |z - z'| \cos \beta]\} \times \sin \beta \, d\beta \, d\varphi \tag{5.31}$$

where Γ_0 is a contour in the complex β plane as specified by the transformation (5.30c), which we shall now consider in detail.

First, let us consider the case when k is real, $k = k'$. Equation (5.30c) then implies

$$\lambda' + i\lambda'' = k'(\sin \beta' \cosh \beta'' + i \cos \beta' \sinh \beta'') \tag{5.32a}$$

or

$$\lambda' = k' \sin \beta' \cosh \beta'', \qquad \lambda'' = k' \cos \beta' \sinh \beta''. \tag{5.32b}$$

It is easy to show that the mapping of (5.30c) transforms the quadrants of the λ plane in parallel strips of width $\pi/2$ radians, as in Fig. 5.5, where the numbers in the brackets denote the corresponding quadrants in the λ plane as shown in Fig. 5.4. The path of integration in (5.30b) that ran from $\lambda = 0$ to $\lambda = \infty$ in the right half plane in Fig. 5.4 is now transformed into the path Γ_0 as shown in Fig. 5.5.

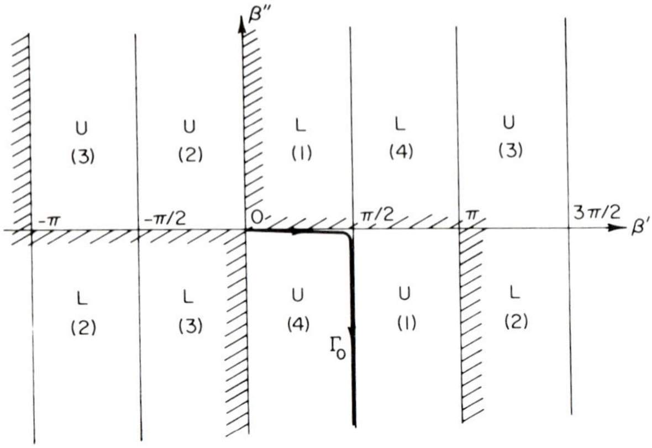

Fig. 5.5. The β plane obtained by the transformation $\lambda = k \sin \beta$ when k is real.

The requirement $\operatorname{Im}(k^2 - \lambda^2)^{1/2} > 0$ of (5.28a) in the language of the transformation $\lambda = k \sin \beta$ of (5.30c) now amounts to satisfying the condition $\operatorname{Im}(k \cos \beta) > 0$ on the path of integration. In the present case k is considered to be real, so we must have

$$\sin \beta' \sinh \beta'' < 0. \tag{5.33}$$

The strips of the complex β plane on which the inequality (5.33) is satisfied are denoted by U (upper Riemann sheet), the others by L (lower Riemann sheet). It is seen that the path Γ_0 lies totally on the upper Riemann sheet.

In the case when k is complex, $k = k' + ik''$, the transformation is more complicated. In this case we have

$$\begin{aligned} \lambda' &= k' \sin \beta' \cosh \beta'' - k'' \cos \beta' \sinh \beta'', \\ \lambda'' &= k'' \sin \beta' \cosh \beta'' + k' \cos \beta' \sinh \beta''. \end{aligned} \tag{5.34a}$$

The real axis of the λ plane ($\lambda'' = 0$) transforms as $\tan \beta' = -(k'/k'') \tanh \beta''$, from which we obtain the asymptotes

$$\beta' \xrightarrow[\beta'' \to \pm\infty]{} \mp \arctan \frac{k'}{k''}, \qquad \lambda'' = 0, \tag{5.34b}$$

which in the case $k'' = 0$ is in complete agreement with the previous lossless case. The imaginary axis of the λ plane ($\lambda' = 0$) transforms as $\tan \beta' = (k''/k') \tanh \beta''$, which has the asymptotes

$$\beta' \xrightarrow[\beta'' \to \pm\infty]{} \pm \arctan \frac{k''}{k'}, \qquad \lambda' = 0, \tag{5.34c}$$

again in agreement with the lossless case when k'' is allowed to vanish. Moreover, in the vicinity of the real axis $\beta'' \to 0$ we have $(\beta' \pm m\pi) \sim -(k'/k'')\beta''$ for $\lambda'' = 0$ and $(\beta' \pm m\pi) \sim (k''/k')\beta''$ for $\lambda' = 0$, denoting that the mapping of the $\lambda'' = 0$ and $\lambda' = 0$ axes is conformal as it should be and the slope is a function of the magnitude k''. Finally, the lines $\text{Im}(k \cos \beta) = 0$ can be easily determined by noting that $k \cos \beta = k \sin(\beta \pm \pi/2)$, which indicates that the lines $\text{Im}(k \cos \beta) = 0$ will be identical to the lines $\lambda'' = 0$, previously determined, except that they will be shifted by ± 90 degrees from the origin. Collecting all of the previous information, we find that the β plane becomes as shown in Fig. 5.6.

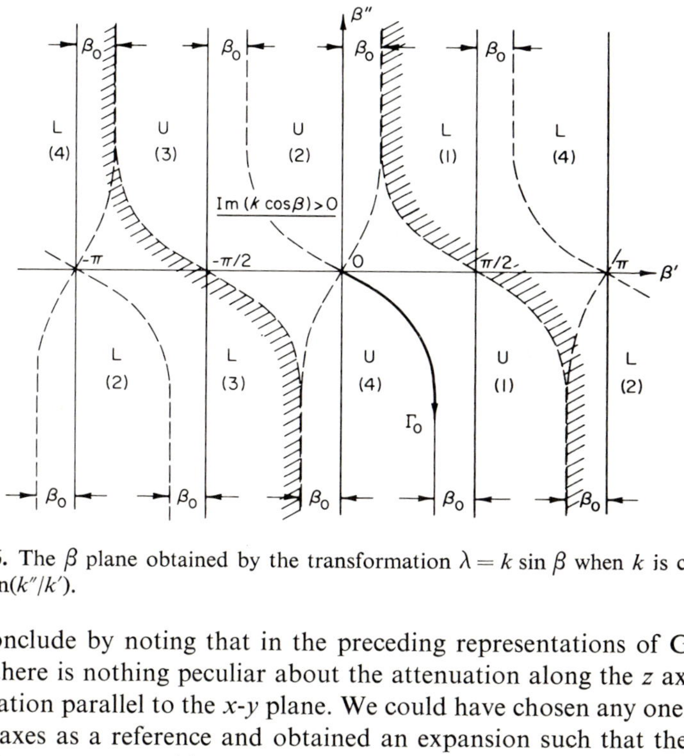

Fig. 5.6. The β plane obtained by the transformation $\lambda = k \sin \beta$ when k is complex ($\beta_0 = \arctan(k''/k')$).

We conclude by noting that in the preceding representations of Green's function there is nothing peculiar about the attenuation along the z axis and no attenuation parallel to the x-y plane. We could have chosen any one of the cartesian axes as a reference and obtained an expansion such that the component plane waves would propagate unattenuated in the direction normal to the reference axis.

To conclude this section, we list the most often used integral representations of the function $\exp(ikR)/R$, which follow immediately from the previously found results.

$$\frac{\exp(ikR)}{R} = i \int_0^\infty J_0(\lambda\rho) \exp[i(k^2 - \lambda^2)^{1/2}|z - z'|] \cdot \frac{\lambda \, d\lambda}{(k^2 - \lambda^2)^{1/2}} \tag{5.35a}$$

$$\frac{\exp(ikR)}{R} = \frac{ik}{2}\int_{\Gamma_1} H_0^{(1)}(k\rho \sin\beta)\exp(ik\cos\beta\,|z-z'|)\sin\beta\,d\beta \tag{5.35b}$$

$$\frac{\exp(ikR)}{R} = \frac{i}{2\pi}\int_0^{\infty}\int_{-\pi}^{\pi} \times \exp\{i[\lambda\cos\varphi(x-x') + \lambda\sin\varphi(y-y') + (k^2-\lambda^2)^{1/2}|z-z'|]\} \times \frac{\lambda\,d\lambda\,d\varphi}{(k^2-\lambda^2)^{1/2}} \tag{5.35c}$$

$$\frac{\exp(ikR)}{R} = \frac{ik}{\pi}\int_{\Gamma_0}\int_{-\pi}^{\pi} \times \exp\{ik[(x-x')\sin\beta\cos\varphi + (y-y')\sin\beta\sin\varphi + |z-z'|\cos\beta]\}\sin\beta\,d\beta\,d\varphi. \tag{5.35d}$$

5.1.2 Green's Function for Cylindrical Waves in Free Space

In this case the fields do not vary along one of the axes, which we shall assume to be the z axis, then $\partial/\partial z = 0$. The Green function then satisfies the two-dimensional wave equation

$$(\nabla_{xy}^2 + k^2)G(\boldsymbol{\rho}, \boldsymbol{\rho}') = -\delta(\boldsymbol{\rho} - \boldsymbol{\rho}') \tag{5.36}$$

where $\boldsymbol{\rho}$ is a vector in the x-y plane from the origin to the field point and $\boldsymbol{\rho}'$ is a vector from the origin to the source point as shown in Fig. 5.7.

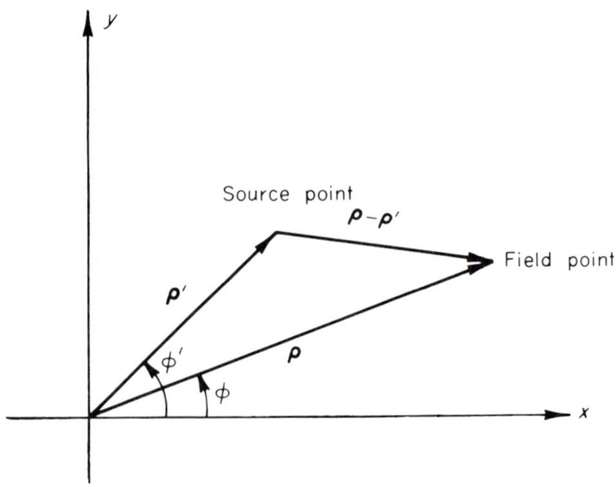

Fig. 5.7. The vectors $\boldsymbol{\rho}$ and $\boldsymbol{\rho}'$ in the x-y plane.

Since the region is infinite in the x-y plane we can use the two-dimensional Fourier transform

$$\tilde{\tilde{F}}(\boldsymbol{\alpha}) = (2\pi)^{-1} \iint_{-\infty}^{\infty} F(\boldsymbol{\rho}) \exp(-i\boldsymbol{\alpha} \cdot \boldsymbol{\rho})\, d\boldsymbol{\rho} \tag{5.37a}$$

and its inverse

$$\tilde{\tilde{F}}(\boldsymbol{\rho}) = (2\pi)^{-1} \iint_{-\infty}^{\infty} \tilde{\tilde{F}}(\boldsymbol{\alpha}) \exp(i\boldsymbol{\alpha} \cdot \boldsymbol{\rho})\, d\boldsymbol{\alpha} \tag{5.37b}$$

where in this case $\boldsymbol{\alpha} = \mathbf{a}_1 \alpha_1 + \mathbf{a}_2 \alpha_2$, $d\boldsymbol{\rho} = dx\, dy$, and $d\boldsymbol{\alpha} = d\alpha_1\, d\alpha_2$. We can express (5.36) in the transform space by multiplying by

$$(2\pi)^{-1} \exp(-i\boldsymbol{\alpha} \cdot \boldsymbol{\rho})\, d\boldsymbol{\rho}$$

and integrating from $-\infty$ to $+\infty$. The right side yields at once $-(2\pi)^{-1}$ $\exp(-i\boldsymbol{\alpha} \cdot \boldsymbol{\rho}')$ while the left side must be integrated by parts. It is easy to show that in the case that the condition (5.13) is satisfied, the transformed equation (5.36) becomes

$$\tilde{\tilde{G}}(\alpha, \boldsymbol{\rho}') = (2\pi)^{-1} \frac{\exp(-i\boldsymbol{\alpha} \cdot \boldsymbol{\rho}')}{\alpha^2 - k^2}. \tag{5.38}$$

Now we invert, using (5.37b), and obtain

$$G(\boldsymbol{\rho}, \boldsymbol{\rho}') = (2\pi)^{-2} \iint_{-\infty}^{\infty} \frac{\exp[i\boldsymbol{\alpha} \cdot (\boldsymbol{\rho} - \boldsymbol{\rho}')]}{\alpha^2 - k^2}\, d\boldsymbol{\alpha}, \tag{5.39}$$

which is a formal solution to the problem.

We can express $G(\boldsymbol{\rho}, \boldsymbol{\rho}')$ in terms of a well-known function as follows. First, we put

$$\begin{aligned} \alpha_1 &= \lambda \cos \beta, \qquad & x - x' &= |\boldsymbol{\rho} - \boldsymbol{\rho}'| \cos \varphi, \\ \alpha_2 &= \lambda \sin \beta, \qquad & y - y' &= |\boldsymbol{\rho} - \boldsymbol{\rho}'| \sin \varphi, \end{aligned} \tag{5.40a}$$

and obtain

$$\begin{aligned} G(\boldsymbol{\rho}, \boldsymbol{\rho}') &= (2\pi)^{-2} \int_0^{\infty} \frac{\lambda\, d\lambda}{\lambda^2 - k^2} \int_{-\pi}^{\pi} \exp[i\lambda |\boldsymbol{\rho} - \boldsymbol{\rho}'| \cos(\beta - \varphi)]\, d\beta \\ &= (4\pi)^{-1} \int_{\infty \exp(i\pi)}^{\infty} \frac{H_0^{(1)}(\lambda |\boldsymbol{\rho} - \boldsymbol{\rho}'|)}{\lambda^2 - k^2}\, \lambda\, d\lambda, \end{aligned} \tag{5.40b}$$

where we used (5.24) and the relation connecting Bessel and Hankel functions. The last integral can be evaluated by allowing λ to take on complex values

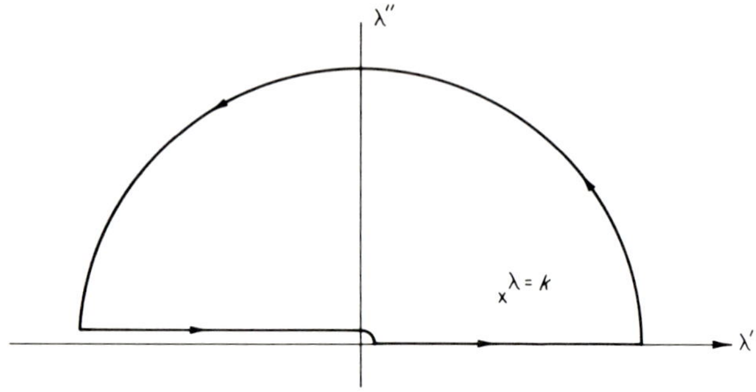

Fig. 5.8. Contour for evaluation of the integral in (5.40b).

and integrating along the contour of Fig. 5.8. in the λ plane. It is easy to show then that the contribution from the large and the small semicircular arcs vanishes in the limit as its radius becomes infinite and zero, respectively, and the integral is equal to $2\pi i$ times the residue at the pole $\lambda = +k$. Thus,

$$G(\boldsymbol{\rho}, \boldsymbol{\rho}') = \frac{i}{4} H_0^{(1)}(k\,|\boldsymbol{\rho} - \boldsymbol{\rho}'|) \tag{5.41}$$

and it follows that $G(\boldsymbol{\rho}, \boldsymbol{\rho}')$ satisfies the radiation condition (5.13) by virtue of the asymptotic behavior of the Hankel function for large values of the argument.

An alternate representation of $G(\boldsymbol{\rho}, \boldsymbol{\rho}')$ can be found by starting from (5.39) with the substitution $\tau = (k^2 - \alpha_1{}^2)^{1/2}$, which gives

$$G(\boldsymbol{\rho}, \boldsymbol{\rho}') = (2\pi)^{-2} \int_{-\infty}^{\infty} d\alpha_1 \exp[i\alpha_1(x - x')] \int_{-\infty}^{\infty} \frac{\exp[i\alpha_2(y - y')]}{\alpha_2{}^2 - \tau^2}\, d\alpha_2 . \tag{5.42}$$

The second integral on the right is identical to that in (5.22a). Thus, we have

$$G(\boldsymbol{\rho}, \boldsymbol{\rho}') = \frac{i}{4\pi} \int_{-\infty}^{\infty} \frac{\exp\{i[\alpha_1(x - x') + (k^2 - \alpha_1{}^2)^{1/2}|y - y'|]\}}{(k^2 - \alpha_1{}^2)^{1/2}}\, d\alpha_1. \tag{5.43}$$

Furthermore, if we put $\alpha_1 = k \sin \beta$, we obtain

$$G(\boldsymbol{\rho}, \boldsymbol{\rho}') = \frac{i}{4\pi} \int_{\Gamma_1} \exp\{ik[\sin \beta(x - x') + \cos \beta\,|y - y'|]\}\, d\beta, \tag{5.44}$$

where Γ_1 is a path in the complex β plane as shown in Fig. 5.9.

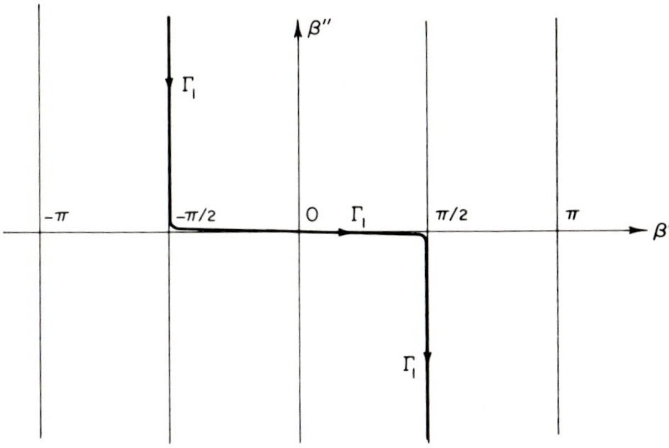

Fig. 5.9. The path Γ_1 in the complex β plane.

We can compare the results of (5.41), (5.43), and (5.44) and thus obtain the well-known integral representations of the Hankel functions

$$H_0^{(1)}(k\,|\boldsymbol{\rho} - \boldsymbol{\rho}'|) = \frac{1}{\pi}\int_{-\infty}^{\infty} \frac{\exp\{i[\alpha_1(x - x') + (k^2 - \alpha_1{}^2)^{1/2}|y - y'|]\}}{(k^2 - \alpha_1{}^2)^{1/2}}\, d\alpha_1 \tag{5.45a}$$

$$H_0^{(1)}(k\,|\boldsymbol{\rho} - \boldsymbol{\rho}'|) = \frac{1}{\pi}\int_{\Gamma_1} \exp\{ik[\sin\beta(x - x') + \cos\beta\,|y - y'|]\}\, d\beta. \tag{5.45b}$$

5.2 Spectral Representation of Dipoles and Line Sources

5.2.1 Electric and Magnetic Dipoles

Suppose that the prescribed current densities appearing in the inhomogeneous term of (5.2a) and (5.3a) are in the form of dipoles defined by

$$\mathbf{p}^{(e)} = I^{(e)}\,\mathbf{dl}\; e^{-i\omega t}, \tag{5.46a}$$

$$\mathbf{p}^{(m)} = I^{(m)}\,\mathbf{dl}\; e^{-i\omega t}, \tag{5.46b}$$

where **dl** is a small line element. If these dipoles are situated in the position $\mathbf{r} = \mathbf{r}'$, then the corresponding current densities can be written

$$\mathbf{J}^{(e)} = \mathbf{p}^{(e)}\,\delta(\mathbf{r} - \mathbf{r}'), \tag{5.47a}$$

$$\mathbf{J}^{(m)} = \mathbf{p}^{(m)}\,\delta(\mathbf{r} - \mathbf{r}'). \tag{5.47b}$$

We can, without loss of generality, assume that the dipoles are oriented in the positive z direction, in which case Eqs. (5.2a) and (5.3a) become

$$(\nabla^2 + k^2)\Pi_z^{(e)} = -\frac{i\mu p^{(e)}}{\omega\varepsilon_0 n^2}\,\delta(\mathbf{r} - \mathbf{r}'), \tag{5.48a}$$

$$(\nabla^2 + k^2)\Pi_z^{(m)} = -\frac{ip^{(m)}}{\omega\mu_0\mu}\,\delta(\mathbf{r}-\mathbf{r}'). \tag{5.48b}$$

Apart from a constant, these differential equations are identical to the one satisfied by the Green function in (5.5). Thus, we obtain immediately, using (5.19),

$$\Pi_z^{(e)} = \frac{i\mu p^{(e)}}{4\pi\omega\varepsilon_0 n^2}\frac{e^{ikR}}{R}, \tag{5.49a}$$

$$\Pi_z^{(m)} = \frac{ip^{(m)}}{4\pi\omega\mu_0\mu}\frac{e^{ikR}}{R}; \tag{5.49b}$$

furthermore, we can use the various integral representations of the function $\exp(ikR)/R$ listed in (5.35) to obtain a spectral representation of the solution. The field components are found by using (5.2b) and (5.2c) in connection with (5.49a) for the electric case and (5.3b) and (5.3c) in connection with (5.49b) for the magnetic case.

5.2.2 Electric and Magnetic Line Sources

In the present case, the prescribed current densities appearing in (5.2a) and (5.3a) are assumed to be in the form of thin current filaments running parallel, say, to the z axis and situated at the point $x = x'$ and $y = y'$ in the x-y plane .The current densities can then be written

$$\mathbf{J}^{(e)} = \mathbf{l}_z I^{(e)}\,\delta(\boldsymbol{\rho}-\boldsymbol{\rho}'), \tag{5.50a}$$

$$\mathbf{J}^{(m)} = \mathbf{l}_z I^{(m)}\,\delta(\boldsymbol{\rho}-\boldsymbol{\rho}'), \tag{5.50b}$$

in which case Eqs. (5.2a) and (5.3a) become

$$(\nabla_{xy}^2 + k^2)\Pi_z^{(e)} = -\frac{i\mu I^{(e)}}{\omega\varepsilon_0 n^2}\,\delta(\boldsymbol{\rho}-\boldsymbol{\rho}'), \tag{5.51a}$$

$$(\nabla_{xy}^2 + k^2)\Pi_z^{(m)} = -\frac{iI^{(m)}}{\omega\mu_0\mu}\,\delta(\boldsymbol{\rho}-\boldsymbol{\rho}'). \tag{5.51b}$$

The foregoing equations are identical to (5.36) except for constants. Thus, using (5.41), we obtain immediately

$$\Pi_z^{(e)} = -\frac{\mu I^{(e)}}{4\omega\varepsilon_0 n^2}\,H_0^{(1)}(k\,|\boldsymbol{\rho}-\boldsymbol{\rho}'|), \tag{5.52a}$$

$$\Pi_z^{(m)} = -\frac{I^{(m)}}{4\omega\mu_0\mu}\,H_0^{(1)}(k\,|\boldsymbol{\rho}-\boldsymbol{\rho}'|), \tag{5.52b}$$

and the field components can be obtained by using (5.2b) and (5.2c) in connection with (5.52a) for the electric case and (5.3b) and (5.3c) in connection with (5.52b) for the magnetic case.

5.3 Asymptotic Evaluation of Integrals

To extract useful information from rather complicated functions we have on several occasions used the asymptotic form, i.e., the value that a function approaches when its argument becomes extremely large. In cases encountered thus far, the finding of an appropriate asymptotic formula did not present a serious problem since we dealt with well-known functions whose asymptotic forms are readily available. In what follows, however, we shall encounter certain definite integrals of a complicated nature that cannot be evaluated exactly. In such cases it will prove useful to know the generally applicable procedures for finding an asymptotic value of an integral when some parameter is large.

5.3.1 The Method of Stationary Phase

This method was formulated late in the nineteenth century and is usually credited to either Kelvin, who used it in connection with a problem in hydrodynamics, or Stokes, who used it to extract asymptotic values from the Airy integral.

The method is applicable to integrals of the type

$$I(x) = \int_a^b F(\xi) \exp(ixf(\xi))\, d\xi, \tag{5.53}$$

where $F(\xi)$ and $f(\xi)$ are real functions and ξ is a real variable. The parameter x is also real and assumed to be large. The function $F(\xi)$ is assumed to vary slowly with ξ in the interval $a \le \xi \le b$. Since x in the exponential is large, it is clear that the integrand will fluctuate rapidly as a function of ξ from positive to negative values except at some value of $\xi = \xi_0$ where the derivative $f'(\xi_0) = 0$, $a \le \xi_0 \le b$, and the phase is constant. It is now argued that at points other than the neighborhood of $\xi = \xi_0$ the contributions to the integral will approximately cancel each other. Thus, the integrand can be approximated by its value in the neighborhood of $\xi = \xi_0$ and the limits can be extended to run from $-\infty$ to $+\infty$.

In the neighborhood of the point $\xi = \xi_0$ the function $F(\xi)$ can be taken out from under the integral sign with the value $F(\xi_0)$. The function $f(\xi)$ can can be expanded in the Taylor series:

$$f(\xi) = f(\xi_0) + \tfrac{1}{2} f''(\xi_0)(\xi - \xi_0)^2 + \cdots, \tag{5.54a}$$

and the integral in (5.53) becomes

$$I(x) \sim F(\xi_0)\exp(ixf(\xi_0)) \int_{-\infty}^{\infty} \exp\left[\frac{ixf''(\xi_0)(\xi - \xi_0)^2}{2}\right] d\xi. \qquad (5.54b)$$

The integral in (5.54b) can be evaluated by noting that

$$\int_{-\infty}^{\infty} \exp(-\rho s^2)\, ds = \left(\frac{\pi}{\rho}\right)^{1/2}, \qquad \rho > 0, \qquad (5.55)$$

to obtain

$$I(x) \sim F(\xi_0)\left(\frac{2\pi}{x\,|f''(\xi_0)|}\right)^{1/2} \exp\left[i\left(xf(\xi_0) \pm \frac{\pi}{4}\right)\right] \qquad (5.56)$$

where the + or − is appropriate for $f''(\xi_0) > 0$ or $f''(\xi_0) < 0$, respectively. When the stationary point $\xi = \xi_0$ coincides with one of the integral limits, the formula (5.56) must be multiplied by $\frac{1}{2}$.

5.3.2 The Method of Steepest Descents (Saddle-Point Method)

The method of stationary phase just outlined makes no provisions for calculating the higher-order terms of the resulting asymptotic expansion. A method that is more general, inasmuch as it permits us to find (at least in theory) any order term of the expansion, is the method of steepest descents or saddle-point method.

The method of steepest descents deals with the approximate evaluation of integrals of the form

$$I(\rho) = \int_C e^{\rho f(\zeta)} F(\zeta)\, d\zeta \qquad (5.57)$$

for large values of ρ. The functions $f(\zeta)$ and $F(\zeta)$ are any arbitrary analytic functions of the complex variable ζ and C is a path of integration in the complex ζ plane. The method was first introduced by Debye in 1909 for obtaining the well-known asymptotic expansion of the Hankel function. It was later improved by Ott (1942) and van der Waerden (1951), and generalized by Oberhettinger (1959). This method is inherently connected with the so-called Watson lemma (see e.g., Jeffreys and Jeffreys, 1950).

The general philosphy of the method of steepest descents is as follows. Within certain limits, the path of integration in the complex ζ plane may be altered without affecting the value of the integral. Thus, we attempt to select a path in the complex ζ plane in such a way as to intercept most of the total value of the integral in a relatively short portion of the path. Following that, we replace the integrand by another, simpler function, which sufficiently closely approximates the original integrand over the important parts of the path. The behavior of this new function outside of the important part of the path is of little concern.

Without loss of generality, we can consider ρ to be real and positive. The complex function $f(\zeta)$ can be separated into the real and imaginary parts as follows.

$$f(\zeta) = f_r(\zeta) + if_i(\zeta) \tag{5.58a}$$

and the exponent in (5.57) becomes

$$\rho f(\zeta) = \rho f_r(\zeta) + i\rho f_i(\zeta). \tag{5.58b}$$

The desired path of integration will satisfy the requirements stated previously if it is chosen in such a manner that the function $f_r(\zeta)$ exhibits a maximum at some point and then decreases as rapidly as possible with the increasing distance from that point. Now we put

$$\zeta = \eta + i\xi, \tag{5.59}$$

it follows from the Cauchy–Riemann condition that

$$\frac{df}{d\zeta} = \frac{\partial f_r}{\partial \eta} + i\frac{\partial f_i}{\partial \eta} = -i\frac{\partial f_r}{\partial \xi} + \frac{\partial f_i}{\partial \xi} \tag{5.60a}$$

or

$$\frac{\partial f_r}{\partial \eta} = \frac{\partial f_i}{\partial \xi}, \qquad \frac{\partial f_r}{\partial \xi} = -\frac{\partial f_i}{\partial \eta}. \tag{5.60b}$$

Thus, if f_r has an extremum at a certain point in the complex ζ plane, then $\partial f_r/\partial \eta = 0$ and $\partial f_r/\partial \xi = 0$ and it follows that f_i must be a constant at that point. Thus, the lines of most rapid increase (or decrease) of one part of the complex function $f(\zeta)$ are lines of constant value of the other. Since the path of integration must pass along the line of most rapid increase and decrease of f_r it must coincide with the line $f_i =$ constant, which is a line of constant phase. The point of the path at which f_r is an extremum is called the *saddle point* (not a maximum) and the derivative of f_r must be zero at this point. Since f_i is a constant on this path, then its derivative must also be zero, and consequently

$$\frac{df}{d\zeta} = 0 \tag{5.61}$$

at the saddle point.

Thus, the most advantageous path of integration must go through the saddle point determined by (5.61) and must leave this point along the line of the most rapid decrease of the function f_r, which coincides with the line

f_i = constant. We shall henceforth call this path *the path of steepest descents*. If, in addition, the parameter ρ is real and positive, then the exponential in (5.57) will decrease rapidly with the distance from the saddle point and a relatively small portion of the path of integration, including the saddle point, will contribute significantly to the value of the integral.

Suppose that (5.61) has been solved and the saddle point has been determined to be at the point $\zeta = \zeta_0$. Then it follows that the path of integration will be determined from

$$f(\zeta) = f(\zeta_0) - s^2, \tag{5.62}$$

where s is real and $-\infty \leq s \leq +\infty$ and the saddle point corresponds to the point $s = 0$. Now using (5.58a) and (5.62), we obtain

$$f_r(\zeta) = f_r(\zeta_0) - s^2, \tag{5.63a}$$

$$f_i(\zeta) = f_i(\zeta_0). \tag{5.63b}$$

It follows from (5.63) that the imaginary part of $f(\zeta)$ remains constant on the new path of integration, while the real part attains a maximum at $s = 0$ and decreases as s becomes different from zero on either side of the saddle point.

It is illuminating to inquire into the geometrical aspects of the transformation (5.62). To this end we let s be complex, $s = s' + is''$; then by (5.62)

$$s'^2 - s''^2 = f_r(\zeta_0) - f_r(\zeta), \tag{5.64a}$$

$$2s's'' = f_i(\zeta_0) - f_i(\zeta). \tag{5.64b}$$

We recognize (5.64a) as a family of hyperbolas about the s' and s'' axes. Moreover, we transform (5.64b) by rotating the axes by $\pi/4$ radians according to the transformation

$$\begin{pmatrix} s' \\ s'' \end{pmatrix} = \begin{pmatrix} \cos \pi/4 & -\sin \pi/4 \\ \sin \pi/4 & \cos \pi/4 \end{pmatrix} \begin{pmatrix} \sigma' \\ \sigma'' \end{pmatrix} \tag{5.65a}$$

and consequently obtain

$$2s's'' = \sigma'^2 - \sigma''^2, \tag{5.65b}$$

which shows that (5.64b) represents a family of hyperbolas about the axes σ' and σ''. The situation is depicted in Fig. 5.10. Now since

$$f(\zeta) = (f_r(\zeta_0) - s'^2 + s''^2) + i(f_i(\zeta_0) - 2s's''), \tag{5.66}$$

then

$$f(\zeta) \xrightarrow[\substack{s''=0 \\ s' \to \pm\infty}]{} -\infty + if_i(\zeta_0) \tag{5.67a}$$

and

$$f(\zeta) \xrightarrow[\substack{s'=0 \\ s'' \to \pm\infty}]{} +\infty + if_i(\zeta_0), \tag{5.67b}$$

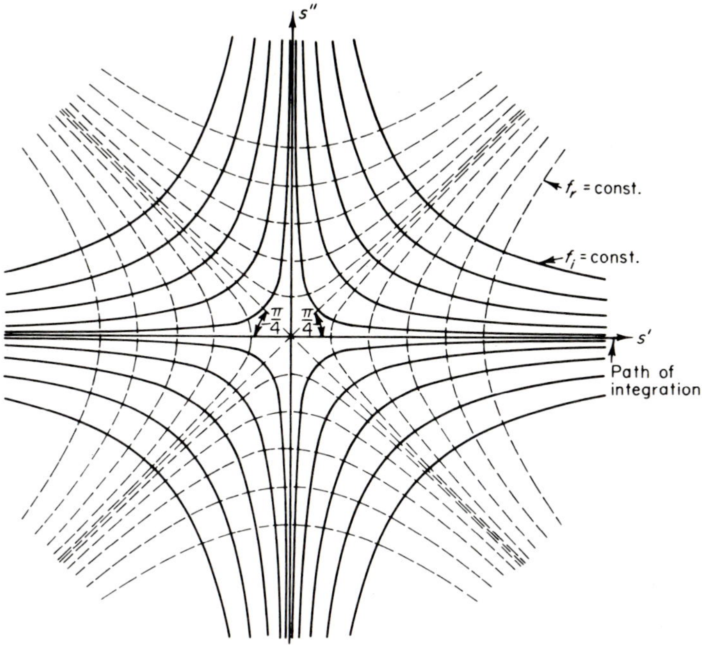

Fig. 5.10. The complex s plane.

from which it follows that the exponential in (5.57) decreases most rapidly going away from the saddle point along the real axis, $s'' = 0$, in the complex s plane. The same exponential increases most rapidly when leaving the saddle point along the imaginary axis, $s' = 0$.

We can draw the relief of $f_r(s', s'') = f_r(\zeta_0) - s'^2 + s''^2$ in the complex s plane by erecting line segments proportional to the value of $f_r(s', s'')$ at a given point. Then taking a cut through the real axis s', the map of values of $f_r(s', s'')$ would look as shown in Fig. 5.11. We note that in the neighborhood of the point $s = 0$ the relief of $f_r(s', s'')$ has the form of a saddle, since on both sides of $s = 0$ along the real axis $f_r(s', s'')$ decreases and in the perpendicular direction along the imaginary axis it increases. Proceeding along the real axis from left to right, we first go upward in a valley having its lowest points on the real axis and rising upward to the left and right. We then reach the highest of the lowest points in the valley and we again descend into a valley.

The integral

$$I = \int_C e^{\rho f(\zeta)} F(\zeta)\, d\zeta \qquad [5.57]$$

taken over some arbitrary path C can, by Cauchy's theorem, be replaced by an integral with the same integrand taken over some new path, provided that

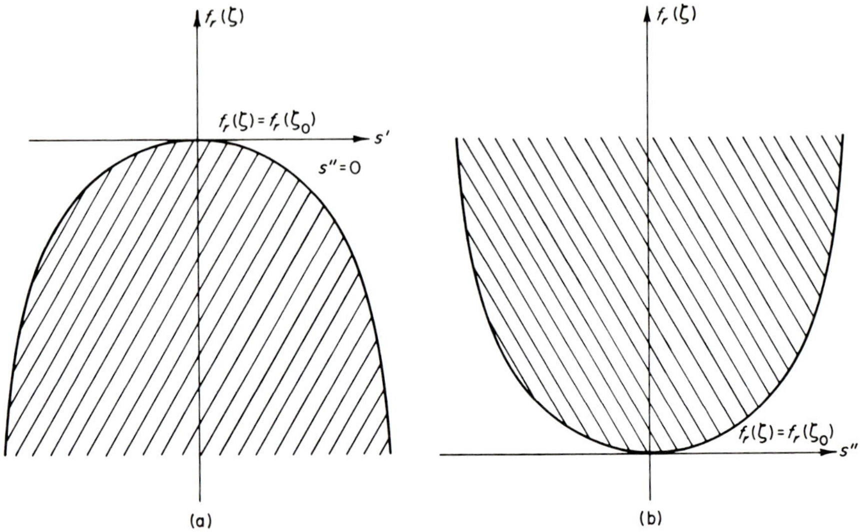

Fig. 5.11. Relief of the function $f_r(\zeta)$ as seen from (a) the cut along the s' axis; (b) the cut along the s'' axis.

all of the singularities situated between the old and the new path are properly accounted for. Thus, when crossing a pole, we must add the residue at that pole, and if we encounter a branch point, we must add the integral over the edges of an appropriate cut made to make the function single valued. Thus, we can write

$$I = I_{SD} + I_{SI} \tag{5.68}$$

where I_{SD} denotes the integral along the path of the steepest descents and I_{SI} denotes integrals around the singularities.

We now focus our attention on the contribution from integration along the line of the steepest descents. Substituting the results of (5.62) into (5.57), we obtain

$$I_{SD} = \exp(\rho f(\zeta_0)) \int_{-\infty}^{\infty} \exp(-\rho s^2)\Phi(s)\, ds, \tag{5.69a}$$

where

$$\Phi(s) = F(\zeta)\frac{d\zeta}{ds}. \tag{5.69b}$$

Now if ρ is large, then the integrand in (5.69a) will fall off rapidly with an increasing s, the distance from the saddle point. Thus, only small values of s

will contribute significantly and, therefore, we can expand $\Phi(s)$ in a Taylor series about the saddle point $s = 0$. Consequently, we write

$$\Phi(s) = \Phi(0) + \Phi'(0)s + \tfrac{1}{2}\Phi''(0)s^2 + \cdots \tag{5.70}$$

and obtain

$$I_{\text{SD}} = \exp(\rho f(\zeta_0)) \times \left[\Phi(0)\int_{-\infty}^{\infty} \exp(-\rho s^2)\,ds + \tfrac{1}{2}\Phi''(0)\int_{-\infty}^{\infty} s^2 \exp(-\rho s^2)\,ds + \cdots\right] \tag{5.71}$$

since the integrals with the odd power of s vanish. Now we note the well-known result

$$\int_{-\infty}^{\infty} \exp(-\rho s^2)s^{2m}\,ds = \frac{\sqrt{\pi}\,(2m)!}{m!\,2^{2m}} \cdot \frac{1}{\rho^{m+1/2}}, \tag{5.72}$$

which when substituted into (5.71) gives

$$I_{\text{SD}} = \left(\frac{\pi}{\rho}\right)^{1/2} \exp(\rho f(\zeta_0))\left[\Phi(0) + \frac{1}{4\rho}\Phi''(0) + \cdots\right]. \tag{5.73}$$

We must now seek the expansion of the function $\Phi(s)$ directly in terms of the integrand in (5.57). To this end we first expand the function $f(\zeta)$ in a Taylor series about the saddle point $\zeta = \zeta_0$. Setting $\zeta - \zeta_0 = x$, we obtain

$$\begin{aligned} f(\zeta) &= f(\zeta_0) + \tfrac{1}{2}f''(\zeta_0)x^2 + \tfrac{1}{6}f'''(\zeta_0)x^3 + \tfrac{1}{24}f^{\text{iv}}(\zeta_0)x^4 + \cdots \\ &= f(\zeta_0) - s^2, \end{aligned} \tag{5.74a}$$

which gives

$$-s^2 = -Ax^2 + Bx^3 + Cx^4 + \cdots, \tag{5.74b}$$

where

$$A = -\tfrac{1}{2}f''(\zeta_0), \qquad B = \tfrac{1}{6}f'''(\zeta_0), \qquad C = \tfrac{1}{24}f^{\text{iv}}(\zeta_0). \tag{5.74c}$$

We must now invert the power series (5.74b) to find the powers of x in terms of the powers of s. To this end we put

$$x = a_0 s(1 + a_1 s + a_2 s^2 + a_3 s^3 + \cdots) \tag{5.75a}$$

and obtain

$$x^2 = a_0{}^2 s^2[1 + 2a_1 s + (2a_2 + a_1{}^2)s^2 + \cdots], \tag{5.75b}$$

$$x^3 = a_0{}^3 s^3[1 + 3a_1 s + 3(a_1{}^2 + a_2)s^2 + \cdots], \tag{5.75c}$$

$$x^4 = a_0{}^4 s^4[1 + 4a_1 s + 2(2a_2 + 3a_1{}^2)s^2 + \cdots]. \tag{5.75d}$$

Substituting the foregoing results back into (5.74b) and comparing the terms of equal powers of s, we obtain

$$a_0 = \frac{1}{\sqrt{A}}, \qquad a_1 = \frac{B}{2A^{3/2}}, \qquad a_2 = \frac{C}{2A^2} + \frac{5}{8}\frac{B^2}{A^3}. \tag{5.76}$$

Next we expand the function $F(\zeta)$ in a Taylor series about the point $\zeta = \zeta_0$, setting as before $\zeta - \zeta_0 = x$. We obtain

$$F(\zeta) = F(\zeta_0)(1 + Px + Qx^2 + \cdots), \tag{5.77a}$$

where

$$P = \frac{F'(\zeta_0)}{F(\zeta_0)}, \qquad Q = \frac{F''(\zeta_0)}{2F(\zeta_0)}. \tag{5.77b}$$

Now noting

$$\frac{dx}{ds} = \frac{d\zeta}{ds} = \frac{1}{\sqrt{A}} + \frac{B}{A^2}s + \frac{3}{\sqrt{A}}\left(\frac{C}{2A^2} + \frac{5}{8}\frac{B^2}{A^3}\right)s^2 + \cdots, \tag{5.78}$$

we obtain finally

$$\Phi(s) = \frac{F(\zeta_0)}{\sqrt{A}}\left[1 + \left(\frac{P}{A^{1/2}} + \frac{B}{A^{3/2}}\right)s + \left(\frac{Q}{A} + \frac{15B^2}{8A^3} + \frac{3BP}{2A^2} + \frac{3C}{2A^2}\right)s^2 + \cdots\right]. \tag{5.79}$$

It is now easy to find that

$$\Phi(0) = \left(\frac{2}{-f''(\zeta_0)}\right)^{1/2} F(\zeta_0), \tag{5.80a}$$

$$\Phi''(0) = 2\Phi(0)\left[\frac{f'''}{(f'')^2}\cdot\frac{F'}{F} + \frac{1}{4}\frac{f^{\text{iv}}}{(f'')^2} - \frac{5}{12}\frac{(f''')^2}{(f'')^3} - \frac{F''}{Ff''}\right], \tag{5.80b}$$

which, when substituted back into (5.73) gives finally

$$I_{\text{SD}} = e^{\rho f(\zeta_0)}\left(\frac{2\pi}{-\rho f''(\zeta_0)}\right)^{1/2} F(\zeta_0)$$
$$\times \left\{1 + \frac{1}{2\rho}\left[\frac{f'''}{(f'')^2}\frac{F'}{F} + \frac{1}{4}\frac{f^{\text{iv}}}{(f'')^2} - \frac{5}{12}\frac{(f''')^2}{(f'')^3} - \frac{F''}{Ff''}\right] + \cdots\right\}. \tag{5.81}$$

Note that the result of this analysis is a series in inverse powers of ρ. It will follow from (5.72) that this series is divergent for a fixed ρ as the number of terms increases. Such a series is called an asymptotic series due to the definition by Poincaré. We shall now go briefly into the meaning of the asymptotic series.

A divergent series

$$f(z) = A_0 + \frac{A_1}{z} + \frac{A_2}{z^2} + \cdots + \frac{A_n}{z^n} + \cdots \tag{5.82}$$

in which the sum of the first $(n+1)$ terms is $S_n(z)$,

$$S_n(z) = \sum_{i=0}^{n} A_i z^{-i}, \tag{5.83}$$

is said to be an asymptotic expansion of a function $f(z)$ for a given range of argument z if the expression

$$R_n(z) = z^n\{f(z) - S_n(z)\} \tag{5.84}$$

satisfies the condition

$$\lim_{|z|\to\infty} R_n(z) = 0, \qquad n \text{ fixed} \tag{5.85a}$$

even though

$$\lim_{|n|\to\infty} |R_n(z)| = \infty, \qquad z \text{ fixed.} \tag{5.85b}$$

When this is the case, we can make

$$|R_n(z)| = |z^n\{f(z) - S_n(z)\}| < \varepsilon, \tag{5.85c}$$

where ε is arbitrarily small, by taking $|z|$ sufficiently large. This definition is due to Poincaré. We state here, without proof, several important consequences of this definition.

(a) Asymptotic expansions can be multiplied unconditionally.
(b) Asymptotic expansions can be integrated unconditionally.
(c) An asymptotic expansion of a function is unique.
(d) One asymptotic expansion may represent several functions.
(e) Asymptotic expansions can be divided providing the divisor contains at least one nonzero coefficient.

5.3.3 Fourier-Type Integrals

Of special interest in future analysis will be an integral resulting from the inverse Fourier transform that can be reduced to the form

$$I(kr) = \int_{\Gamma_1} F(\beta) \exp[ikr\cos(\beta - \theta)]\, d\beta, \tag{5.86}$$

where Γ_1 is a path of integration in the complex β plane as shown in Fig. 5.9. When the product kr is large, this kind of integral lends itself to approximate evaluation by the method of steepest descents.

By analogy with previous analysis we put

$$f(\beta) = ik \cos(\beta - \theta) \tag{5.87}$$

from which it will follow immediately that the saddle point corresponding to $f'(\beta) = 0$ will be located at the point $\beta = \theta$ on the real axis in the complex β plane. To find the path of the steepest descents passing through the saddle point we set

$$f(\beta) = f(\theta) - s^2, \tag{5.88}$$

where s is a real variable. It will follow from (5.87) and (5.88) that $\cos(\beta - \theta) = 1 + is^2/k$ and since in the vicinity of the saddle point $(\beta - \theta)$ is small, we can expand $\cos(\beta - \theta)$ and retain only the first two terms, obtaining

$$(\beta' - \theta)^2 - \beta''^2 = -\frac{2k''}{|k|^2} s^2, \tag{5.89a}$$

$$\beta''(\beta' - \theta) = -\frac{k'}{|k|^2} s^2. \tag{5.89b}$$

Let us first examine the consequence of the transformation (5.88) in the neighborhood of the saddle point when k is real. In this case, $\beta''(\beta' - \theta) = -s^2/k$ and $\beta' - \theta = \pm\beta''$. The ambiguity of sign in the latter expression can be resolved by noting that when $\beta' < \theta$, the $\beta'' > 0$; and when $\beta' > \theta$, then $\beta'' < 0$, as follows from the former expression. Thus, we have in this case

$$\beta'' = -(\beta' - \theta), \tag{5.90}$$

which represents a straight line intersecting the axis of the reals at the point $\beta' = 0$ at an angle $-\pi/4$.

If k is complex, $k = k' + ik''$, then we can solve the equation in (5.89) simultaneously and obtain

$$\beta'' = -\frac{k'' + |k|}{k'}(\beta' - \theta), \tag{5.91}$$

where we choose the negative sign to agree with (5.90) when $k'' \to 0$. Note that in this case the angle of intersection is larger than $\pi/4$ and it increases with increasing k''.

Now let us investigate the asymptotes of the steepest-descent path as defined in (5.88) when $\beta'' \to \pm\infty$. It readily follows from (5.87) and (5.88) by separating the real and the imaginary parts that $\cos(\beta' - \theta)\cosh\beta'' = 1 + k''s^2/|k|^2$ and $\sin(\beta' - \theta)\sinh\beta'' = -k's^2/|k|^2$ or alternately

$$\tan(\beta' - \theta)\tanh\beta'' = -\frac{k's^2}{|k|^2 + k''s^2} \xrightarrow[s \to \pm\infty]{} -\frac{k'}{k''}. \tag{5.92}$$

We conclude immediately that as $\beta'' \to +\infty$, then $\beta' \to \theta - \arctan(k'/k'')$, and as $\beta'' \to -\infty$, then $\beta' \to \theta + \arctan(k'/k'')$. The new path of integration, which we shall henceforth denote by Γ, the path of steepest descents, is shown in Fig. 5.12.

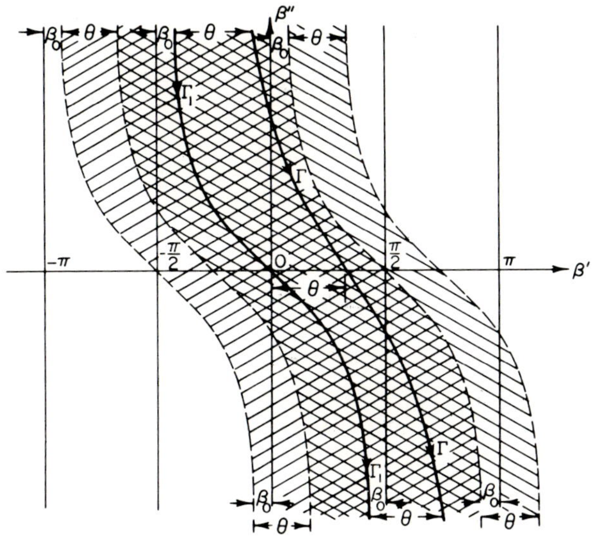

Fig. 5.12. Transformation $\lambda = k \sin \beta$ showing the paths Γ_1 and Γ and the branch cuts $\mathrm{Im}(k \cos \beta) = 0$. Inside the shaded region ///// $\mathrm{Im}(k \cos \beta) > 0$, whereas inside the region \\\\ $\mathrm{Im}(k \cos(\beta - \theta)) > 0$. ($\beta_0 = \arctan(k''/k')$.)

We are now ready to evaluate the contribution to the integral in (5.86) due to the integration along the path of the steepest descents. We write

$$I_{\mathrm{SD}} = \int_\Gamma F(\beta) \exp[ikr \cos(\beta - \theta)] \, d\beta = \exp(ikr) \int_{-\infty}^{\infty} \Phi(s) \exp(-rs^2) \, ds, \tag{5.93a}$$

where now

$$\Phi(s) = F(\beta) \frac{d\beta}{ds}. \tag{5.93b}$$

Now using the results of (5.81) together with those of (5.87), we obtain finally

$$I_{\mathrm{SD}} = \exp\left[i\left(kr - \frac{\pi}{4}\right)\right]\left(\frac{2\pi}{kr}\right)^{1/2} F(\theta)\left[1 - \frac{i}{2kr}\left(\frac{1}{4} + \frac{F''(\theta)}{F(\theta)}\right) + \cdots\right]. \tag{5.94}$$

5.3.4 Pole near Saddle Point

A complication often arises when the function $F(\beta)$ has a singularity near the saddle point of which the simplest version may be a simple pole. If the function $F(\beta)$ has a simple pole, say at $\beta = \beta_P$, that is near the $\beta = \theta$, then the radius of convergence of the expansion of $F(\beta)$ in powers of s will be limited by the distance from the saddle point to the nearest pole. It follows then that instead of using a Taylor series, we may use the Laurent series for the expansion of $F(\beta)$. Thus, we can write

$$F(\beta) = \frac{F_1(\beta_P)}{\beta - \beta_P} + F_2(\beta), \tag{5.95}$$

where $F_1(\beta_P)$ is the residue of the function $F(\beta)$ at the pole $\beta = \beta_P$ and $F_2(\beta)$ is analytic, everywhere about $\beta = \beta_P$, and can therefore be expanded in a Taylor series about the saddle point $\beta = \theta$. Instead of (5.93a) we thus have

$$I_{\mathrm{SD}} = I_{\mathrm{SD}}^{(1)} + I_{\mathrm{SD}}^{(2)}, \tag{5.96a}$$

where

$$I_{\mathrm{SD}}^{(1)} = \exp(ikr) \int_{-\infty}^{\infty} \Phi_1(s) \exp(-rs^2)\, ds, \tag{5.96b}$$

$$I_{\mathrm{SD}}^{(2)} = \exp(ikr) \int_{-\infty}^{\infty} \Phi_2(s) \exp(-rs^2)\, ds \tag{5.96c}$$

and

$$\Phi_1(s) = \frac{F_1(\beta_P)}{\beta - \beta_P} \frac{d\beta}{ds}, \tag{5.96d}$$

$$\Phi_2(s) = \left[F(\beta) - \frac{F_1(\beta_P)}{\beta - \beta_P}\right] \frac{d\beta}{ds}. \tag{5.96e}$$

Now we shall deal with each one of the foregoing integrals separately. The integral $I_{SD}^{(1)}$ can be written

$$I_{\mathrm{SD}}^{(1)} = F_1(\beta_P) \exp(ikr) \int_{-\infty}^{\infty} \left(\frac{d\beta/ds}{\beta - \beta_P}\right) \exp(-rs^2)\, ds \tag{5.97}$$

and we must seek an expansion of the quantity in the brace in terms of the powers of s. Noting that $\beta - \beta_P = (\beta - \theta) + (\theta - \beta_P)$ and by (5.87) and (5.88) $\beta - \theta \sim (2/ik)^{1/2} s$, we obtain to the first order in s

$$I_{\mathrm{SD}}^{(1)} = F_1(\beta_P) L(kr) \exp(ikr), \tag{5.98a}$$

where

$$L(kr) = 2a\sqrt{r} \int_0^{\infty} \frac{\exp(-t^2)}{ra^2 - t^2}\, dt, \tag{5.98b}$$

$$a = \left(\frac{ik}{2}\right)^{1/2} (\theta - \beta_P) \tag{5.98c}$$

with $t = \sqrt{r}s$.

The integral $L(kr)$ in (5.98b) can be evaluated exactly in terms of the incomplete error function defined by the integral

$$\operatorname{erfc}(-iz) = \frac{2iz \exp(z^2)}{\pi} \int_0^\infty \frac{\exp(-t^2)}{z^2 - t^2}\, dt, \qquad \operatorname{Im}(z) > 0 \tag{5.99a}$$

and as $z \to \infty$, $|\arg z| < \pi/4$, we have

$$\operatorname{erfc}(-iz) \sim \frac{i \exp(z^2)}{z\sqrt{\pi}} \left[1 + \sum_{m=1}^{\infty} \frac{1 \cdot 3 \cdots (2m-1)}{(2z^2)^m}\right]. \tag{5.99b}$$

Thus, we obtain

$$L(kr) = -i\pi \exp(-ra^2) \operatorname{erfc}(-ia\sqrt{r}) \tag{5.100a}$$

$$\sim \frac{\sqrt{\pi}}{a\sqrt{r}} \left[1 + \sum_{m=1}^{\infty} \frac{1 \cdot 3 \cdots (2m-1)}{(2a^2r)^m}\right]. \tag{5.100b}$$

Substituting back into (5.97) and using (5.98a), we finally secure

$$I_{\text{SD}}^{(1)} = \frac{F_1(\beta_P)}{(\theta - \beta_P)} \left(\frac{2\pi}{ikr}\right)^{1/2} \exp(ikr) \left\{1 + \sum_{m=1}^{\infty} \frac{1 \cdot 3 \cdots (2m-1)}{[ikr(\theta - \beta_P)^2]^m}\right\}. \tag{5.101}$$

The second integral

$$I_{\text{SD}}^{(2)} = \exp(ikr) \int_{-\infty}^{\infty} \exp(-rs^2) \left[F_2(\beta) \frac{d\beta}{ds}\right] ds \qquad [5.96c]$$

can be dealt with in a routine manner since the function in the square brackets has no singularities near the saddle point. Thus, it can be expanded in a Taylor series as follows.

$$F_2(\beta) = \sum_0^{\infty} A_n(\beta - \theta)^n, \tag{5.102a}$$

where

$$A_n = \frac{1}{n!} \frac{d^n}{d\beta^n} \left[F(\beta) - \frac{F_1(\beta_P)}{\beta - \beta_P}\right]_{\beta = \theta}. \tag{5.102b}$$

Finally, combining the results of (5.101) and (5.102), we obtain in this case

$$I_{\text{SD}} = \left(\frac{2\pi}{kr}\right)^{1/2} \exp\left[i\left(kr - \frac{\pi}{4}\right)\right] \left\{F(\theta) + \frac{F_1(\beta_P)}{\theta - \beta_P} \sum_{m=1}^{\infty} \frac{1 \cdot 3 \cdot 5 \cdot \cdots \cdot (2m-1)}{[ikr(\theta - \beta_P)^2]^m}\right\}. \tag{5.103}$$

Problems

5.1 Show that (1.56), (5.1a), and (5.2) are consistent with Maxwell's equations in the form

$$\nabla \times \mathbf{E} = i\omega\mu_0 \mu \mathbf{H},$$

$$\nabla \times \mathbf{H} = -i\omega\varepsilon_0 \varepsilon\left(1 + \frac{i\sigma}{\omega\varepsilon_0 \varepsilon}\right)\mathbf{E} + \mathbf{J}^{(e)},$$

$$\nabla \cdot \mathbf{J}^{(e)} = i\omega\rho^{(e)},$$

$$\nabla \cdot \mathbf{E} = \frac{\mu\rho^{(e)}}{\varepsilon_0 n^2},$$

$$\nabla \cdot \mathbf{H} = 0,$$

while (1.57), (5.1b), and (5.3) are consistent with

$$\nabla \times \mathbf{E} = i\omega\mu_0 \mu \mathbf{H} - \mathbf{J}^{(m)},$$

$$\nabla \times \mathbf{H} = -i\omega\varepsilon_0 \varepsilon\left(1 + \frac{i\sigma}{\omega\varepsilon_0 \varepsilon}\right)\mathbf{E},$$

$$\nabla \cdot \mathbf{H} = \frac{\rho^{(m)}}{\mu_0 \mu},$$

$$\nabla \cdot \mathbf{J}^{(m)} = i\omega\rho^{(m)},$$

$$\nabla \cdot \mathbf{E} = 0,$$

where $\mathbf{J}^{(m)}$ and $\rho^{(m)}$ are fictitious magnetic currents and charges, respectively.

5.2 Show that the Green function $G(\mathbf{r}, \mathbf{r}')$ satisfying the differential equation (5.5.) and the homogeneous boundary conditions (5.8) is symmetrical:

$$G(\mathbf{r}, \mathbf{r}') = G(\mathbf{r}', \mathbf{r}).$$

5.3 Derive (5.41) using the Fourier–Bessel transforms given by

$$\tilde{f}_n(\lambda) = \int_0^\infty f_n(\rho) J_n(\lambda\rho)\rho \, d\rho,$$

$$f_n(\rho) = \int_0^\infty \tilde{f}_n(\lambda) J_n(\lambda\rho)\lambda \, d\lambda.$$

5.4 Starting with (5.40), derive (5.41). Is the indentation about the point $\lambda = 0$ in Fig. 5.8 necessary?

5.5 Show that the cylindrical Green function $G(\boldsymbol{\rho}, \boldsymbol{\rho}')$ can be obtained from the spherical Green function $G(\mathbf{r}, \mathbf{r}')$ by "integrating out" one of the variables. Give a physical motivation for this mathematical manipulation.

5.6 Show that in the case when the stationary point coincides with one of the limits of the integral in (5.53), the formula in (5.56) must be divided by 2.

5.7 Show that the Sommerfeld integral representations of the Hankel functions of the first and second kind,

$$H_\nu^{(1)}(x) = \frac{\exp(-i\nu\pi/2)}{\pi} \int_{C_1} \exp(ix\cos\beta + i\nu\beta)\, d\beta,$$

$$H_\nu^{(2)}(x) = \frac{\exp(-i\nu\pi/2)}{\pi} \int_{C_2} \exp(ix\cos\beta + i\nu\beta)\, d\beta,$$

where C_1 and C_2 are shown in Fig. 5.13 and x is assumed to be real

Fig. 5.13.

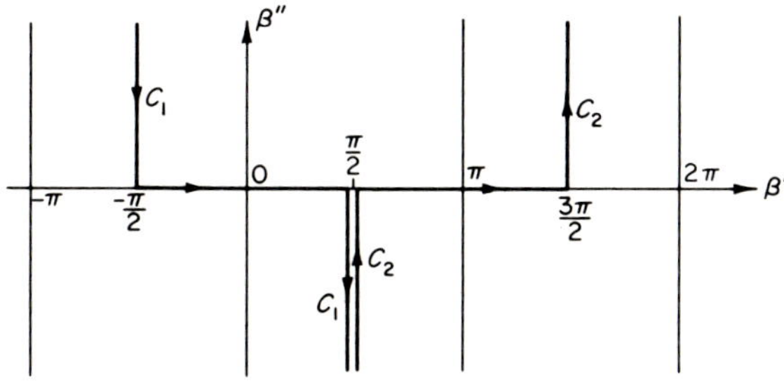

and positive, are solutions to the Bessel differential equation

$$\left[\frac{d^2}{dx^2} + \frac{1}{x}\frac{d}{dx} + \left(1 - \frac{\nu^2}{x^2}\right)\right]\mathscr{C}_\nu = 0.$$

5.8 Using the contour integral representation for $H_\nu^{(1)}(x)$ and $H_\nu^{(2)}(x)$ as found in Problem 5.7 find the first three terms of the asymptotic expansion of $H_\nu^{(1)}(x)$ and $H_\nu^{(2)}(x)$ when x is real and $x \to \infty$. State all of the necessary restrictions.

5.9 Apply the method of steepest descents to the factorial function

$$n! = \int_0^\infty u^n e^{-u}\, du$$

to obtain the Stirling formula valid for a large n

$$n! \sim (2\pi)^{1/2} n^{n+1/2} e^{-n}.$$

Hint: Use the transformation $u = nv$.

References

Abramowitz, M., and Stegun, I. A. (1964). "Handbook of Mathematical Functions." U.S. Govt. Printing Office, Washington, D.C. (*Natl. Bur. Std. U.S. Appl. Math. Ser.* **55**).

Arfken, G. (1966). "Mathematical Methods for Physicists." Academic Press, New York.

Baños, A., Jr. (1966). "Dipole Radiation in the Presence of a Conducting Half-Space." Pergamon, Oxford.

Brekhovskikh, L. M. (1960). "Waves in Layered Media." Academic Press, New York.

Carrier, G. F., Krook, M., and Pearson, C. E. (1966). "Function of a Complex Variable." McGraw-Hill, New York.

Clemmow, P. C. (1966). "Plane Wave Spectrum Representation of Electromagnetic Fields." Pergamon, Oxford.

Felsen, L. B., and Marcuvitz, N. (1961). "Model Analysis and Synthesis of Electromagnetic Fields." *Polytech. Inst. Brooklyn Rept.* **PIBMRI-841-60** (July).

Jeffreys, H., and Jeffreys, B. S. (1950). "Methods of Mathematical Physics," Cambridge Univ. Press, London and New York.

Oberhettinger, F. (1959). On a modification of Watson's lemma, *Natl. Bur. Std. J. Res.* **63B**, 15–17.

Ott, H. (1942). Reflexion und Brechung von Kugelwellen: Effekte 2. Ordnung, *Ann. Physik* **41**, 443–467.

van der Waerden, B. L. (1951). On the method of saddle-points, *Appl. Sci. Res.* **B2**, 33–46.

Watson, G. N. (1962). "Theory of Bessel Functions." Cambridge Univ. Press, London and New York.

CHAPTER 6

FIELD OF A DIPOLE IN A STRATIFIED MEDIUM

In this chapter we consider in detail the simplest problem of radiation from a discrete source in the presence of a separation boundary. We begin with a completely general formulation of the problem of a vertical electric dipole situated in one of the two arbitrary homogeneous and isotropic media separated by a planar boundary. Subsequently, we shall treat various cases of the two media in greater detail, including the classical problem of a vertical dipole above a conducting earth (Sommerfeld problem).

Finally, we shall look into the wave excitation in a three-layer problem, which will lead us to the discussion of electromagnetic wave propagation in the earth ionosphere wave guide.

6.1 Vertical Electric Dipole above an Interface

Consider a region of two adjoining nonmagnetic media separated by a plane infinite interface that shall be assumed to be the x-y plane. A vertical electric dipole of moment $p^{(e)}$ is situated in the upper medium at $z = h$ and is oriented in the positive z direction as shown in Fig. 6.1. The problem can be formulated conveniently in terms of the Hertzian vector of the electric type derived previously. Since the electric dipole is oriented in the z direction, it may be possible to formulate the solution in terms of a single corresponding component of the Hertzian vector $\Pi_z^{(e)}$. Thus, on the basis of the results of Chapter 1 we write

$$(\nabla^2 + k_1{}^2)\Pi_1 = \frac{ip}{\omega\varepsilon_0\, n_1{}^2}\, \delta(x)\, \delta(y)\, \delta(z - h), \tag{6.1a}$$

$$(\nabla^2 + k_2{}^2)\Pi_2 = 0, \tag{6.1b}$$

where $k_1 = k_0 n_1$, $k_2 = k_0 n_2$, $\mu_1 = \mu_2 = 1$, and the refractive index n is defined in (1.51). In writing (6.1) we dropped the subscript z and the super-

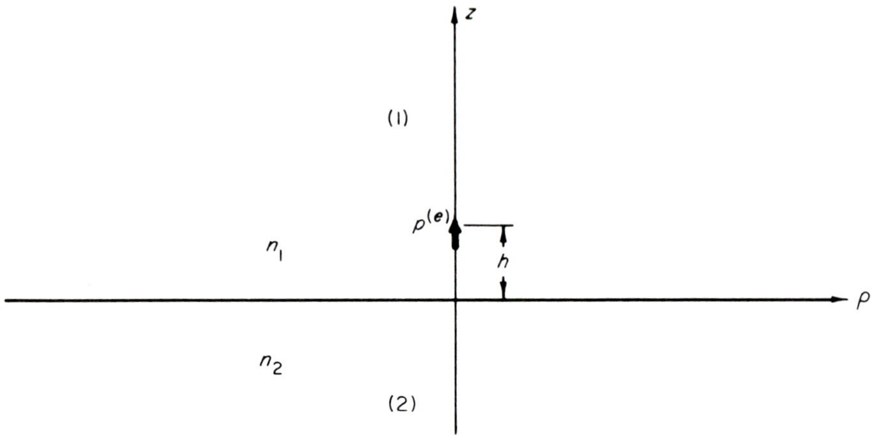

Fig. 6.1. A vertical electric dipole above an interface of two media.

script (e) without risk of confusion. We note that the electric and magnetic fields can be derived from

$$\mathbf{H} = -i\omega\varepsilon_0 n^2\left(\mathbf{1}_x \frac{\partial \Pi}{\partial y} - \mathbf{1}_y \frac{\partial \Pi}{\partial x}\right), \tag{6.2a}$$

$$\mathbf{E} = \mathbf{1}_x\left(\frac{\partial^2 \Pi}{\partial x\, \partial z}\right) + \mathbf{1}_y\left(\frac{\partial^2 \Pi}{\partial y\, \partial z}\right) + \mathbf{1}_z\left(k^2 + \frac{\partial^2}{\partial z^2}\right)\Pi, \tag{6.2b}$$

which follows immediately from (1.67).

At the interface $z = 0$, the tangential electric and magnetic field components must be continuous, which in terms of the Hertzian vector components can be stated as follows.

$$n_1^{\,2} \frac{\partial \Pi_1}{\partial y} = n_2^{\,2} \frac{\partial \Pi_2}{\partial y}, \tag{6.3a}$$

$$n_1^{\,2} \frac{\partial \Pi_1}{\partial x} = n_2^{\,2} \frac{\partial \Pi_2}{\partial x}, \tag{6.3b}$$

$$\frac{\partial}{\partial x}\left(\frac{\partial \Pi_1}{\partial z}\right) = \frac{\partial}{\partial x}\left(\frac{\partial \Pi_2}{\partial z}\right), \tag{6.3c}$$

$$\frac{\partial}{\partial y}\left(\frac{\partial \Pi_1}{\partial z}\right) = \frac{\partial}{\partial y}\left(\frac{\partial \Pi_2}{\partial z}\right). \tag{6.3d}$$

Since these boundary conditions must hold at $z = 0$ for all x and y, it follows that the x and y dependence of the fields on either side of the interface must be the same. Note that the foregoing statement implies first that the fields must be symmetric about the z axis and second that Snell's law must be satisfied. Thus the boundary conditions can be stated concisely as

$$n_1^2\Pi_1 = n_2^2\Pi_2, \tag{6.4a}$$

$$\frac{\partial\Pi_1}{\partial z} = \frac{\partial\Pi_2}{\partial z}, \tag{6.4b}$$

and we must now construct solutions satisfying the differential equations (6.1) and the boundary conditions (6.4).

In Chapter 5 we found the particular solution to the nonhomogeneous equation (6.1). It is clear, however, that in the present problem we shall not be able to satisfy the boundary conditions with that solution alone. We can, however, attempt to find an appropriate solution of the form

$$\Pi_1 = \Pi_1^{(p)} + \Pi_1^{(c)} \tag{6.5}$$

where $\Pi_1^{(p)}$ will denote the particular and $\Pi_1^{(c)}$ will denote the complementary solution of (6.1a). The particular solution, as was just mentioned, we already found in Chapter 5. Thus, by (5.49a) and (5.22b) we write

$$\Pi_1^{(p)} = \frac{ip}{4\pi\omega\varepsilon_0 n_1^2}\cdot\frac{\exp(ik_1r_1)}{r_1} \tag{6.6a}$$

$$\Pi_1^{(p)} = -\frac{p}{8\pi^2\omega\varepsilon_0 n_1^2}\iint_{-\infty}^{\infty}\frac{\exp\{i[\alpha_1 x + \alpha_2 y + \tau_1|z-h|]\}}{\tau_1}\,d\alpha_1\,d\alpha_2, \tag{6.6b}$$

where in this case

$$r_1 = [x^2 + y^2 + (z-h)^2]^{1/2} = [\rho^2 + (z-h)^2]^{1/2}, \tag{6.6c}$$

$$\tau_1 = [k_1^2 - (\alpha_1^2 + \alpha_2^2)]^{1/2}. \tag{6.6d}$$

The complementary solution must satisfy the homogeneous differential equation

$$(\nabla^2 + k_1^2)\Pi_1^{(c)} = 0 \tag{6.7}$$

and it can be constructed in a desired form by first transforming (6.7) using a two-dimensional Fourier transform on the x and y coordinates and subsequently solving the ordinary differential equation with respect to the remaining z variable. We obtain in this case

$$\tilde{\tilde{\Pi}}_1^{(c)} = A(\alpha_1, \alpha_2)\exp(i\tau_1 z) + B(\alpha_1, \alpha_2)\exp(-i\tau_1 z). \tag{6.8}$$

Consistent with the convention $\mathrm{Im}(\tau_1) > 0$, we must set $B(\alpha_1, \alpha_2) = 0$ in (6.8) so that fields will remain finite as $z \to \infty$. Inverting and adding to (6.6b), we obtain a complete integral representation for the fields in the upper half space as follows.

$$\Pi_1 = -\frac{p}{8\pi^2\omega\varepsilon_0 n_1{}^2}\iint_{-\infty}^{\infty}\left[\frac{\exp(i\tau_1|z-h|)}{\tau_1} + R(\alpha_1, \alpha_2)\exp(i\tau_1 z)\right]$$
$$\times \exp[i(\alpha_1 x + \alpha_2 y)]\, d\alpha_1\, d\alpha_2 . \tag{6.9}$$

Since the problem of finding the integral representation of the fields in the lower region is analogous to finding the complementary integral in the upper region, we can write by inspection

$$\Pi_2 = -\frac{p}{8\pi^2\omega\varepsilon_0 n_1{}^2}\iint_{-\infty}^{\infty} T(\alpha_1, \alpha_2)\exp(-i\tau_2 z)\exp[i(\alpha_1 x + \alpha_2 y)]\, d\alpha_1\, d\alpha_2 , \tag{6.10}$$

where

$$\tau_2 = [k_2{}^2 - (\alpha_1{}^2 + \alpha_2{}^2)]^{1/2} \tag{6.11}$$

and since in this case the negative exponential solution for the z coordinate must be retained. The factor $-p/8\pi^2\omega\varepsilon_0 n_1{}^2$ has been inserted in front of the integral in (6.10) for future convenience.

Applying the boundary conditions at $z = 0$ as specified by (6.4), we obtain two simultaneous equations.

$$\frac{\exp(i\tau_1 h)}{\tau_1} + R = n_{21}^2 T, \tag{6.12a}$$

$$\frac{\exp(i\tau_1 h)}{\tau_1} - R = \frac{\tau_2}{\tau_1} T, \tag{6.12b}$$

$$n_{21} = \frac{n_2}{n_1}, \qquad n_{12} = \frac{n_1}{n_2}, \tag{6.12c}$$

which can be readily solved to give

$$R = \frac{n_{21}^2\tau_1 - \tau_2}{n_{21}^2\tau_1 + \tau_2}\,\frac{\exp(i\tau_1 h)}{\tau_1}, \tag{6.13a}$$

$$T = \frac{2\exp(i\tau_1 h)}{n_{21}^2\tau_1 + \tau_2} . \tag{6.13b}$$

Substitution of the last results into (6.9) and (6.10) concludes the formal solution to the problem.

Before leaving this section, it will be instructive to inquire into the physical interpretation of these results. To this end we note that R in (6.13a) can be written

$$R = \frac{\exp(i\tau_1 h)}{\tau_1} - \frac{2\tau_2}{n_{21}^2\tau_1 + \tau_2}\,\frac{\exp(i\tau_1 h)}{\tau_1}, \tag{6.14}$$

and upon substitution into (6.9) gives

$$\Pi_1 = \frac{ip}{4\pi\omega\varepsilon_0 n_1{}^2}\left[\frac{\exp(ik_1 r_1)}{r_1} + \frac{\exp(ik_1 r_2)}{r_2} + I_1\right], \tag{6.15a}$$

where

$$r_2 = [\rho^2 + (z + h)^{1/2}] \tag{6.15b}$$

and

$$I_1 = -\frac{i}{\pi}\iint_{-\infty}^{\infty} \frac{\tau_2}{\tau_1(n_{21}^2\tau_1 + \tau_2)} \exp\{i[\alpha_1 x + \alpha_2 y + \tau_1(z + h)]\}\, d\alpha_1\, d\alpha_2, \tag{6.16a}$$

$$= -i\int_{-\infty}^{\infty} \frac{\tau_2 H_0^{(1)}(\lambda\rho)}{\tau_1(n_{21}^2\tau_1 + \tau_2)} \exp[i\tau_1(z + h)]\lambda\, d\lambda, \tag{6.16b}$$

and in order to arrive at (6.15) we used (5.22b) and (5.19), and to obtain (6.16b) we used the transformations in (5.23a). Note that in the case $n_2 \gg n_1$, the contribution of the integral I_1 in (6.15a) vanishes and the solution approaches the case of a vertical electric dipole above a perfectly conducting half space as shown in Fig. 6.2. Thus the role of the integral I_1 in (6.15a) can be interpreted as one of a correction factor that must be added in the case when the lower half space has a finite conductivity or a finite value of the electric permittivity.

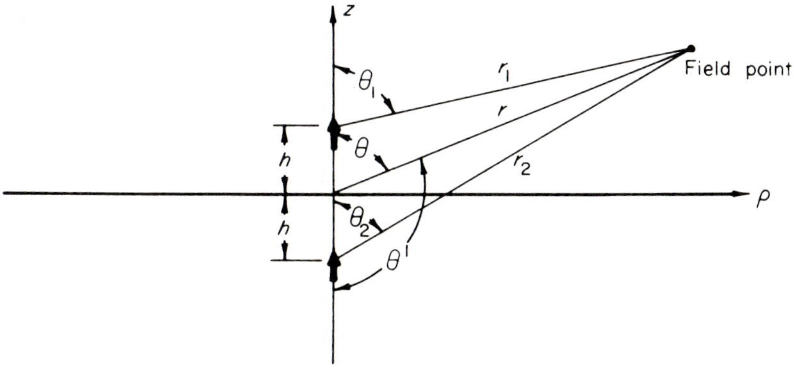

Fig. 6.2. Problem of a vertical electric dipole above a perfectly conducting half space.

6.1.1 Analysis of the Reflected Field

According to the previous results, the field in the upper half space consists of the direct field of the dipole situated on the z axis at $z = h$, its image situated at $z = -h$ and the correction term contained in the integral I_1 in (6.16). It follows that the reflected (complementary) field in the upper half space can be written

$$\Pi_1^{(c)} = \frac{ip}{4\pi\omega\varepsilon_0 n_1{}^2}\left(\frac{\exp(ik_1 r_2)}{r_2} + I_1\right), \tag{6.17}$$

and hence the problem of determining the reflected field amounts to the evaluation of the integral I_1 given by (6.16).

In what follows it will be convenient to regard λ in (6.16b) as a complex variable rather than a real one. Moreover, in order to evaluate (6.16b) by contour integration, we must close the path of integration with a large semicircle in the upper half-plane in such a way that the contribution from the integration along the semicircle will vanish as $|\lambda|$ approaches infinity. To ensure this convergence, we must inquire into the nature of the integrand in (6.16b). (Note that we could not close the contour in the lower half plane due to the branch cut of $H_0^{(1)}(\lambda\rho)$ along the negative real axis as shown in Fig. 5.4a.)

We note that the said integrand contains two radicals $\tau_1 = (k_1{}^2 - \lambda^2)^{1/2}$ and $\tau_2 = (k_2{}^2 - \lambda^2)^{1/2}$, which are double-valued functions of the complex variable λ. In order that the integral I_1 properly represent a field solution, we must ensure that it will be single valued everywhere on the path of integration. This can be accomplished by connecting the points $\tau_1 = 0$, $\pm\infty$ and $\tau_2 = 0$, $\pm\infty$ by means of lines, called branch cuts, that cannot be crossed by the path of integration. The details of the branch cuts are arbitrary as far as the single-valuedness of the integrand is concerned; the vanishing of the contribution from the integration along the semicircular arc is another matter, however.

In the integrand under consideration we can "extract" the asymptotic behaviour of the Hankel function for large values of the argument by defining

$$\breve{H}_\nu^{(1)}(\lambda\rho) = H_\nu^{(1)}(\lambda\rho)\exp\left[-i\left(\lambda\rho - \frac{\nu\pi}{2} - \frac{\pi}{4}\right)\right], \tag{6.18a}$$

$$\breve{H}_\nu^{(1)}(\lambda\rho) \xrightarrow[|\lambda\rho|\to\infty]{} \left(\frac{2}{\pi\lambda\rho}\right)^{1/2}. \tag{6.18b}$$

Using (6.18a) in (6.16b), we obtain

$$I_1 = -\exp\left(\frac{i\pi}{4}\right)\int_{-\infty}^{\infty} \frac{\tau_2 \breve{H}_0^{(1)}(\lambda\rho)}{\tau_1(n_{21}^2\tau_1 + \tau_2)} \exp\{i[\lambda\rho + \tau_1(z+h)]\}\lambda\, d\lambda. \tag{6.19}$$

Now let us focus our attention on the exponent in the integrand. We can write

$$\tau_1 = [(k_1 + \lambda)(k_1 - \lambda)]^{1/2} = M \exp\left[\frac{i(\varphi_1 + \varphi_2)}{2}\right] \tag{6.20}$$

where M is a positive real number and $\varphi_1 = \arctan[(k_1'' + \lambda'')/(k_1' + \lambda')]$ and $\varphi_2 = \arctan[(k_1'' - \lambda'')/(k_1' - \lambda')]$. We note that on the line $\lambda'' = k_1''$, $\varphi_2 = 0$; whereas on the line $\lambda'' = -k_1''$, $\varphi_1 = 0$, as shown in Fig. 6.3. Now let us put

$$\begin{aligned} z + h &= r_2 \cos\theta_2, \\ \rho &= r_2 \sin\theta_2, \qquad 0 \le \theta_2 \le \pi/2, \end{aligned} \tag{6.21}$$

in agreement with the diagram in Fig. 6.2. Hence, the exponent becomes

$$\begin{aligned} i[\lambda\rho + \tau_1(z + h)] &= ir_2[\lambda' \sin\theta_2 + M \cos\theta_2 \cos(\varphi_1 + \varphi_2)/2] \\ &\quad - r_2[\lambda'' \sin\theta_2 + M \cos\theta_2 \sin(\varphi_1 + \varphi_2)/2] \end{aligned} \tag{6.22}$$

and it follows that the integral in (6.19) will converge in the upper half plane of the complex λ plane if the quantity

$$\lambda'' \sin\theta_2 + M \cos\theta_2 \sin(\varphi_1 + \varphi_2)/2 > 0. \tag{6.23}$$

Now in the upper half plane $\lambda'' > 0$; moreover, since $0 \le \theta_2 \le \pi/2$, the first term in (6.23), as well as part of the second term $M \cos\theta_2$, will always be positive. Thus, the convergence will certainly be ensured if we require that $\sin(\varphi_1 + \varphi_2)/2 > 0$, or equivalently

$$0 < (\varphi_1 + \varphi_2) < 2\pi. \tag{6.24}$$

Consider Fig. 6.3, where three possible configurations of the branch cut

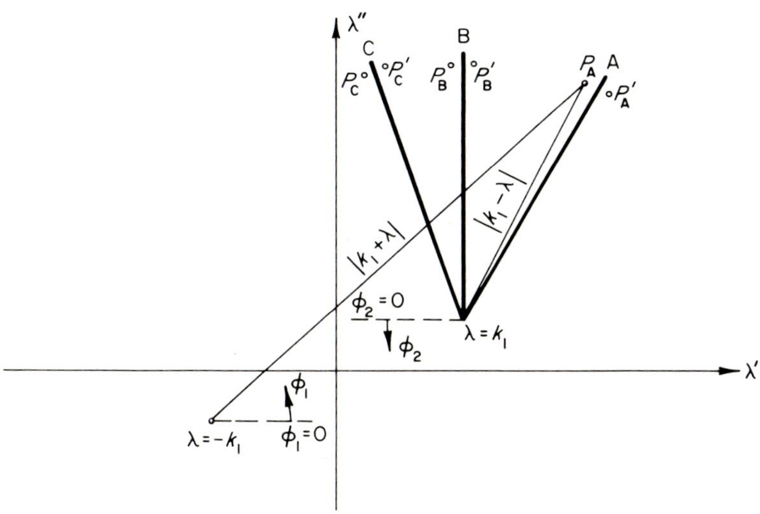

Fig. 6.3. A proper location of the branch cut.

connecting the points $\lambda = k_1$, ∞ are shown. First, let us look at the configuration A. On the right side of the branch cut facing the branch point, at the point P_A the phase of $k_1 + \lambda$ is $0 < \varphi_1 < \pi/2$ and the phase of $k_1 - \lambda$ is $-\pi < \varphi_2 < -\pi/2$; hence the sum $-\pi < \varphi_1 + \varphi_2 < 0$, which is in violation of (6.24). However, on the opposite side of the branch cut at the point P_A', $\pi < \varphi_1 + \varphi_2 < 2\pi$; consequently, (6.24) is satisfied at that point. Evidently, in order to improve the phase angle on the right side of the branch cut, we must swing the branch cut toward the imaginary axis.

Now consider the branch cut configuration B in Fig. 6.3. At the point P_B, the phases are $0 < \varphi_1 < \pi/2$ and $-\pi/2 < \varphi_2 < 0$; hence $-\pi/2 < \varphi_1 + \varphi_2 < \pi/2$, which seems to violate (6.24). However, as λ'' approaches infinity, φ_1 approaches $\pi/2$ and φ_2 approaches $-\pi/2$; hence $\varphi_1 + \varphi_2$ approaches zero. In fact it can be shown (problem) that in this case

$$\tan(\varphi_1 + \varphi_2) = \frac{2[k_1' k_1'' + \lambda''(\delta - k_1')]}{\lambda''^2 + \delta(2k_1' - \delta) - k_1''^2}, \tag{6.25}$$

where δ is the distance from the point P_B to the branch cut. It is clear that the right side of (6.25) approaches zero as λ'' approaches infinity and hence $\varphi_1 + \varphi_2$ approaches zero as stated earlier. At the point P_B' we have $0 < \varphi_1 < \pi/2$ and $\pi < \varphi_2 < 3\pi/2$ and, consequently, $\pi < \varphi_1 + \varphi_2 < 2\pi$ which again is acceptable. Thus, the branch cut configuration B satisfies the requirement of convergence.

Finally, consider the branch cut configuration C. At the point P_C, the sum $\varphi_1 + \varphi_2$ is well within the bounds of (6.24). However, at the point P_C', the sum $\varphi_1 + \varphi_2$ exceeds 2π, which violates the condition of (6.24). Thus, the branch cut configuration C is incompatible with the convergence requirement.

We could also choose a branch cut in such a manner that

$$\operatorname{Im}(k_1^2 - \lambda^2)^{1/2} = 0 \tag{6.26a}$$

on the branch cut. It will follow from (6.20) that in this case we must have $\varphi_2 = -\varphi_1$ or $\varphi_2 = 2\pi - \varphi_1$, and moreover

$$\lambda'\lambda'' = k_1' k_1'' \tag{6.26b}$$

on the branch cut. Equation (6.26b) represents a set of two hyperbolas in the λ'-λ'' plane running through the points $(\pm k_1', \pm k_1'')$ as shown in Fig. 6.4. It is easy to show (problem) that on the right side of the branch cut facing the origin

$$(k_1^2 - \lambda^2)^{1/2} = M \exp(i\alpha) \quad \text{(right)}, \tag{6.27a}$$

whereas on the left side

$$(k_1^2 - \lambda^2)^{1/2} = M \exp[i(\pi - \alpha)] \quad \text{(left)}, \tag{6.27b}$$

where α is an arbitrary small and positive angle. In fact, it is easy to show that on a large semicircle in the upper λ plane $0 < \varphi_1 + \varphi_2 < \pi$ on the right side of the branch cut facing the origin, and $\pi < \varphi_1 + \varphi_2 < 2\pi$ on the left.

There is, in this case, no specific requirement as to how the other branch points, $\lambda = \pm k_2$, should be connected with infinity as far as the convergence of the integral in (6.19) is concerned. It will, however, prove convenient to construct the corresponding branch cut in a manner similar to the other branch cut, which will ensure the convergence of the integral representation for the fields in the lower region as given by (6.10).

In addition to the branch point singularities just discussed the integrand in (6.19) admits the possibility of a pole at the point where the denominator vanishes; that is, when

$$n_2^2(k_1^2 - \lambda^2)^{1/2} + n_1^2(k_2^2 - \lambda^2)^{1/2} = 0, \tag{6.28a}$$

which amounts to

$$\lambda = \frac{\pm n_1 n_2 k_0}{(n_1^2 + n_2^2)^{1/2}}. \tag{6.28b}$$

The exact locations of these poles are not easy to determine in this general case. However, we can still show that the poles will be situated in the first and third quadrants of the λ plane. To do this we proceed as follows. We put

$$n_1 = N_1 \exp(i\theta_1), \qquad n_2 = N_2 \exp(i\theta_2), \tag{6.29a}$$

and note that for the usual physical media $0 \leq \theta_{1,2} \leq \pi/4$. Then

$$\begin{aligned}\frac{\lambda}{k_0} &= \frac{\pm N_1 N_2 \exp[i(\theta_1 + \theta_2)]}{[N_1^2 \exp(i2\theta_1) + N_2^2 \exp(i2\theta_2)]^{1/2}} \\ &= \pm N \exp[i(\theta_1 + \theta_2 - \theta)]\end{aligned} \tag{6.29b}$$

and it will follow that $0 < \theta_1 + \theta_2 - \theta < \pi/2$, so that the poles will be located in the first and third quadrants, as stated before.

Next, the location of the poles should be examined more closely inasmuch as they affect the evaluation of the field integral (6.19). In what follows it will be convenient to talk about the problem in terms of a four-sheeted Riemann surface, which comes about from two possible choices of sign in front of the two radicals in the integrand of (6.19). We shall designate these sheets as in Table 6.1. Now to ensure the proper behavior of the fields at large distances

Table 6.1

Sheet	$\mathrm{Im}(k_1^2 - \lambda^2)^{1/2}$	$\mathrm{Im}(k_2^2 - \lambda^2)^{1/2}$
I	+	+
II	+	−
III	−	+
IV	+	−

from the source, we must make sure that the integration is carried out on sheet I. This will be satisfied if the path of integration never crosses any one of the branch cuts in Fig. 6.4. Furthermore, we must know whether the pole found in (6.28b) lies on sheet I and must be taken into account when carrying out the integration.

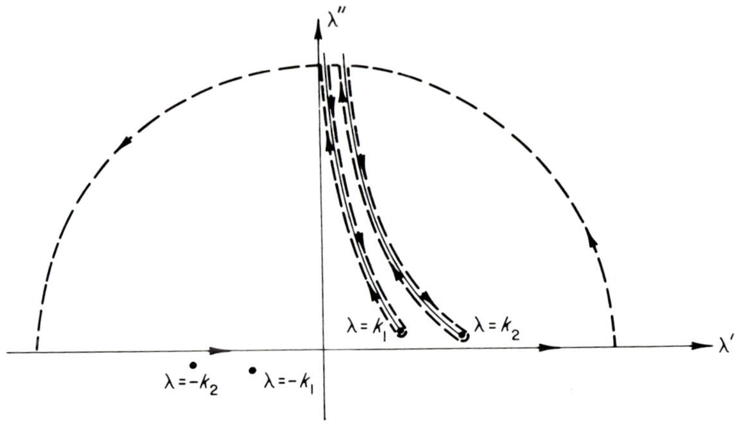

Fig. 6.4. Path of integration in the complex λ plane.

It appears that in the present general formulation of the problem the task of determining the exact pole location is extremely difficult. We shall postpone it, then, until specific media are considered.

The evaluation of the integral in (6.19) is greatly simplified if we make the transformations

$$\lambda = k_1 \sin\beta, \qquad z + h = r_2 \cos\theta_2, \qquad \rho = r_2 \sin\theta_2, \tag{6.30a}$$

to obtain

$$I_1 = -k_1 \exp\left(\frac{i\pi}{4}\right) \int_{\Gamma_1} \frac{(n_{21}^2 - \sin^2\beta)^{1/2} \breve{H}_0^{(1)}(k_1 \rho \sin\beta) \sin\beta}{n_{21}^2 \cos\beta + (n_{21}^2 - \sin^2\beta)^{1/2}}$$
$$\times \exp[ik_1 r_2 \cos(\beta - \theta_2)]\, d\beta, \tag{6.30b}$$

where Γ_1 is a path in the complex β plane as shown in Fig. 5.12.

Let us now inquire into the consequences of the transformation $\lambda = k_1 \sin\beta$ on the branch points, branch cuts, and poles. The location of the branch points $\lambda = \pm k_1$ in the λ plane is now transformed into $\sin\beta = \pm 1$ in the β plane. Thus, these branch points are now situated at $\beta = \pm\pi/2$, $\pm 3\pi/2$, and so on. The branch cuts $\text{Im}({k_1}^2 - \lambda^2)^{1/2} = 0$ are now transformed into $\text{Im}(k_1 \cos\beta) = 0$ and they must begin at the branch points $\beta = \pm\,\pi/2$. Since $\text{Im}(k_1 \cos\beta) = \text{Im}(k_1 \sin(\pi/2 \pm \beta)) = 0$ it will follow that these lines run parallel to the lines $\mathrm{Im}(k_1 \sin \beta) = 0$, the Γ_1 path, except that they will be shifted from the Γ_1 path by a distance of $\pi/2$ along the real axis, as shown in Figs. 5.6. and 5.12.

The remaining branch points $\lambda = \pm k_2$ are now transformed into $\sin \beta = \pm n_{21}$, which has solutions

$$\beta_B = -i \ln\{\pm[in_{21} \pm (1 - n_{21}^2)^{1/2}]\}. \tag{6.31}$$

The branch cuts $\mathrm{Im}({k_2}^2 - \lambda^2)^{1/2} = 0$ now transform into

$$\mathrm{Im}({n_2}^2 - {n_1}^2 \sin^2 \beta)^{1/2} = 0,$$

which is impossible to trace in the general case.

Finally, the location of the pole given by (6.27b) for the λ plane will now be given by

$$\sin \beta_P = \frac{\pm n_{21}}{(1 + n_{21}^2)^{1/2}}, \tag{6.32}$$

which again is impossible to trace in general, but must be examined closely in connection with specific problems to follow.

6.1.2 Analysis of the Transmitted Field

Now let us consider the field in the lower half space as given in (6.10) and (6.13b). We write for convenience

$$\Pi_2 = -\frac{p}{4\pi\omega\varepsilon_0 {n_1}^2} I_2, \tag{6.33a}$$

where

$$I_2 = \frac{1}{\pi} \iint_{-\infty}^{\infty} \frac{\exp(i\tau_1 h) \exp(-i\tau_2 z)}{n_{21}^2 \tau_1 + \tau_2} \exp[i(\alpha_1 x + \alpha_2 y)]\, d\alpha_1\, d\alpha_2, \tag{6.33b}$$

$$I_2 = \int_{-\infty}^{\infty} \frac{\exp(i\tau_1 h) H_0^{(1)}(\lambda\rho)}{n_{21}^2 \tau_1 + \tau_2} \exp(-i\tau_2 z)\lambda\, d\lambda. \tag{6.33c}$$

As before, we extract the asymptotic behavior of the Hankel function using (6.18a) and write

$$I_2 = \exp\left(-\frac{i\pi}{4}\right) \int_{-\infty}^{\infty} \frac{\exp(i\tau_1 h) \breve{H}_0^{(1)}(\lambda\rho)}{n_{21}^2 \tau_1 + \tau_2} \exp[i(\lambda\rho - \tau_2 z)]\lambda\, d\lambda. \tag{6.34}$$

It is easy to show that with the choice of the branch cut

$$\mathrm{Im}(\tau_2) = \mathrm{Im}({k_2}^2 - \lambda^2)^{1/2} = 0$$

as in Fig. 6.4, the contribution from the semicircular arc will vanish as $|\lambda| \to \infty$. Thus, the same contour can be used for evaluation I_2 as was used for I_1.

Now using the transformations

$$\lambda = k_2 \sin\beta, \qquad \rho = r\sin\theta', \qquad -z = r\cos\theta', \qquad 0 \le \theta' \le \pi/2, \tag{6.35}$$

we obtain

$$I_2 = \frac{k_0 \exp[-(i\pi/4)]}{n_2} \int_{\Gamma_1} \frac{\exp[ik_2(n_{12}^2 - \sin^2\beta)^{1/2} h] \breve{H}_0^{(1)}(k_2 \rho \sin\beta)}{n_{12}^2 \cos\beta + (n_{12}^2 - \sin^2\beta)^{1/2}} \times \sin\beta\cos\beta \exp[ik_2 r\cos(\beta - \theta')]\, d\beta. \tag{6.36}$$

We note that the only essential difference between (6.36) and (6.30) is the interchange of the roles of n_1 and n_2. Thus, all of the transformations developed for the previous case apply with the interchange of n_1 and n_2.

6.2 Vertical Electric Dipole above a Conducting Earth (Sommerfeld Problem)

We shall now use the formalism developed in the preceding section to consider the classical problem of a vertical electric dipole above a conducting earth. The problem was formally solved by Sommerfeld in 1909 but subsequently a great deal of effort was expended by many authors in interpretation of the results. A comprehensive résumé of the problem can be found in the monograph by Baños (1966).

Before we attempt to evaluate the various integrals it is essential to estimate the range of values of the relevant parameters. Since the upper medium in this case is air, we set $n_1 = 1$. The lower medium is earth or sea water, which can be represented by the index of refraction

$$n_2 = n = (\varepsilon_r + i\sigma/\omega\varepsilon_0)^{1/2}, \tag{6.37}$$

where ε_r is the relative dielectric constant of the earth or sea water, σ is the conductivity, and the relative permeability is assumed to be equal to unity. To get an idea of the range of values of the refractive index, let us consider the so-called good earth and sea water. The generally accepted values for earth are $\sigma = 12 \times 10^{-3}$ mho per meter, $\varepsilon_r = 15$; and for sea water, $\sigma = 4$ mhos per meter, $\varepsilon_r = 81$. Based on these values, the tabulation of values of the refractive index as a function of frequency is given in Table 6.2. It will be noted

Table 6.2

VARIATION OF THE REFRACTIVE INDEX OF EARTH AND SEA WATER WITH FREQUENCY[a]

f	Earth		Sea water	
	n^2	n	n^2	n
1 kHz	$15 + i2.16 \times 10^5$	$328(1 + i)$	$81 + i7.2 \times 10^7$	$6000(1 + i)$
10 kHz	$15 + i2.16 \times 10^4$	$104(1 + i)$	$81 + i7.2 \times 10^6$	$1900(1 + i)$
100 kHz	$15 + i2.16 \times 10^3$	$32.8(1 + i)$	$81 + i7.2 \times 10^5$	$600(1 + i)$
1 MHz	$15 + i2.16 \times 10^2$	$10.4(1 + i)$	$81 + i7.2 \times 10^4$	$190(1 + i)$
10 MHz	$15 + i21.6$	$6.02 + i3.14$	$81 + i7.2 \times 10^3$	$60(1 + i)$
100 MHz	$15 + i2.16$	$3.87 + i0.28$	$81 + i720$	$60 + i53.4$

[a] Earth, $\sigma = 12 \times 10^{-3}$ mho/meter, $\varepsilon_r = 15$; sea water, $\sigma = 4$ mhos/meter, $\varepsilon_r = 81$.

that the displacement current is negligible if we limit ourselves to frequencies below about 1 megacycle per second (MHz), in which case the results should be valid in the very low (3–30 kHz) and low (30–300 kHz) frequency ranges.

6.2.1 FIELDS IN THE AIR

Let us now examine the reflected field in the air (upper half space). Setting $n_1 = \mu_1 = 1$ and $n_2 = n$ in (6.30b), we write

$$I_1 = -k_0 \exp\left(\frac{i\pi}{4}\right) \int_{\Gamma_1} \frac{(n^2 - \sin^2 \beta)^{1/2} \breve{H}_0^{(1)}(k_0 \rho \sin \beta) \sin \beta}{n^2 \cos \beta + (n^2 - \sin^2 \beta)^{1/2}} \times \exp[ik_0 r_2 \cos(\beta - \theta_2)]\, d\beta. \tag{6.38}$$

When the refractive index is large, then according to (6.31) the possible location of branch points will be given approximately by

$$\beta_B = \pm \ln(\pm i2n). \tag{6.39a}$$

In the present case, $n = |n| \exp(i\pi/4)$ and the following locations are possible.

$$\beta_{B1} = \pi/4 + i \ln(2|n|), \tag{6.39b}$$

$$\beta_{B2} = -\pi/4 - i \ln(2|n|), \tag{6.39c}$$

$$\beta_{B3} = 3\pi/4 - i \ln(2|n|), \tag{6.39d}$$

$$\beta_{B4} = -3\pi/4 + i \ln (2|n|). \tag{6.39e}$$

Similarly, the possible location of poles (6.32) can be approximated by

$$\pm\sin\beta_P = 1 - \frac{1}{2n^2} \sim \cos\left(\frac{\pi}{2} \mp \beta_P\right), \tag{6.40a}$$

which gives

$$\beta_P = \pm\left(\frac{\pi}{2} \pm \frac{1}{n}\right), \tag{6.40b}$$

where all combinations of signs are possible. Thus for $n = n'(1 + i)$

$$\beta_{P1} = \left(\frac{\pi}{2} - \frac{1}{2n'}\right) + \frac{i}{2n'}, \tag{6.40c}$$

$$\beta_{P2} = \left(\frac{\pi}{2} + \frac{1}{2n'}\right) - \frac{i}{2n'}, \tag{6.40d}$$

$$\beta_{P3} = -\left(\frac{\pi}{2} - \frac{1}{2n'}\right) - \frac{i}{2n'}, \tag{6.40e}$$

$$\beta_{P4} = -\left(\frac{\pi}{2} + \frac{1}{2n'}\right) + \frac{i}{2n'}. \tag{6.40f}$$

The complex β plane showing possible locations of branch points and poles is shown in Fig. 6.5. Note that of the four possible locations of the branch

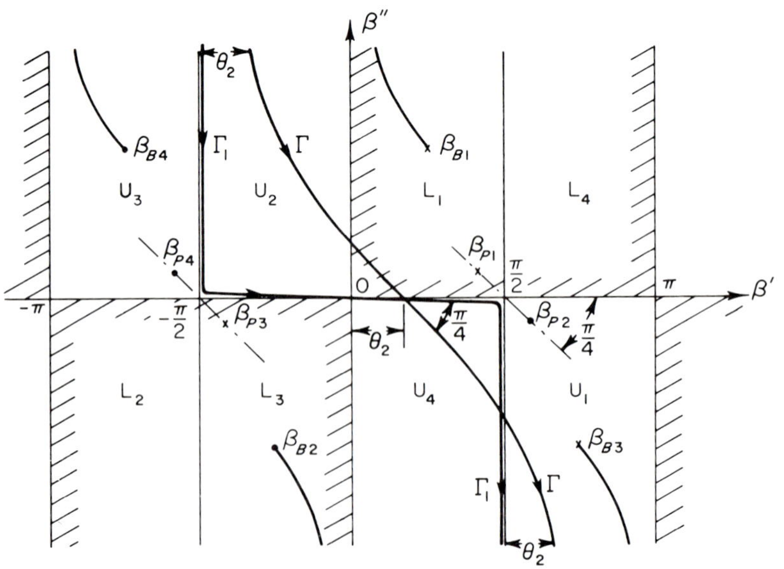

Fig. 6.5. The complex β plane showing possible locations of branch points and poles when $n = n'(1 + i)$ and $|n| \gg 1$.

points, only two are situated on the upper sheet on which the integration must be performed. Similarly, only two of the four possible poles are situated on the upper sheet. It is also important to note that none of the upper-sheet poles is situated between the original path of integration Γ_1 and the path of steepest descents Γ. In particular, the pole β_{P2} is on a line having a slope $-\pi/4$ with respect to the horizontal axis; thus, it will never be intercepted by the path of steepest descents, which also has a slope of $-\pi/4$ on the real axis, but increases in the negative direction in the lower half plane. However, due to the proximity of the pole β_{P2} with respect to the point $\beta = \pi/2$, special precautions must be taken in evaluation of the field integral (6.38).

The presence of the branch point β_{B3} should ordinarily be taken into account in deforming the path Γ_1 into the path Γ. It will be seen in the following section that such a deformation entails an addition of an integral along the borders of the relevant branch cut to the integral along the path of the steepest descents. In the present case the location of the branch point is such that its imaginary part is proportional to $\ln(2|n|)$, which for a large $|n|$ is a very large number. Because of this it will follow that the contribution from the integration along the borders of the branch cut is a fast-decreasing exponential that can be neglected by comparison with the contribution from the saddle-point integration.

We are now ready to proceed with the evaluation of the integral (6.38). Since the deformation of the contour Γ_1 into the path of steepest descents Γ can be readily accomplished, we can use (5.94) for the case $\theta_2 < \pi/2$ to obtain

$$I_1 = -\left[\frac{2(n^2 - \sin^2\theta_2)^{1/2}}{n^2\cos\theta_2 + (n^2 - \sin^2\theta_2)^{1/2}} + \frac{1}{2ik_0 r_2}\,G(\theta_2) + \cdots\right]\frac{\exp(ik_0 r_2)}{r_2}, \tag{6.41a}$$

where

$$G(\theta_2) = \frac{\begin{aligned}2n^2(n^2-1)[2n^2(n^2-1) + n^2\cos^2\theta_2(3-\cos^2\theta_2)\\ + \cos\theta(n^2-\sin^2\theta_2)^{1/2}(2n^2-\sin^2\theta_2)]\end{aligned}}{(n^2-\sin^2\theta_2)^{3/2}[n^2\cos\theta_2 + (n^2-\sin^2\theta_2)^{1/2}]^3}. \tag{6.41b}$$

The complementary (reflected) field in the upper half space can now be found by using (6.17). We obtain

$$\Pi_1^{(c)} = \frac{ip}{4\pi\omega\varepsilon_0}\left[\frac{n^2\cos\theta_2 - (n^2-\sin^2\theta_2)^{1/2}}{n^2\cos\theta_2 + (n^2-\sin^2\theta_2)^{1/2}} - \frac{1}{2ik_0 r_2}\,G(\theta_2) + \cdots\right] \times \frac{\exp(ik_0 r_2)}{r_2}. \tag{6.42}$$

When the conductivity of the earth is large, $|n| \gg 1$, we obtain for the total field in the upper half space

$$\Pi_1 = \frac{ip}{4\pi\omega\varepsilon_0}\left\{\frac{\exp(ik_0 r_1)}{r_1} + \left[1 - \frac{2}{n\cos\theta_2 + 1} - \frac{2n^2}{ik_0 r_2}\left(\frac{1}{n\cos\theta_2 + 1}\right)^3 + \cdots\right] \times \frac{\exp(ik_0 r_2)}{r_2}\right\}. \quad (6.43)$$

Note that as $n \to \infty$, Π_1 goes properly into the form of a field due to a vertical electric dipole above a perfectly conducting plane. However, when $\theta_2 \sim \pi/2$, we obtain

$$\Pi_1 = \frac{ip}{4\pi\omega\varepsilon_0}\left[\frac{\exp(ik_0 r_1)}{r_1} - \frac{\exp(ik_0 r_2)}{r_2} - \frac{2n^2}{ik_0}\cdot\frac{\exp(ik_0 r_2)}{r_2^{\,2}} + \cdots\right]. \quad (6.44)$$

Here we observe that the sum of the first two terms may be smaller than the third term. As a matter of fact, if both the transmitter and the receiver are on the ground, then $h = 0$ and $r_1 = r_2$, and consequently the field will be solely determined by the second- and higher-order terms. The reason for this poor convergence in the vicinity of $\theta_2 = \pi/2$ is that the effect of the pole at $\beta = \beta_{P2}$ now becomes important. This can be properly taken into account right from the start by using the modified saddle-point technique outlined in Chapter 5.

6.2.2 Fields near the Interface

As we pointed out earlier, when the observation point is near the interface, a straightforward application of the saddle-point method of integration is not convenient since the second- and higher-order terms become important. Even though a mechanism is available for generation of the higher-order terms, the procedure is very tedious since it entails higher-order derivatives of the integrand. This can be avoided by application of the modified method of steepest descents outlined in Chapter 5.

As the first step toward application of the modified method of steepest descents, let us write $R(\alpha_1, \alpha_2)$ in (6.13a) as follows.

$$R = -\frac{\exp(i\tau_1 h)}{\tau_1} + \frac{2n^2\exp(i\tau_1 h)}{n^2\tau_1 + \tau_2}, \quad (6.45)$$

which upon substitution into (6.9) gives

$$\Pi_1 = \frac{ip}{4\pi\omega\varepsilon_0}\left[\frac{\exp(ik_0 r_1)}{r_1} - \frac{\exp(ik_0 r_2)}{r_2} + I_1'\right], \quad (6.46a)$$

where

$$I_1' = in^2 \int_{-\infty}^{\infty} \frac{\breve{H}_0^{(1)}(\lambda\rho)\exp[i\tau_1(z+h)]}{n^2\tau_1+\tau_2} \lambda\, d\lambda. \tag{6.46b}$$

The reason for splitting R in two terms according to (6.45) now becomes clear. When the dipole and the observation point are both at the interface, a case that is of great practical importance, then $r_1 = r_2$ and the first two terms in (6.46a) cancel each other and the field is solely determined by the integral I_1'. Moreover, from comparison of (6.44) and (6.46a), it follows that the third term in (6.44) should be just the leading term of the asymptotic expansion of I_1' in (6.46b) valid in the vicinity of the interface.

Using the transformation (6.30a) with $k_1 = k_0$, we readily obtain

$$I_1' = k_0 n^2 \exp\left(\frac{i\pi}{4}\right) \int_{\Gamma_1} \frac{\breve{H}_0^{(1)}(k_0\rho\sin\beta)\sin\beta\cos\beta}{n^2\cos\beta + (n^2-\sin^2\beta)^{1/2}} \exp[ik_0 r_2\cos(\beta-\theta_2)]\, d\beta. \tag{6.47}$$

The presence of the troublesome pole in (6.47) can be more clearly displayed if we note the identity

$$[n^2\cos\beta + (n^2-\sin^2\beta)^{1/2}]^{-1} = (n^4-1)^{-1} \frac{(n^2-\sin^2\beta)^{1/2} - n^2\cos\beta}{\sin(\beta+\beta_P)\sin(\beta-\beta_P)}, \tag{6.48}$$

where, according to (6.40b),

$$\beta_P \sim \frac{\pi}{2} + \frac{1}{n}. \tag{6.49}$$

Reference now is made to the result in (5.103). We note that in this case

$$F(\theta_2) = \left(\frac{2ik_0}{\pi r_2}\right)^{1/2} \frac{n^2\cos\theta_2}{n^2\cos\theta_2 + (n^2-\sin^2\theta_2)^{1/2}}, \tag{6.50a}$$

$$F(\theta_2) = \sim n\left(\frac{2ik_0}{\pi r_2}\right)^{1/2} \cos\theta_2, \tag{6.50b}$$

and

$$F_1(\beta_P) = \frac{k_0 n^2 \exp(i\pi/4)}{2(n^4-1)} [(n^2-\sin^2\beta_P)^{1/2} - n^2\cos\beta_P]\breve{H}_0^{(1)}(k_0\rho\sin\beta_P), \tag{6.51a}$$

$$F_1(\beta_P) = \sim \frac{1}{n}\left(\frac{2ik_0}{\pi r_2\sin\theta_2}\right)^{1/2}. \tag{6.51b}$$

Combining (6.50) and (6.51) according to the prescription in (5.103), we obtain

$$I_1' = \frac{2\exp(ik_0 r_2)}{r_2} \times \left\{ n\cos\theta_2 + (\sin\theta_2)^{-1/2}\cdot[n(\theta_2-\beta_P)]^{-1} \sum_{m=1}^{\infty} \frac{1\cdot 3\ \cdots\ (2m-1)}{[ik_0 r_2(\theta_2-\beta_P)^2]^m} \right\}, \tag{6.52}$$

which upon substitution into (6.46a) gives the total field in the upper half space appropriate to observation points near the interface.

A more explicit expression can be obtained by noting that near the interface $r_{1,2} \sim \rho + (z \mp h)^2/2\rho$, $\cos\theta_2 \sim (z+h)/\rho$, $\sin\theta_2 \sim 1$, and $\theta_2 - \beta_P \sim -1/n$: thus

$$\Pi_1 = \frac{ip}{2\pi\omega\varepsilon_0}\frac{\exp(ik_0\rho)}{\rho}\left[-i\ {}_0hz + \frac{n(z+h)}{\rho} - \sum_{m=1}^{\infty}\frac{1\cdot 3\ \cdots\ (2m-1)}{(ik_0\rho/n^2)^m}\right]. \tag{6.53}$$

Finally, if the transmitter and receiver are both on the ground, then $z = h = 0$ and we obtain simply

$$\Pi_1 = -\frac{ip}{2\pi\omega\varepsilon_0}\frac{\exp(ik_0\rho)}{\rho}\sum_{m=1}^{\infty}\frac{1\cdot 3\cdot 5\ \cdots\ (2m-1)}{(ik_0\rho/n^2)^m}. \tag{6.54}$$

The factor $ik_0\rho/n^2$ is equivalent, for large n, to $2\rho_s$ where $\rho_s = ik_0\rho(n^2-1)/2n^4$ is the *numerical distance* as defined by Bremmer (1958). It expresses a normalized distance between the transmitter and the receiver on the ground. It is clear that the expansion (6.54) is valid only for a sufficiently large distance, i.e., $k_0\rho \gg |n|^2$. Thus, for example, at a frequency of 1MHz, $|n|^2 \sim 4\times 10^4$, $k_0 \sim 2\times 10^{-2}$, and it would follow that $\rho \gg 2000$. This is a rather severe limitation on the results and it is apparent that a more careful evaluation of the integrals is needed for closer ranges. (This is examined by Baños (1966).)

Before leaving this section, it will be instructive to look more closely at the vertical dependence of the fields near the interface. For simplicity, let us assume that the dipole is at the interface so that $h = 0$. Retaining only the leading, ρ^{-2}, term of the expansion in (6.53), we obtain

$$\Pi_1 \sim -\frac{pn^2\exp(ik_0\rho)}{2\pi\omega\varepsilon_0 k_0\rho^2}\left(1 - \frac{ik_0 z}{n}\right)$$

$$\sim -\frac{pn^2}{2\pi\omega\varepsilon_0 k_0\rho^2}\exp\left[ik_0\left(\rho - \frac{z}{n}\right)\right] \tag{6.55}$$

where the last approximation is valid if $|k_0 z/n| \ll 1$. If $n = n'(1 + i)$ is substituted into (6.55), we find that

$$\Pi_1 \sim \exp\left(-\frac{k_0 z}{2n'}\right) \cdot \exp\left[ik_0\left(\rho - \frac{z}{2n'}\right)\right], \tag{6.56}$$

which is in the form of a surface wave inasmuch as it attains a maximum at $z = 0$ and it decays exponentially in the vertical direction. The vertical attenuation is, however, slow for large n'.

Let us now look into the field components. Noting that in cylindrical coordinates

$$\begin{aligned} E_\rho &= \frac{\partial}{\partial\rho}\left(\frac{\partial\Pi_1}{\partial z}\right), \\ E_z &= \left(k_0{}^2 + \frac{\partial^2}{\partial z^2}\right)\Pi_1, \\ H_\varphi &= i\omega\varepsilon_0 \frac{\partial\Pi_1}{\partial\rho}, \end{aligned} \tag{6.57}$$

while all of the other field components are zero. It follows from (6.55) that to the order $0(\rho^{-2})$, $\partial\Pi_1/\partial\rho = ik_0\Pi_1$, $\partial\Pi_1/\partial z = -ik_0\Pi_1/n$ and consequently

$$\begin{aligned} E_\rho &= \frac{k_0{}^2}{n}\Pi_1, \\ E_z &= k_0{}^2\Pi_1, \\ H_\varphi &= -\omega\varepsilon_0 k_0 \Pi_1. \end{aligned} \tag{6.58}$$

It follows that the radial electric field component is small in comparison with the vertical component and vanishes for infinite conductivity, as it should. The angle that the electric field makes with the normal to the interface is called the "tilt" angle and is equal to

$$\alpha_t = \arctan(1/n). \tag{6.59}$$

It is interesting to inquire how the energy is transmitted across the interface. The situation is depicted in Fig. 6.6. As mentioned earlier, the total electric field at the interface is tilted with respect to the normal by the angle $\alpha_t = \arctan(1/n)$. The magnetic field H_φ is parallel to the interface and normal to the electric field. Thus, it follows that the direction of the Poynting vector will be as shown in Fig. 6.6, that is, at an angle α_t from the interface. As the conductivity increases, the tilt angle decreases and in the limit of infinite conductivity, the direction of power flow will be parallel to the interface and consequently there will be no power transfer to the lower half space.

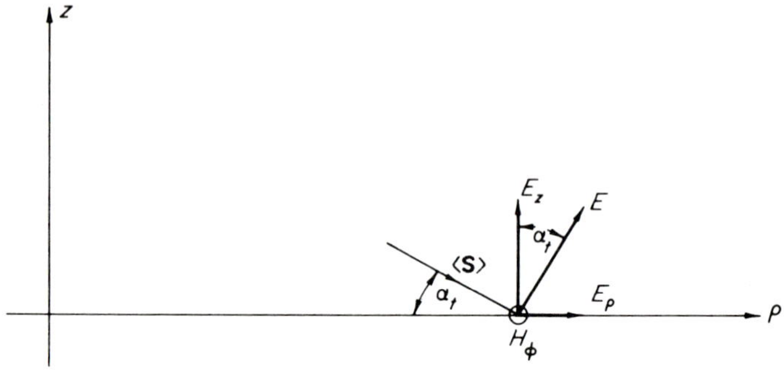

Fig. 6.6. Fields and power flow at the interface.

6.2.3 Fields in the Ground

Let us now examine the fields transmitted in the lower half space (ground) given by (6.33a). We have in this case

$$\Pi_2 = -\frac{p}{4\pi\omega\varepsilon_0} I_2, \tag{6.60a}$$

where

$$I_2 = k_2 \exp\left(\frac{-i\pi}{4}\right) \int_{\Gamma_1} \frac{\exp[ik_0 h(1 - n^2 \sin^2 \beta)^{1/2}]\breve{H}_0^{(1)}(k_2 \rho \sin \beta)}{\cos \beta + n(1 - n^2 \sin^2 \beta)^{1/2}}$$
$$\times \sin \beta \cos \beta \exp[ik_2 r \cos(\beta - \theta')]\, d\beta. \tag{6.60b}$$

In anticipation of the deformation of the contour Γ_1 to the path of the steepest descents Γ, we must investigate the singularities of the integrand in the complex β plane.

Evidently, there are branch points where the radical vanishes, i.e., $\sin \beta = \pm 1/n$. For a large $n = n'(1 + i)$ the location of these branch points will be given by

$$\beta_{B1} = \frac{1 - i}{2n'}, \tag{6.61a}$$

$$\beta_{B2} = -\frac{1 + i}{2n'}. \tag{6.61b}$$

The associated branch cuts will satisfy $(1 - n^2 \sin^2 \beta)^{1/2} = a$ where a is a real number. This will amount to

$$\sin\beta = \pm\frac{(1-a^2)^{1/2}}{|n|}\exp\left(-\frac{i\pi}{4}\right), \qquad a \le 1,$$

$$= \pm\frac{(a^2-1)^{1/2}}{|n|}\exp\left(\frac{i\pi}{4}\right), \qquad a \ge 1, \tag{6.62}$$

on the branch cuts. It follows from (6.62) that both branch cuts have slopes of $-\pi/4$ with respect to the β' axis. They approach the origin $\beta = 0$ and then for a large β'' they approach the asymptotes $\beta = \pm\pi/4$ as shown in Fig. 6.7. The asymptotic behavior of the branch cuts follows from the fact that on the branch cuts $\tan\beta' = \tanh\beta''$; thus, as $\beta'' \to \infty$, $\beta' \to \pi/4$ and as $\beta'' \to -\infty$, $\beta' \to -\pi/4$.

Next, let us investigate the possibility of pole singularities. It follows from (6.60b) that the denominator of the integrand will vanish when $\sin\beta = \pm$

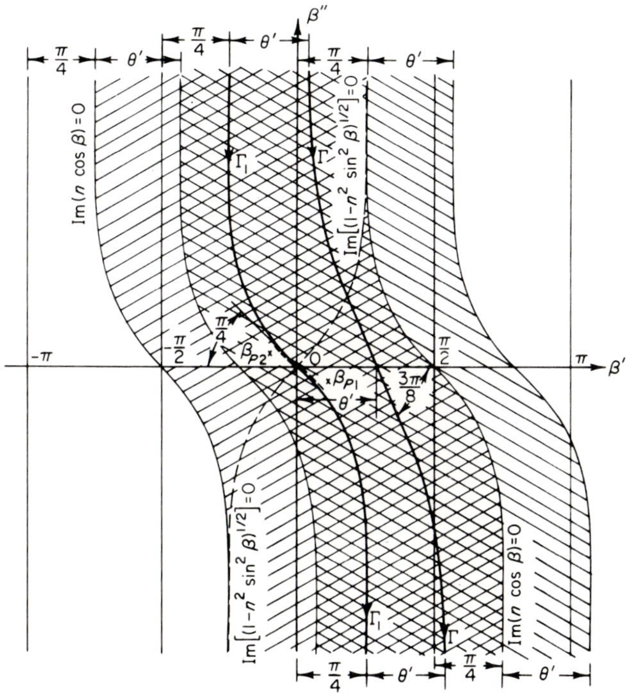

Fig. 6.7. The complex β plane for evaluation of the fields in the lower half space when $|n| \gg 1$. Inside the shaded region ///// $\mathrm{Im}(n\cos\beta) > 0$, while inside the region \\\\\ $\mathrm{Im}(n\cos(\beta-\theta')) > 0$.

$(1+n^2)^{-1/2}$, which for a large $n = n'(1+i)$ amounts to

$$\beta_{P1,2} = (2n')^{-1}\{[1 + (2|n|^2)^{-1}] - i[1 - (2|n|^2)^{-1}]\}. \tag{6.63}$$

The location of these two poles is shown in Fig. 6.7.

The trace of the lines $\mathrm{Im}(n \cos \beta) = 0$ is the same as in Fig. 5.12 for the special case of $\beta_0 = \pi/4$. It is important to note that the lines $\mathrm{Im}(n \cos \beta) = 0$ intersect the β' axis at an angle $-\pi/4$ radians. The trace of the lines $\mathrm{Im}(n \cos (\beta - \theta'))$ is shifted to the right by the angle θ' from the lines $\mathrm{Im}(n \cos \beta) = 0$.

Finally, the trace of the line of steepest descents can be obtained from (5.88) through (5.92). Since $n' = n''$ in the present case it follows from (5.91) that the angle of intersection with β' axis will be $-3\pi/8$ radians. Moreover, from (5.92) it will follow that the path of steepest descents will approach the asymptotes $\beta' - \theta' = \pm\pi/4$ as $\beta'' \to \pm\infty$. The complete picture is depicted in Fig. 6.7.

We note that in the present case the deformation of the contour Γ_1 into the contour Γ is more complicated than before due to the presence of the branch cut $\mathrm{Im}[(1 - n^2 \sin^2 \beta)^{1/2}] = 0$, which cannot be crossed if we are to remain on sheet I. We note from Fig. 6.8a that for some angle $\theta' < \theta_B$ the path of steepest descents will cross the branch cut twice. As we follow the path Γ from $-\infty$ on sheet I, we cross the branch cut for the first time and thus find ourselves on sheet III. We then cross the branch cut once more, thus reentering sheet I, and continue to $+\infty$. The convergence of the field integral at the end points is therefore ensured.

Now consider the situation when $\theta' > \theta_B$. Here we cross the branch cut only once, finding ourselves on sheet III without apparent possibility of returning onto sheet I if we follow the path Γ. It is clear that in this case we must modify the path of integration in order to ensure the convergence of the integral at the endpoints. This can be done as shown in Fig. 6.8b. Instead of crossing the branch cut as we approach it following the path Γ, we follow the borders of the branch cut until we find ourselves on its opposite side. Here we cross it once; then, following the path, we cross the branch cut once more, thus reentering sheet I. Moreover; for $\theta' > \theta_P$ we must also deform the path to include the pole at $\beta = \beta_{P1}$. Consequently, the complete contribution to the field integral can now be written

$$I_2 = k_2 \exp\left(-\frac{i\pi}{4}\right)[I_{SD} + I_{BC}\, u(\theta' - \theta_B) + 2\pi i \,\mathrm{Res}(\beta = \beta_{P1}) u(\theta' - \theta_P)], \tag{6.64a}$$

where

$$I_{SD} = \int_\Gamma F(\beta) \exp[ik_2 r \cos(\beta - \theta')]\, d\beta, \tag{6.64b}$$

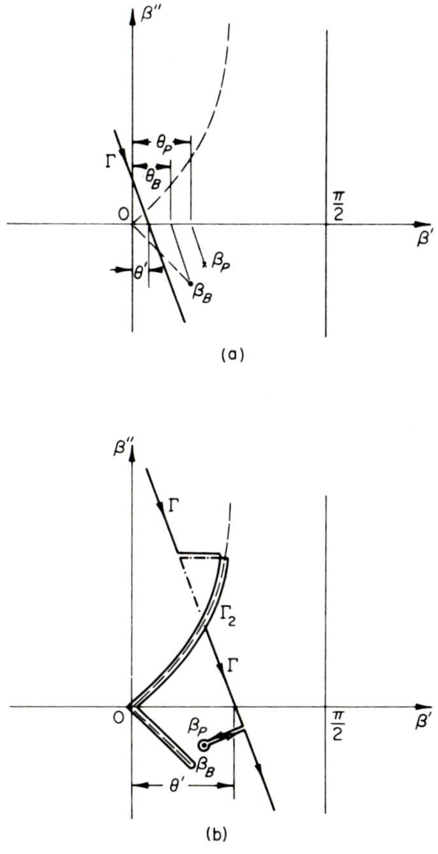

Fig. 6.8. Integration paths for evaluation of the field integrals in the lower half space: (a) integration path for $\theta' \ll \theta_B \ll \theta_P$, (b) for $\theta' > \theta_P > \theta_B$.

$$I_{BC} = \int_{\Gamma_2} F(\beta) \exp[ik_2 r \cos(\beta - \theta')] \, d\beta, \tag{6.64c}$$

and we shall now evaluate each of the foregoing contributions separately.

The integral I_{SD} can be evaluated using (5.94). The result is

$$I_{\text{SD}} = \frac{2\exp[i(k_2 r - \pi/4)]}{k_2 r} \frac{\cos\theta' \exp[i\ {}_0 h(1 - n^2 \sin^2\theta')^{1/2}]}{\cos\theta' + n(1 - n^2\sin^2\theta')^{1/2}} \times \left[1 + \frac{G(\theta')}{2ik_2 r} + \cdots\right], \tag{6.65a}$$

where

$$G(\theta') = \frac{n(n^2-1)[n\cos\theta'(1+\sin^2\theta')(1-n^2\sin^2\theta')^{1/2} + \cos^2\theta' + (\cos^2\theta' - 2\sin^2\theta')(1-n^2\sin^2\theta')]}{\cos\theta'(1-n^2\sin^2\theta')^{3/2}[\cos\theta' + n(1-n^2\sin^2\theta')^{1/2}]^2}. \tag{6.65b}$$

Note that the second-order term $G(\theta')$ would become infinite at the branch point $\sin\theta' = 1/n$ if n were real. This is the consequence of the fact that the saddle-point method of integration would not be valid near the branch point, since the function $F(\beta)$ would not be slowly varying there. However, since n is complex, no severe difficulty arises. Moreover, the result (6.65a) would not be valid in the neighborhood of the pole $\sin\theta' = (1+n^2)^{-1/2}$ for a similar reason.

Now when $\theta' = \pi/2$, the leading term of (6.65a) becomes

$$I_{\rm SD}\left(\theta' = \frac{\pi}{2}\right) = \frac{i2\exp[i(k_2\rho + k_0 h(1-n^2)^{1/2} - \pi/4)]}{n^2(1-n^2)(k_2\rho)^2}, \tag{6.66}$$

which, indeed, is small on account of n being large and complex. By comparing (6.66) with the leading term of (6.54), we conclude that (6.66) is not the important part of the field at the interface. The important contribution to the field at the interface is the one resulting from the integration about the branch cut, as we shall see next.

Thus, we now focus our attention on the integral I_{BC} in (6.64a). This integral can be evaluated in the complex β plane, but it will prove more convenient to make the transformation

$$\sin\beta = \sigma/n, \tag{6.67}$$

whereupon the integral (6.64c) becomes

$$I_{BC} = n^{-1}\int_{C_1+C_2+C_3} Q(\sigma)\exp\{ik_0[\rho\sigma - z(n^2-\sigma^2)^{1/2}]\}\,d\sigma, \tag{6.68a}$$

where

$$Q(\sigma) = \frac{\exp[ik_0 h(1-\sigma^2)^{1/2}]\breve{H}_0^{(1)}(k_0\rho\sigma)}{(n^2-\sigma^2)^{1/2} + n^2(1-\sigma^2)^{1/2}}, \tag{6.68b}$$

and the new path of integration in the complex σ plane is shown in Fig. 6.9. Note that the pertinent branch point is now situated at $\sigma = \sigma_B = \pm 1$ and the pole at $\sigma = \sigma_P$ where

$$\sigma_P = 1 - \frac{3}{8|n|^4} + \frac{i}{2|n|^2}, \tag{6.69}$$

and it is important to note that the pole is to the left of a vertical line passing

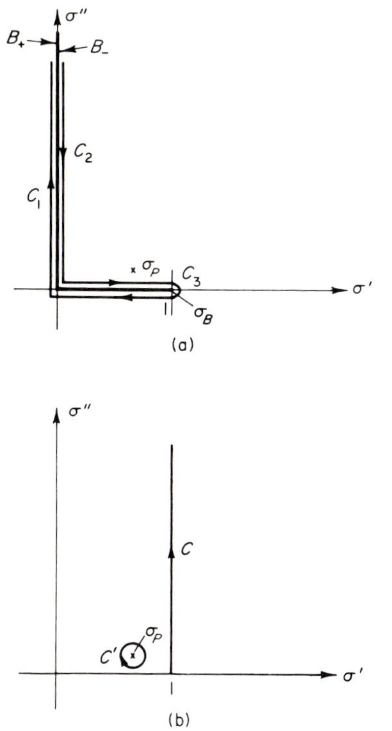

Fig. 6.9. Contours C_1, C_2, C, and C'.

through the branch point $\sigma = 1$.

To evaluate the contour integral in (6.68a) let us first dispose of its contribution along the small semicircle C_3 encircling the branch point $\sigma = 1$. On C_3 we can set $\sigma = 1 - \varepsilon \exp(-i\theta)$, $-\pi/2 \leq \theta \leq \pi/2$, where ε is the radius of the semicircle. Substituting this change of variable into the integrand of (6.68a), we find that (problem)

$$\int_{C_3} \underset{\varepsilon \to 0}{\longrightarrow} 0. \tag{6.70}$$

Hence, we are left with the contributions from C_1 and C_2 only.

Now the branch cut B in Fig. 6.9a is a line on which $\operatorname{Im}(1 - \sigma^2)^{1/2} = 0$ and according to (6.27) the real part of $(1 - \sigma^2)^{1/2}$ is positive on its right side B_+ and negative on its left side B_- facing the branch point. Hence, if Q_+ and Q_- denote the value of Q in (6.68b) on B_+ and B_-, respectively, then we can write

$$I_{BC} = n^{-1}\Bigg(\int_{B_+} Q_+(\sigma) \exp\{ik_0[\rho\sigma - z(n^2 - \sigma^2)^{1/2}]\}\, d\sigma$$

$$+ \int_{B_-} Q_-(\sigma) \exp\{ik_0[\rho\sigma - z(n^2 - \sigma^2)^{1/2}]\}\, d\sigma\Bigg), \tag{6.71a}$$

where

$$Q_{\pm}(\sigma) = \frac{\check{H}_0^{(1)}(k_0 \rho\sigma) \exp[\pm ik_0 h(1-\sigma^2)^{1/2}]}{(n^2-\sigma^2)^{1/2} \pm n^2(1-\sigma^2)^{1/2}}. \tag{6.71b}$$

Finally, inasmuch as the integral along B_- is just the negative of the same integral along B_+, we obtain

$$I_{BC} = n^{-1} \int_{B_+} P(\sigma) \exp\{ik_0[\rho\sigma - z(n^2-\sigma^2)^{1/2}]\}\, d\sigma, \tag{6.72a}$$

where

$$P(\sigma) = \frac{2\check{H}_0^{(1)}(k_0\rho\sigma)}{(n^4-1)(\sigma^2-\sigma_P{}^2)}$$

$$\times \{i(n^2-\sigma^2)^{1/2} \sin[k_0 h(1-\sigma^2)^{1/2}] - n^2(1-\sigma^2)^{1/2} \cos[k_0 h(1-\sigma^2)^{1/2}]\} \tag{6.72b}$$

since $\sigma_P{}^2 = n^2/(1+n^2)$.

For reasons that will become apparent later, we now replace the path B_+ by the equivalent paths $C + C'$ as shown in Fig. 6.9b. Thus, we write

$$I_{BC} = \frac{1}{n} I_{BC}^{(c)} - \frac{2\pi i}{n} \operatorname{Res}(\sigma = \sigma_P), \tag{6.73}$$

where $I_{BC}^{(c)}$ is the integral in (6.72a) evaluated along the path C in Fig. 6.9b. The residue at the pole $\sigma = \sigma_P$ is easy to find but we must note that $(1-\sigma_P{}^2)^{1/2} = -(1+n^2)^{-1/2}$ so that $\operatorname{Im}(1-\sigma^2)^{1/2} > 0$ is satisfied everywhere. Thus, we find

$$\operatorname{Res}(\sigma = \sigma_P) \sim \frac{\check{H}_0^{(1)}(k_0\rho)}{n^3} \exp\left[ik_0\left(\rho - nz - \frac{h}{n}\right)\right]. \tag{6.74}$$

The integral $I_{BC}^{(c)}$ can be evaluated by the method of steepest descents. We put

$$\sigma = 1 + is^2 \tag{6.75a}$$

and calculate

$$(1-\sigma^2)^{1/2} = -\sqrt{2} \exp\left(-\frac{i\pi}{4}\right) s\left(1 + \frac{is^2}{4} + \cdots\right), \tag{6.75b}$$

$$(n^2-\sigma^2)^{1/2} = n\left(1 - \frac{is^2}{n^2} + \cdots\right). \tag{6.75c}$$

Hence, the exponent in the integrand of (6.72a) becomes

$$ik_0[\rho\sigma - z(n^2 - \sigma^2)^{1/2}] = ik_0(\rho - zn) - k_0(\rho + z/n)s^2 + \cdots, \tag{6.76}$$

and it follows that the value of the integral (6.72a) will be significant only in the neighborhood of the point $s = 0$ providing that $\rho > |z|/2n'$, which does not seem to be a serious limitation. The function $P(\sigma)$ in (6.72b) can now be expanded in a power series of s of which the leading term is

$$P(\sigma) = \frac{4\exp(-i\pi/4)\exp(-ik_0 h/n)}{(\pi k_0 \rho)^{1/2}} s + \cdots \tag{6.77}$$

and the integral $I_{BC}^{(c)}$ can be evaluated approximately to give

$$I_{BC}^{(c)} \sim \frac{2\exp(i\pi/4)}{(k_0\rho)^2}\exp\left[ik_0\left(\rho - zn - \frac{h}{n}\right)\right]. \tag{6.78}$$

Combining the results of (6.78) and (6.74) according to (6.73), we obtain for the integral I_{BC} in (6.72a)

$$I_{BC} = \frac{2}{n}\left[\frac{\exp(i\pi/4)}{(k_0\rho)^2} - \pi i\frac{\check{H}_0^{(1)}(\ _0\rho)}{n^3}\right]\exp\left[ik_0\left(\rho - nz - \frac{h}{n}\right)\right]. \tag{6.79}$$

The remaining task is the evaluation of the residue term in (6.64a) at the pole $\beta = \beta_{P1}$ given by (6.63). We find in this case

$$\text{Res}(\beta = \beta_{P1}) = \frac{\check{H}_0^{(1)}(\ _0\rho)}{n^4}\exp\left[ik_0\left(\rho - nz - \frac{h}{n}\right)\right], \tag{6.80}$$

which completes the evaluation of the integral I_2 in (6.64a).

Of particular interest are the fields near the interface $\theta' \sim \pi/2$. Adding the various contributions in this case according to (6.64a), we find that the residue term (6.80) exactly cancels the residue term in (6.79) for $\theta' > \theta_P > \theta_B$. Thus, we obtain in this case

$$I_2 = 2k_0\exp\left(-\frac{i\pi}{4}\right)\left\{\frac{\cos\theta'\exp[i\ _0 h(1 - n^2\sin^2\theta')^{1/2}]}{\cos\theta' + n(1 - n^2\sin^2\theta')^{1/2}}\left[1 + \frac{G(\theta')}{2ik_2 r} + \cdots\right]\right.$$
$$\left.\times\frac{\exp[i(k_2 r - \pi/4)]}{k_0 r} + \frac{\exp(i\pi/4)}{(k_0\rho)^2}\exp\left[ik_0\left(\rho - nz - \frac{h}{n}\right)\right]\right\}. \tag{6.81}$$

Note that the first term in (6.81) will be extremely small at large distances because of the presence of the exponential dependence

$$\exp(ik_2 r) = \exp(-k_0 n'r + ik_0 n'r)$$

when n' is large. Thus, the most important contribution to I_2 will come from the second term and consequently Π_2 in (6.60a) can be written

$$\Pi_2 \sim -\frac{p}{2\pi\omega\varepsilon_0 k_0 \rho^2} \exp\left[i\ {}_0\left(\rho - nz - \frac{h}{n}\right)\right]. \tag{6.82}$$

Compare (6.82) with the corresponding expression for the upper half space given in (6.54). Note that (6.82) is just the leading term of the series that is needed to satisfy the boundary condition (6.4a) at the interface $z = 0$. As a matter of fact, if we set $z = h = 0$ in (6.72b), we can easily expand the function $P(\sigma)$ in a power series of s, finding all of the terms as follows.

$$P(\sigma) = \frac{4\exp(-i\pi/4)}{(\pi k_0 \rho)^{1/2}} \sum_0^\infty (-i2n^2)^m s^{2m+1}, \tag{6.83}$$

which upon substitution in (6.72) can be readily integrated. Collecting the results, we obtain in this case

$$\Pi_2 = -\frac{ip}{2\pi\omega\varepsilon_0 n^2} \frac{\exp(ik_0\rho)}{\rho} \sum_1^\infty \frac{1\cdot 3\cdot 5\ \cdots\ (2m-1)}{(ik_0\rho/n^2)^m}, \tag{6.84}$$

which by virtue of (6.4a) and (6.54) is seen to match the fields on the other side of the interface term by term.

Let us now return to the leading term of Π_2 in (6.82) to find the form of the fields near the interface. The fields in the lower region are given by

$$\begin{aligned} E_{\rho 2} &= \frac{\partial}{\partial\rho}\frac{\partial \Pi_2}{\partial z}, \\ E_{z2} &= \left(k_2^{\,2} + \frac{\partial^2}{\partial z^2}\right)\Pi_2, \\ H_{\varphi 2} &= i\omega\varepsilon_0 n^2 \frac{\partial \Pi_2}{\partial \rho}. \end{aligned} \tag{6.85}$$

Thus, using (6.82), we obtain for the leading terms

$$\begin{aligned} E_{\rho 2} &= k_2^{\,2}\Pi_2, \\ E_{z2} &= 0, \\ H_{\varphi 2} &= -\omega\varepsilon_0 k_2^{\,2}\Pi_2. \end{aligned} \tag{6.86}$$

Note that since $E_{z2} = 0$, the field in the lower region cannot propagate in the radial direction. Evidently, since only $E_{\rho 2}$ and $H_{\varphi 2}$ are present, then the direction of the Poynting vector must be normal to the interface as shown in Fig. 6.10, where we also superposed the pertinent results concerning the power flow in the upper region. We also note that the ratio of the electric and

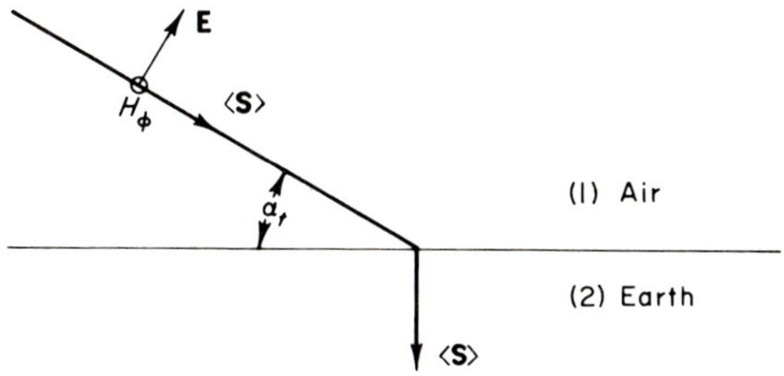

Fig. 6.10. Directions of the power flow from the air into the earth ($\alpha_t = \arctan n^{-1}$).

magnetic field components in the lower region is

$$\frac{E_{\rho 2}}{-H_{\varphi 2}} = \left(\frac{\mu_0}{\varepsilon_0 n^2}\right)^{1/2} = Z_2, \tag{6.87}$$

which is the intrinsic impedance of the earth. The last expression justifies the use of impedance (Leontovitch) boundary conditions in this case, as could have been concluded earlier on the basis of the results of Section 2.7.

The vertical dependence of the fields above and below the ground deserves attention. According to (6.56), above the ground the fields decay as $\exp(-k_0 z/2n')$, which is slow since n' is large. In the ground, however, according to (6.82), the fields will decay as $\exp(-k_0 n'|z|)$, which is very rapid for large n'. It is to be noted that there is no exponential decay of fields in the radial direction.

6.3 Vertical Electric Dipole above a Slightly Conducting Plasma

As a second application of the formal solutions obtained in the earlier sections of this chapter we now look into the problem of a dielectric half space with an index of refraction $|n| < 1$, the upper half space being air. The condition $|n| < 1$ can be achieved by a slightly conducting plasma or by two dielectric media other than plasmas in which n_1 is real and n_2 is slightly conducting and the ratio $n_2/n_1 < 1$.

In the case of a plasma in absence of a magnetostatic field and when $\omega \gg \nu$, then from (3.37a) the effective index of refraction can be written

$$n = \left[1 - \left(\frac{\omega_p}{\omega}\right)^2\right]^{1/2} + \frac{i\nu\omega_p{}^2}{2\omega^3[1-(\omega_p/\omega)^2]^{1/2}} \tag{6.88}$$

and we note that in this case $n'' < n' < 1$.

6.3.1 Fields in the Air

Let us now examine the reflected field in the air in this case. Putting $n_1 = 1$, $n_{21} = n$ in (6.30), we get

$$I_1 = -k_0 \exp\left(\frac{i\pi}{4}\right) \int_{\Gamma_1} \frac{(n^2 - \sin^2\beta)^{1/2} \breve{H}_0^{(1)}(k_0 \rho \sin\beta) \sin\beta}{n^2 \cos\beta + (n^2 - \sin^2\beta)^{1/2}} \times \exp[ik_0 r_2 \cos(\beta - \theta_2)]\, d\beta. \tag{6.89}$$

The poles of the integrand are still given by $\sin\beta_P = \pm n(1+n^2)^{-1/2}$; in this case, however, since $1 > n' > n''$ we obtain approximately

$$\sin\beta_{P1,2} \sim \pm[n'(1 - n'^2/2) + in''(1 - 3n'^2/2)]. \tag{6.90}$$

In addition to the foregoing pole singularities there will be branch point singularities where $\sin\beta_B = \pm n$. On the branch cut $\mathrm{Im}(n^2 - \sin^2\beta)^{1/2} = 0$ from which it follows that

$$\sin 2\beta' \sinh 2\beta'' = 4n'n'' \tag{6.91}$$

on the branch cut. Equation (6.91) is the equation of a hyperbola with the asymptotes given by

$$\beta' \xrightarrow[\beta'' \to \infty]{} \left(\frac{n'n''}{4}\right) \exp(-2\beta''),$$
$$\beta' \xrightarrow[\beta'' \to -\infty]{} \left(\frac{n'n''}{4}\right) \exp(-2|\beta''|). \tag{6.92}$$

Collecting the foregoing information, the branch cuts will appear as shown in Fig. 6.11.

In what follows it will be important to know the location of the poles β_{P1} and β_{P2} with respect to the branch cuts. Using the results of (6.90), we easily ascertain that at the poles

$$\sinh 2\beta_P'' = \frac{1}{(1+n'^2)^2} \frac{4n'n''}{\sin 2\beta_P'}, \tag{6.93a}$$

whereas on the branch cut at $\beta = \beta_P' + i\beta''$

$$\sinh 2\beta'' = \frac{4n'n''}{\sin 2\beta_P'}. \tag{6.93b}$$

Since β'' is greater on the branch cut than at the pole (at the same β' location), we conclude that the poles lie below their respective branch cuts, as shown in Fig. 6.11. Finally, on the basis of the analysis of Chapter 5, it follows that the

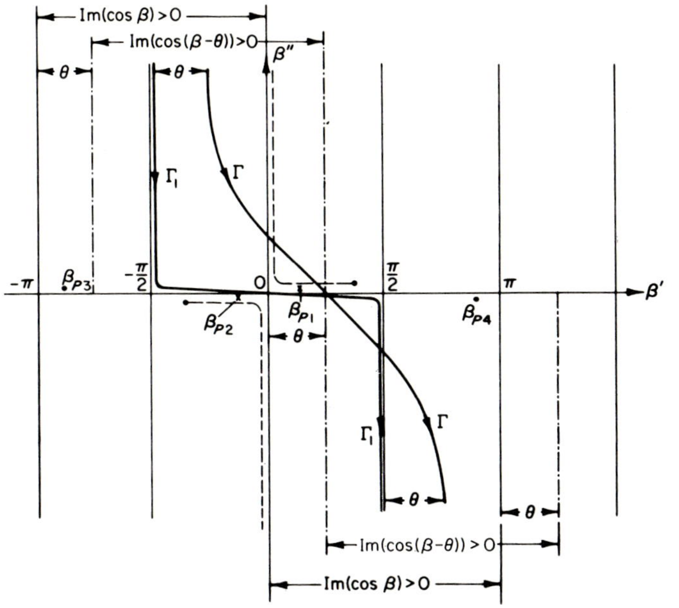

Fig. 6.11. The complex β plane for evaluation of the reflected field when $|n| < 1$.

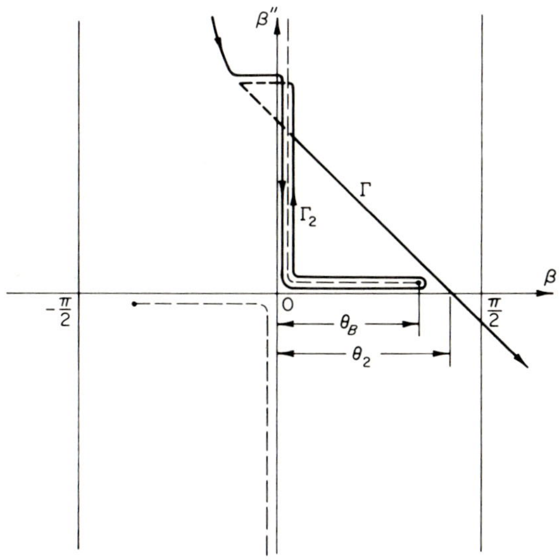

Fig. 6.12. Modified path of integration to account for the presence of the branch cut.

path of steepest descents will intersect the axis at an angle of $-\pi/4$ radians.

Looking over Fig. 6.11, we note that the deformation of the original path of integration into the path of steepest descents presents no particular difficulties when $\theta_2 < \theta_B$. In this case the path Γ crosses the branch cut twice and thus it returns on sheet I. However, when $\theta_2 > \theta_B$, the path crosses the branch cut $\mathrm{Im}(n^2 - \sin^2 \beta)^{1/2} = 0$ only once; thus it enters sheet II without apparent possibility of reentering sheet I if continued along the path Γ. As in the previous section we must here supplement the path Γ with the integral along the borders of the branch cut as shown in Fig. 6.12. Thus the field integral (6.89) can be written

$$I_1 = -k_0 \exp\left(\frac{i\pi}{4}\right)[I_{\mathrm{SD}} + u(\theta_2 - \theta_{\mathrm{B}})I_{BC}], \tag{6.94a}$$

where

$$I_{\mathrm{SD}} = \int_{\Gamma} F(\beta) \exp[ik_0 r_2 \cos(\beta - \theta_2)]\, d\beta \tag{6.94b}$$

and

$$I_{BC} = \int_{\Gamma_2} F(\beta) \exp[ik_0 r_2 \cos(\beta - \theta_2)]\, d\beta, \tag{6.94c}$$

where

$$F(\beta) = \frac{(n^2 - \sin^2 \beta)^{1/2} \check{H}_0^{(1)}(k_0 \rho \sin \beta) \sin \beta}{n^2 \cos \beta + (n^2 - \sin^2 \beta)^{1/2}}. \tag{6.95}$$

The integral I_{SD} in (6.94b) can be evaluated in a routine manner to give

$$I_{\mathrm{SD}} = \frac{\exp[i(k_0 r_2 - \pi/4)]}{k_0 r_2} \times \left[\frac{2(n^2 - \sin^2 \theta_2)^{1/2}}{n^2 \cos \theta_2 + (n^2 - \sin^2 \theta_2)^{1/2}} + \frac{1}{2ik_0 r_2} G(\theta_2) + \cdots\right] \tag{6.96}$$

where $G(\theta_2)$ is given by (6.41b). Note that the second-order term $G(\theta_2)$ becomes infinite at $\sin \theta_2 = \pm n$ when n is real. This is because the saddle-point method of integration is not valid near the branch point, since the function $F(\beta)$ is not slowly varying there. We shall remove this restriction later.

To evaluate the integral I_{BC} we first make the transformation

$$\sin \beta = \sigma \tag{6.97}$$

and the integral now becomes

$$I_{BC} = \int_{C_1 + C_2 + C_3} Q(\sigma) \exp\{ik_0[\rho\sigma + (z + h)(1 - \sigma^2)^{1/2}]\}\, d\sigma, \tag{6.98a}$$

where

$$Q(\sigma) = \frac{\sigma \check{H}_0^{(1)}(k_0 \rho\sigma)(n^2 - \sigma^2)^{1/2}}{(1-\sigma^2)^{1/2}[n^2(1-\sigma^2)^{1/2} + (n^2 - \sigma^2)^{1/2}]} \tag{6.98b}$$

and the paths C_1, C_2, and C_3 are shown in Fig. 6.13a. Note that the pertinent branch point is now at $\sigma = n$ and the pole at $\sigma = n(1+n^2)^{-1/2}$.

To evaluate the contour integral in (6.98a) we first dispose of its contribution along the small semicircle C_3 encircling the branch point at $\sigma = n$. On C_3 we set $\sigma = n - \varepsilon e^{-i\theta}$ where ε is the radius of the semicircle. Substituting this change of variable into the integrand of (6.98a), it will follow that the contribution from C_3 will vanish in the limit as ε approaches zero. Hence, we are left with the contributions from C_1 and C_2 only.

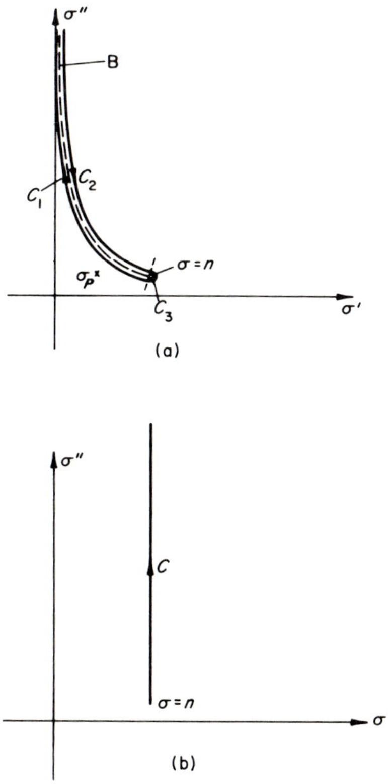

Fig. 6.13. The contours for evaluation of the integral I_{BC} in the case $|n| < 1$.

Now the branch cut in Fig. 6.13a is a line on which $\text{Im}(n^2 - \sigma^2)^{1/2} = 0$ and according to (6.27) the real part of $(n^2 - \sigma^2)^{1/2}$ is positive on its right side B_+ and negative on its left side B_- facing the branch point. Hence, if Q_+ and Q_- denote the values of $Q(\sigma)$ in (6.98b) on B_+ and B_-, respectively, then we can write

$$I_{BC} = \int_{B_+} Q_+(\sigma) \exp\{ik_0[\rho\sigma + (z + h)(1 - \sigma^2)^{1/2}]\}\, d\sigma$$

$$+ \int_{B_-} Q_-(\sigma) \exp\{ik_0[\rho\sigma + (z + h)(1 - \sigma^2)^{1/2}]\}\, d\sigma, \qquad (6.99\text{a})$$

where

$$Q_\pm(\sigma) = \pm \frac{\sigma \breve{H}_0^{(1)}(k_0 \rho\sigma)(n^2 - \sigma^2)^{1/2}}{(1 - \sigma^2)^{1/2}[n^2(1 - \sigma^2)^{1/2} \pm (n^2 - \sigma^2)^{1/2}]}. \qquad (6.99\text{b})$$

Moreover, since the integral along B_- is just the opposite of the same integral along B_+, we obtain

$$I_{BC} = \int_{C_1} P(\sigma) \exp\{ik_0[\rho\sigma + (z + h)(1 - \sigma^2)^{1/2}]\}\, d\sigma, \qquad (6.100\text{a})$$

where

$$P(\sigma) = \frac{2n^2(n^2 - \sigma^2)^{1/2} \breve{H}_0^{(1)}(k_0 \rho\sigma)\sigma}{(1 - n^4)(\sigma^2 - \sigma_P{}^2)} \qquad (6.100\text{b})$$

and $\sigma_P = n(1 + n^2)^{-1/2}$.

To evaluate the integral (6.99a) it will be convenient to deform the path of integration to the path of steepest descents through the point $\sigma = n$, as shown in Fig. 6.13b. It follows from (6.93) that the relevant pole lies below and not above the branch cut. Thus, the deformation to the path of steepest descents through the point $\sigma = n$ can be done straightforwardly. On the new path we set

$$\sigma = n + is^2 \qquad (6.101\text{a})$$

and obtain

$$I_{BC} = \frac{i4n^2 \exp(ik_0 n\rho)}{1 - n^4} \int_0^\infty \frac{\sigma(n^2 - \sigma^2)^{1/2} \breve{H}_0^{(1)}(k_0 \rho\sigma) \exp[ik_0(z + h)(1 - \sigma^2)^{1/2}]}{\sigma^2 - \sigma_P{}^2}$$

$$\times \exp(-k_0 \rho s^2) s\, ds. \qquad (6.101\text{b})$$

Since the integrand falls off rapidly as s increases, only small values of s will be significant. Thus, we can expand the integrand in a power series in s and retain only the first few terms to obtain

$$I_{BC} = -\frac{i8(1+n^2)\exp(-i\pi/4)\exp\{ik_0[n\rho+(1-n^2)^{1/2}(z+h)]\}}{n(1-n^4)(\pi k_0\rho)^{1/2}}$$

$$\times \int_0^\infty s^2 \exp\left\{-k_0\left[\rho - \frac{n(z+h)}{(1-n^2)^{1/2}}\right]s^2 + \frac{ik_0(z+h)}{2(1-n^2)^{3/2}}s^4\right\} ds, \qquad (6.102)$$

where in the exponent of the integrand we retained powers of s up to the fourth, since when $(z+h)/\rho = (1-n^2)^{1/2}/n$, then the coefficient of the second power of s vanishes. It will follow that the integral (6.102) can be evaluated exactly in terms of the parabolic cylinder function defined by†

$$D_{-\nu}(\tau) = \frac{\exp(-\tau^{2/4})}{\Gamma(\nu)}\int_0^\infty t^{\nu-1}\exp\left(-\tau t - \frac{t^2}{2}\right) dt, \qquad (6.103)$$

where ν is a positive integer. Thus, in (6.102) we make the transformation

$$s^2 = (1-n^2)^{3/2}\exp\left(\frac{i\pi}{4}\right)[k_0(z+h)]^{-1/2}\cdot t, \qquad (6.104)$$

which recasts (6.102) into

$$I_{BC} = -\frac{i4\exp(i\pi/8)(1-n^2)^{1/8}}{n(\pi k_0\rho)^{1/2}[k_0(z+h)]^{3/4}}\exp\{ik_0[n\rho+(1-n^2)^{1/2}(z+h)]\}$$

$$\times \int_0^\infty t^{1/2}\exp\left(-\tau t - \frac{t^2}{2}\right) dt, \qquad (6.105a)$$

where

$$\tau = \exp\left(\frac{i\pi}{4}\right)\left[\frac{k_0(1-n^2)^{1/2}}{z+h}\right]^{1/2}[\rho(1-n^2)^{1/2} - n(z+h)]. \qquad (6.105b)$$

From (6.103) it is clear that the integral in (6.105a) is equal to

$$(\sqrt{\pi}/2)\exp(-\tau^2/4)\, D_{-3/2}(\tau);$$

thus

$$I_{BC} = -\frac{2\exp(i5\pi/8)(1-n^2)^{1/8}}{n(k_0\rho)^{1/2}[k_0(z+h)]^{3/4}}\exp\{ik_0[n\rho+(1-n^2)^{1/2}(z+h)]\}$$

$$\times \exp\left(\frac{\tau^2}{4}\right)D_{-3/2}(\tau). \qquad (6.106)$$

† See Abramowitz and Stegun (1964), where integral representation of $U(a, z)$ is given in formula 19.5.3 (p. 687); since $D_{-a-1/2}(z) = U(a, z)$ (formula 19.3.7) our definition in (6.103) follows with the change $\nu = a + \frac{1}{2}$.

When $\tau \to \infty$ and $|\arg \tau| < 3\pi/4$ (which is the present case), $D_{-3/2}(\tau) \sim \tau^{-3/4} \exp(-\tau^2/4)$ and we obtain

$$I_{BC} = -\frac{2 \exp(i\pi/4) \exp\{ik_0[n\rho + (1-n^2)^{1/2}(z+h)]\}}{k_0 n(1-n^2)^{1/4}\sqrt{\rho}\,[\rho(1-n^2)^{1/2} - n(z+h)]^{3/2}} \tag{6.107}$$

valid for $\rho > n(z+h)(1-n^2)^{-1/2}$.

When $\tau \to 0$, $D_{-3/2}(\tau) \sim \sqrt{\pi}2^{5/4}/\Gamma(1/4)$, and we obtain

$$I_{BC} = -\frac{i4\sqrt{\pi}\exp(i\pi/8)}{\Gamma(1/4)n^{3/2}[(1-n^2)k_0 r_2/2]^{1/4}} \frac{\exp(ik_0 r_2)}{k_0 r_2} \tag{6.108}$$

valid in the neighborhood of $\rho \sim n(z+h)(1-n^2)^{-1/2}$.

The contribution I_{BC} resulting from the integration along the borders of the branch cut $\operatorname{Im}(n^2 - \sin^2\beta)^{1/2} = 0$ is the lateral wave encountered earlier in Chapter 2.

If we collect the foregoing results and substitute in (6.5), where we also use (6.94a), we obtain the complete expression for the vector potential in the upper region. Thus, when $\rho > n(z+h)(1-n^2)^{-1/2}$, we obtain for the leading terms

$$\Pi_1 = \frac{ip}{4\pi\omega\varepsilon_0}\left(\frac{\exp(ik_0 r_1)}{r_1} + \left[1 - \frac{2(n^2 - \sin^2\theta_2)^{1/2}}{n^2\cos\theta_2 + (n^2 - \sin^2\theta_2)^{1/2}} + \frac{G(\theta_2)}{2ik_0 r_2}\right]\right.$$
$$\left.\times \frac{\exp(ik_0 r_2)}{r_2} + \frac{i2\exp\{ik_0[n\rho + (1-n^2)^{1/2}(z+h)]\}}{k_0 n(1-n^2)^{1/4}\sqrt{\rho}\,[\rho(1-n^2)^{1/2} - n(z+h)]^{3/2}}\right). \tag{6.109}$$

It is interesting to note that in the case when both the transmitter and the receiver are on the ground, $z = h = 0$, we obtain for the leading terms

$$\Pi_1 = -\frac{p}{2\pi\omega\varepsilon_0 k_0 n(1-n^2)\rho^2}[n^5\exp(ik_0\rho) + \exp(ik_0 n\rho)], \tag{6.110}$$

where the term $n^5 \exp(ik_0\rho)$ is the second-order term of the saddle-point contribution (reflected field) and the term $\exp(ik_0 n\rho)$ is the branch cut contribution (lateral field). When n is lossy, $n'' > 0$, then the lateral field will be exponentially attenuated and the second-order reflected field will predominate for a large ρ.

6.3.2 Fields in the Plasma

According to (6.33a) and (6.36) with $n_1 = 1$ and $n_2 = n$, the fields in the plasma can be found from

$$\Pi_2 = -\frac{p}{4\pi\omega\varepsilon_0} I_2, \tag{6.111a}$$

where

$$I_2 = k_0 n \exp\left(-\frac{i\pi}{4}\right) \int_{\Gamma_1} \times \frac{\exp[ik_0 h(1 - n^2 \sin^2 \beta)^{1/2}] \check{H}_0^{(1)}(k_0 n\rho \sin \beta) \sin \beta \cos \beta}{\cos \beta + n(1 - n^2 \sin^2 \beta)^{1/2}} \times \exp[ik_0 nr \cos(\beta - \theta')]\, d\beta. \tag{6.111b}$$

The branch points will be situated at $\beta = \beta_{B1,2}$ where

$$\sin \beta_{B1,2} = \pm\left(\frac{1}{n'} - \frac{in''}{n'^2}\right) \tag{6.112a}$$

and the poles at $\beta = \beta_{P1,2}$ where

$$\sin \beta_{P1,2} = \pm\left[\frac{1}{(1 + n'^2)^{1/2}} - \frac{in'n''}{(1 + n'^2)^{3/2}}\right]. \tag{6.112b}$$

A brief analysis, similar to the ones carried out before, will reveal that the poles, branch points, and branch cuts will be as shown in Fig. 6.14. Moreover,

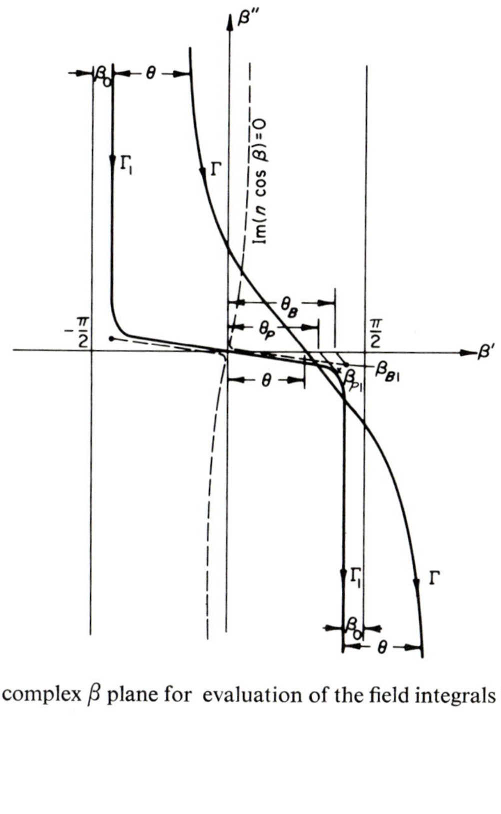

Fig. 6.14. The complex β plane for evaluation of the field integrals in the lower medium when $|n| < 1$.

it can be shown that the pole β_{P1} lies below the branch cut $\mathrm{Im}(n\cos\beta) = 0$ and not above it.

It is clear from what has been said before and on the basis of the diagram in Fig. 6.14 that in this case the integral I_2 can be written

$$I_2 = k_0 n \exp\left(-\frac{i\pi}{4}\right)[I_{SD} + u(\theta' - \theta_P)2\pi i \,\mathrm{Res}(\beta = \beta_{P1}) + u(\theta' - \theta_B)I_{BC}], \tag{6.113}$$

where the various contributions will arise from the saddle point, the pole at $\beta = \beta_{P1}$, and the integration along the borders of the branch cut $\mathrm{Im}(n\cos\beta) = 0$. Since the procedure for evaluation of each of these contributions has already been established, we shall not pursue this problem any further.

6.4 Vertical Electric Dipole in a Three-Layer Medium

The formalism developed in the preceding sections can be easily extended to any number of layers. In this section we consider a three-layer medium with the vertical electric dipole situated in the center layer. The solution to an arbitrary three-layer medium will serve as a prototype for discussion of more specific practical problems, like propagation in the earth-ionosphere wave guide and radiation from a dielectric covered small annular slot antenna.

Consider the structure of Fig. 6.15, consisting of three arbitrary homogeneous media characterized by the indices of refraction n_1, n_2, and n_3 and relative permeabilities $\mu_1 = \mu_2 = \mu_3 = 1$. The problem can be most expeditiously formulated in terms of the z component of the Hertzian vector of the electric type. Let us denote by Π_1, Π_2, and Π_3 the z components of the Hertzian vector in the three respective regions. They must satisfy

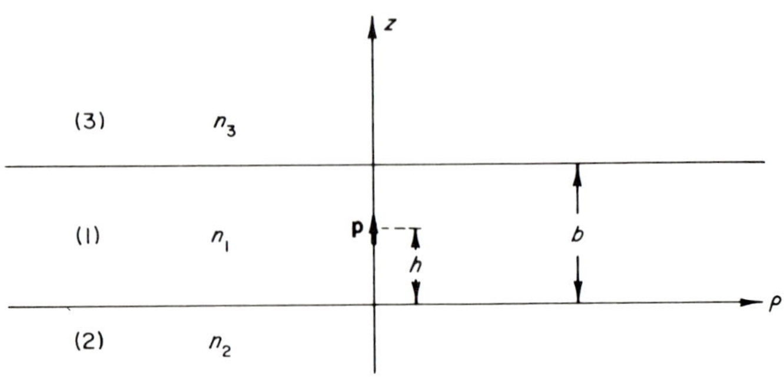

Fig. 6.15. Vertical dipole in a three-layer medium.

$$(\nabla^2 + k_1{}^2)\Pi_1 = \frac{ip}{\omega\varepsilon_0 n_1{}^2}\,\delta(x)\,\delta(y)\,\delta(z-h), \tag{6.114a}$$

$$(\nabla^2 + k_2{}^2)\Pi_2 = 0, \tag{6.114b}$$

$$(\nabla^2 + k_3{}^2)\Pi_3 = 0. \tag{6.114c}$$

The boundary conditions to be satisfied can be stated as follows. At $z = 0$

$$n_1{}^2\Pi_1 = n_2{}^2\Pi_2, \tag{6.115a}$$

$$\frac{\partial \Pi_1}{\partial z} = \frac{\partial \Pi_2}{\partial z}, \tag{6.115b}$$

and at $z = h$

$$n_1{}^2\Pi_1 = n_3{}^2\Pi_3, \tag{6.115c}$$

$$\frac{\partial \Pi_1}{\partial z} = \frac{\partial \Pi_3}{\partial z}, \tag{6.115d}$$

plus the usual requirement that the fields remain finite everywhere.

In region 1, which contains the source, the solution for π_1 must include the particular integral and a suitable complementary solution. Since this region is bounded, we must retain both $A(\alpha_1, \alpha_2)$ and $B(\alpha_1, \alpha_2)$ in (6.8), and consequently write, in place of (6.9),

$$\Pi_1 = -\frac{p}{8\pi^2\omega\varepsilon_0 n_1{}^2}\iint_{-\infty}^{\infty} \times \left[\frac{\exp(i\tau_1|z-h|)}{\tau_1} + A(\alpha_1, \alpha_2)\exp(i\tau_1 z) + B(\alpha_1, \alpha_2)\exp(-i\tau_1 z)\right] \times \exp[i(\alpha_1 x + \alpha_2 y)]\,d\alpha_1\,d\alpha_2. \tag{6.116}$$

In region 2 there must be only a wave traveling in the negative z direction, and in region 3 only a wave traveling in the positive z direction must be allowed. Thus we write

$$\Pi_2 = -\frac{p}{8\pi^2\omega\varepsilon_0 n_1{}^2}\iint_{-\infty}^{\infty} T_{12}(\alpha_1, \alpha_2)\exp(-i\tau_2 z)\exp[i(\alpha_1 x + \alpha_2 y)]\,d\alpha_1\,d\alpha_2, \tag{6.117a}$$

$$\Pi_3 = -\frac{p}{8\pi^2\omega\varepsilon_0 n_1{}^2}\iint_{-\infty}^{\infty} T_{13}(\alpha_1, \alpha_2)\exp[i\tau_3(z-b)]\exp[i(\alpha_1 x + \alpha_2 y)]\,d\alpha_1\,d\alpha_2, \tag{6.117b}$$

where $\tau_j = [k_j{}^2 - (\alpha_1{}^2 + \alpha_2{}^2)]^{1/2}$, $k_j = k_0 n_j$, and $\mathrm{Im}(\tau_j) > 0$.

Applying the boundary conditions (6.115), we obtain a set of four simultaneous equations:

$$\begin{pmatrix} 1 & 1 & -n_{21}^2 & 0 \\ 1 & -1 & \tau_2/\tau_1 & 0 \\ \exp(i\tau_1 b) & \exp(-i\tau_1 b) & 0 & -n_{31}^2 \\ \exp(i\tau_1 b) & -\exp(-i\tau_1 b) & 0 & -\tau_3/\tau_1 \end{pmatrix} \begin{pmatrix} A \\ B \\ T_{12} \\ T_{13} \end{pmatrix} = \begin{pmatrix} -\exp(i\tau_1 h)/\tau_1 \\ \exp(i\tau_1 h)/\tau_1 \\ -\exp[i\tau_1(b-h)]/\tau_1 \\ -\exp[i\tau_1(b-h)]/\tau_1 \end{pmatrix}. \tag{6.118}$$

Hence, we obtain

$$A = R_{12}\left\{\frac{1 + R_{13}\exp[i2\tau_1(b-h)]}{1 - R_{12}R_{13}\exp(i2\tau_1 b)}\right\}\frac{\exp(i\tau_1 h)}{\tau_1}, \tag{6.119a}$$

$$B = R_{13}\exp(i2\tau_1 b)\left[\frac{1 + R_{12}\exp(i2\tau_1 h)}{1 - R_{13}R_{12}\exp(i2\tau_1 b)}\right]\frac{\exp(-i\tau_1 h)}{\tau_1}, \tag{6.119b}$$

$$T_{12} = \frac{2\exp(i\tau_1 h)}{\tau_2 + n_{21}^2\tau_1}\left\{\frac{1 + R_{13}\exp[i2\tau_1(b-h)]}{1 - R_{12}R_{13}\exp(i2\tau_1 b)}\right\}, \tag{6.119c}$$

$$T_{13} = \frac{2\exp[i\tau_1(b-h)]}{\tau_3 + n_{31}^2\tau_1}\left[\frac{1 + R_{12}\exp(i2\tau_1 h)}{1 - R_{12}R_{13}\exp(i2\tau_1 b)}\right], \tag{6.119d}$$

where

$$R_{12} = \frac{n_{21}^2\tau_1 - \tau_2}{n_{21}^2\tau_1 + \tau_2}, \tag{6.120a}$$

$$R_{13} = \frac{n_{31}^2\tau_1 - \tau_3}{n_{31}^2\tau_1 + \tau_3}. \tag{6.120b}$$

It is interesting to note that R_{12} in (6.120a) is the same as R in (6.13a) except for the absence of the factor $\exp(i\tau_1 h)/\tau_1$. Clearly, R_{12} and R_{13} can be considered as generalized reflection coefficients at the upper and lower boundaries, respectively. Moreover, as $b \to \infty$ it will follow that $A \to R$, $B \to 0$, $T_{12} \to T$, $T_{13} \to 0$, and the present solution becomes identical to that obtained in the case of a single interface in Section 6.1.

Since all four coefficients A, B, T_{12}, and T_{13} are even functions of α_1 and α_2 we can employ the transformation (5.23a) and rewrite Π_1, Π_2, and Π_3 as follows.

$$\Pi_1 = \frac{ip}{4\pi\omega\varepsilon_0 n_1^2}\left[\frac{\exp(ik_1 r_1)}{r_1} + I_1\right], \tag{6.121a}$$

$$\Pi_2 = -\frac{p}{4\pi\omega\varepsilon_0 n_1^2} I_2, \tag{6.121b}$$

$$\Pi_3 = -\frac{p}{4\pi\omega\varepsilon_0 n_1^2} I_3, \tag{6.121c}$$

where

$$I_1 = \frac{1}{2}\int_{-\infty}^{\infty} [A(\lambda)\exp(i\tau_1 z) + B(\lambda)\exp(-i\tau_1 z)]H_0^{(1)}(\lambda\rho)\lambda\, d\lambda, \tag{6.122a}$$

$$I_2 = \frac{1}{2}\int_{-\infty}^{\infty} T_{12}(\lambda)\exp(-i\tau_2 z)H_0^{(1)}(\lambda\rho)\lambda\, d\lambda, \tag{6.122b}$$

$$I_3 = \frac{1}{2}\int_{-\infty}^{\infty} T_{13}(\lambda)\exp[i\tau_3(z-b)]H_0^{(1)}(\lambda\rho)\lambda\, d\lambda. \tag{6.122c}$$

This concludes the formal solution of the problem. In order to express the answers in a form suitable for numerical calculations, the various integrals must be evaluated by contour integration. It is clear that in this case it will be important to locate precisely the various singularities of the integrands. Looking over the various integrands, we see that there will be three pairs of branch points at $\lambda = \pm k_1$, $\lambda = \pm k_2$, and $\lambda = \pm k_3$. In addition to these there will be an infinite number of poles where the denominator vanishes. There is no way of finding the location of these poles in a general case and we shall therefore postpone the evaluation of the integrals until consideration is given to a few special limiting cases of interest.

6.4.1 Perfectly Conducting Boundaries

If the media 2 and 3 in Fig. 6.15 are infinitely conducting so that $n_2 = n_3 = \infty$, then $R_{12} = R_{13} = 1$ and the integrals $I_2 = I_3 = 0$. Moreover, in this case

$$A = \frac{i\cos\tau_1(b-h)}{\tau_1\sin\tau_1 b}, \tag{6.123a}$$

$$B = \frac{i \cos \tau_1 h}{\tau_1 \sin \tau_1 b} \exp(i\tau_1 b). \tag{6.123b}$$

Substituting these into (6.116), we obtain

$$\Pi_{1<} = -\frac{ip}{4\pi^2 \omega \varepsilon_0 n_1{}^2} \iint_{-\infty}^{\infty} \frac{\cos \tau_1 z \cos \tau_1 (b-h)}{\tau_1 \sin \tau_1 b} \exp[i(\alpha_1 x + \alpha_2 y)]\, d\alpha_1\, d\alpha_2, \tag{6.124a}$$

$$\Pi_{1>} = -\frac{ip}{4\pi^2 \omega \varepsilon_0 n_1{}^2} \iint_{-\infty}^{\infty} \frac{\cos \tau_1 h \cos \tau_1 (b-z)}{\tau_1 \sin \tau_1 b} \exp[i(\alpha_1 x + \alpha_2 y)]\, d\alpha_1\, d\alpha_2, \tag{6.124b}$$

where $\Pi_{1<}$ is pertinent to the region $z < h$ and $\Pi_{1>}$ to $z > h$. Alternately, using the transformations (5.23a), we obtain

$$\Pi_{1<} = -\frac{ip}{4\pi \omega \varepsilon_0 n_1{}^2} \int_{-\infty}^{\infty} \frac{\cos \tau_1 z \cos \tau_1 (b-h)}{\tau_1 \sin \tau_1 b} H_0^{(1)}(\lambda\rho)\lambda\, d\lambda, \tag{6.125a}$$

$$\Pi_{1>} = -\frac{ip}{4\pi \omega \varepsilon_0 n_1{}^2} \int_{-\infty}^{\infty} \frac{\cos \tau_1 h \cos \tau_1 (b-z)}{\tau_1 \sin \tau_1 b} H_0^{(1)}(\lambda\rho)\lambda\, d\lambda. \tag{6.125b}$$

The integrals in (6.125) can be evaluated in the complex λ plane by closing the contour with a large semicircle in the upper half plane. Setting $\lambda = Re^{i\theta}$ on the semicircle and letting $R \to \infty$, it is readily shown that the absolute value of each one of the integrands in (6.125) will approach $(R/2\pi\rho)^{1/2}$ $\exp\{-R \sin \theta(\rho + |z-h|)\}$; hence it will vanish for $R \to \infty$ everywhere above the real axis.

Each one of the integrands will have poles where $\tau_1 b = \pm m\pi$; that is, where

$$\lambda_m = k_1 \left[1 - \left(\frac{m\pi}{k_1 b}\right)^2\right]^{1/2}. \tag{6.126}$$

In order to find the locus of the poles in (6.126) it will be convenient to assume that medium 1 is slightly lossy so that $k_1 = k_1' + ik_1''$ and $k_1'' < k_1'$. We obtain in this case

$$\lambda_m \sim \left[k_1'^2 - \left(\frac{m\pi}{b}\right)^2\right]^{1/2} + \frac{ik_1' k_1''}{[k_1'^2 - (m\pi/b)^2]^{1/2}}, \qquad m \ll \frac{k_1' b}{\pi}, \tag{6.127a}$$

$$\lambda_m \sim \frac{k_1' k_1''}{m\pi/b} + i\left(\frac{m\pi}{b}\right), \qquad m \gg \frac{k_1' b}{\pi}. \tag{6.127b}$$

The location of these poles is shown in Fig. 6.16. If medium 1 is lossless, so that $k'' = 0$, then the poles for which $m < k_1 b/\pi$ will be situated on the real axis and the poles for which $m > k_1 b/\pi$ will be situated on the imaginary axis. In the latter case $2b/m$ will be the cutoff wavelength for the propagating modes.

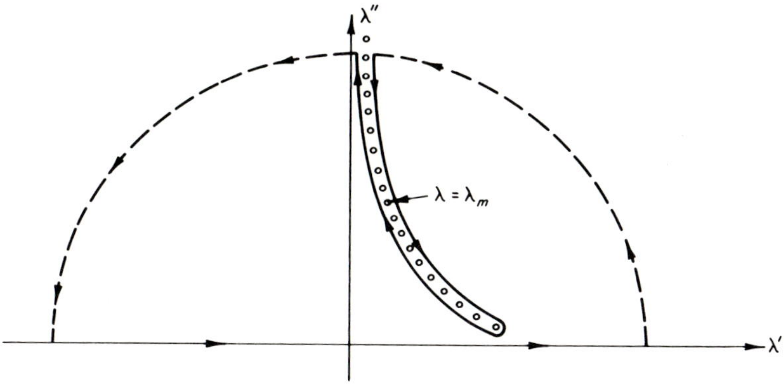

Fig. 6.16. The complex λ plane for evaluation of the integrals in (6.125).

It is clear that each of the integrals in (6.125) must be equal to $2\pi i$ times the sum of residues at the poles $\lambda = \lambda_m$. The branch point $\lambda = k_1$, even though present, has no effect on integration since the integrands in are even functions of τ_1. Noting that

$$\operatorname{Res}\left[\frac{F(\lambda)}{\sin \tau_1 b}\right]_{\lambda=\lambda_m} = \frac{F(\lambda_m)}{d/d\lambda\,(\sin \tau_1 b)\,|_{\lambda=\lambda_m}}, \tag{6.128}$$

we obtain

$$\Pi_1 = -\frac{p}{2\omega\varepsilon_0 {n_1}^2 b} \sum_{-\infty}^{\infty} \cos\frac{m\pi z}{b} \cos\frac{m\pi h}{b} H_0^{(1)}(\lambda_m \rho), \tag{6.129a}$$

$$\Pi_1 = -\frac{p}{\omega\varepsilon_0 {n_1}^2 b}\left[\frac{1}{2} H_0^{(1)}(k_1\rho) + \sum_{1}^{\infty} \cos\frac{m\pi z}{b} \cos\frac{m\pi h}{b} H_0^{(1)}(\lambda_m \rho)\right], \tag{6.129b}$$

where $\lambda_m = \lambda'_m + i\lambda''_m$, $\lambda''_m \geq 0$, is given by (6.126). Note that since $\Pi_{1>}$ can be obtained from $\Pi_{1<}$ by interchanging the roles of z and h, the result in (6.129b) is equally valid for $\Pi_{1<}$ and $\Pi_{1>}$. Equation (6.129b) represents the total field radiated by the dipole and it consists of a cylindrical wave represented by the term $H^{(1)}(k_1 \rho)$ and a summation of all possible modal waves.

At a far distance from the dipole we can replace the Hankel function by its asymptotic form to obtain

$$\Pi_1 = -\frac{p\exp(-i\pi/4)}{\omega\varepsilon_0 n_1{}^2 b}\left(\frac{2}{\pi k_1\rho}\right)^{1/2}\left(\frac{1}{2}\exp(ik_1\rho) + \sum_{m=1}^{\infty}\left[1 - \left(\frac{m\pi}{k_1 b}\right)^2\right]^{-1/4}\right.$$
$$\left.\times \cos\frac{m\pi z}{b}\cos\frac{m\pi h}{b}\exp\{ik_1\rho[1-(m\pi/k_1 b)^2]^{1/2}\}\right). \tag{6.130}$$

6.4.2 Imperfectly Conducting Boundaries

If the media 2 and 3 are conducting but not perfectly, a simple solution can still be obtained. For simplicity we set $n_3 = n_2 \gg n_1$. In this case the reflection coefficients R_{12} and R_{13} in (6.120) can be simplified to

$$R_{12} = R_{13} = R = \frac{n_2\tau_1 - n_1 k_1}{n_2\tau_1 + n_1 k_1}. \tag{6.131}$$

It will be convenient to set

$$\frac{n_1 k_1}{n_2 \tau_1} = i\tan\gamma, \tag{6.132}$$

so that $R = \exp(-i2\gamma)$. We obtain in this case

$$\Pi_{1<} = -\frac{ip}{4\pi\omega\varepsilon_0 n_1{}^2}\int_{-\infty}^{\infty}\frac{\cos(\tau_1 z - \gamma)\cos[\tau_1(b-h)-\gamma]}{\tau_1\sin(\tau_1 b - 2\gamma)}H_0^{(1)}(\lambda\rho)\lambda\,d\lambda, \tag{6.133}$$

and $\Pi_{1>}$ can be obtained from $\Pi_{1<}$ in (6.133) by interchanging the roles of z and h.

The integrand in (6.133) has an infinite number of poles in the first quadrant of the λ plane where

$$\tau_{1m} b - 2\gamma_m = \pm m\pi, \tag{6.134}$$

which, in general, cannot be solved explicitly for λ_m since γ_m is a function of λ_m. The branch point singularity $\lambda = k_1$ and the associated branch cut need not be taken into account in integration since the integrand is an even function of τ_1. Thus we obtain

$$\Pi_1 = -\frac{p}{2\omega\varepsilon_0 n_1{}^2}\sum_{-\infty}^{\infty}\frac{\cos(\tau_{1m}z - \gamma_m)\cos(\tau_{1m}h - \gamma_m)H_0^{(1)}(\lambda_m\rho)}{b - 2ik_1 n_1/n_2\tau_{1m}^2}, \tag{6.135}$$

which as $n_2 \to \infty$ becomes identical with the corresponding expression for perfectly conducting walls in (6.129a).

6.4.3 The Earth-Ionosphere Wave Guide at Extremely Low Frequencies

The prototype solution for a three-layer medium formulated in this section lends itself to the analysis of wave propagation in the earth-ionosphere wave guide.

At frequencies below 10 kilocycles the refractive index of the earth is in excess of 100 (see Table 6.2); hence, to a reasonable approximation, it can be considered a perfect conductor. The lowest region of the ionosphere (E layer) exhibits an electron density of about 750 electrons per cubic centimeter and a collision frequency $\nu \sim 10^7$. Thus, in this case $\omega \ll \nu$ and $\omega_p \ll \nu$ and according to (3.37b) the effective refractive index of the ionosphere becomes

$$n_i = n_3 = (1 + i\omega_p{}^2/\omega\nu)^{1/2}. \tag{6.136}$$

At a frequency of 10 kilocycles, $n_i \sim 1.6 + i1.26$ and it increases with decreasing frequency.

In accordance with the assumptions made, it will follow that $R_{12} = 1$, $T_{12} = 0$, and

$$R_{13} = \exp(-i2\gamma), \tag{6.137a}$$

$$\tan \gamma = -\frac{i\tau_3}{n_3{}^2\tau_1}. \tag{6.137b}$$

Using these, we readily obtain

$$\Pi_{1<} = -\frac{ip}{4\pi\omega\varepsilon_0}\int_{-\infty}^{\infty}\frac{\cos\tau_1 z\cos[\tau_1(b-h)-\gamma]}{\tau_1\sin(\tau_1 b-\gamma)}H_0^{(1)}(\lambda\rho)\lambda\,d\lambda \tag{6.138a}$$

and

$$\Pi_3 = -\frac{ip}{4\pi\omega\varepsilon_0}\int_{-\infty}^{\infty}\frac{\cos\tau_1 h}{\sin(\tau_1 b-\gamma)}H_0^{(1)}(\lambda\rho)\exp[i\tau_2(z-b)]\lambda\,d\lambda, \tag{6.138b}$$

where $\Pi_{1>}$ can be obtained from $\Pi_{1<}$ by interchanging the roles of h and z.

The integral Π_1 in (6.138a) can be evaluated by integrating over the poles given by

$$\tau_{1m} b - \gamma_m = \pm m\pi \tag{6.139}$$

to obtain

$$\Pi_1 = -\frac{p}{2\omega\varepsilon_0}\sum_{-\infty}^{\infty}\frac{\cos\tau_{1m} z\cos\tau_{1m} h H_0^{(1)}(\lambda_m\rho)}{b - ik_0{}^2(n_3{}^2-1)\cos^2\tau_{1m} b/\tau_{1m}^2\tau_{3m}}, \tag{6.140}$$

which in the limit as $n_3 \to \infty$ becomes identical with (6.129a).

Even though the problem has been solved formally, a little can be said about the behavior of fields until the roots of the transcendental equation (6.139) are determined. This is a formidable problem since, in general, an exact solution is not possible. An extensive approximate study of these roots and their effect on terrestrial radio wave propagation has been conducted by Wait (1962).

Problems

6.1 Prove (6.25).

6.2 Prove (6.27a) and (6.27b). Also, show that on a large semicircle in the upper λ plane $0 < \varphi_1 + \varphi_2 < \pi$ and $\pi < \varphi_1 + \varphi_2 < 2\pi$, respectively, on the right and left sides of the branch cut $\lambda'\lambda'' = k_1' k_1''$ facing the origin.

6.3 Prove (6.70).

6.4 An electric dipole of moment p is situated above a perfectly conducting plane coinciding with the x-y plane. The dipole lies in the y-z plane and is tilted with respect to the z axis by an angle γ as shown in Fig. 6.17.
(a) Formulate a rigorous solution to the problem in terms of the components of the Hertzian vector.
(b) Find the image representation of the solution.
(c) When the distance from the dipole is large, find all of the electric and magnetic field components in spherical coordinates.

6.5 An insulated electric current loop of radius a carrying a current $I\exp(-i\omega t)$ is buried below the surface of the earth as shown in Fig. 6.18. The plane of the loop is parallel to the surface of the earth.
(a) Formulate the solution to the problem in terms of the Hertzian vector.
(b) In the case when the radius of the loop is very small, check the solution obtained in (a) with the solution to the equivalent problem of a magnetic dipole.
(c) For the solution obtained in (a), find the asymptotic form of the electric and magnetic fields in the air appropriate to the vlf range in spherical coordinates when the distance from the origin is large.

6.6 Consider an electric dipole located horizontally at a distance h above the interface of two media as shown in Fig. 6.19. Formulate the solution in terms of the Hertzian vector of the electric type.
(a) Show that, in general, one component of the Hertzian vector is not sufficient to satisfy the boundary conditions.
(b) Find the integral representation of the fields in the upper medium

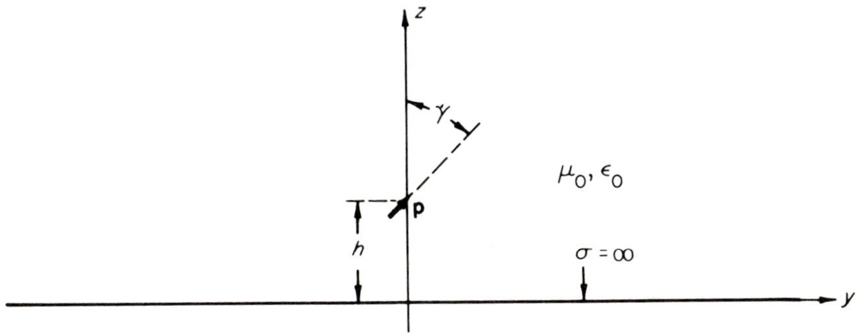

Fig. 6.17.

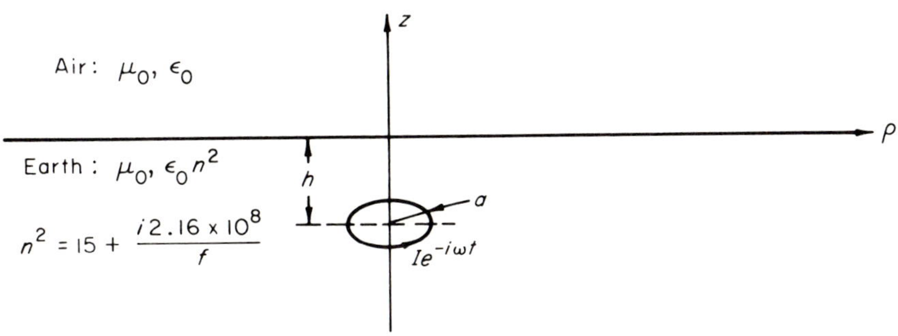

Fig. 6.18.

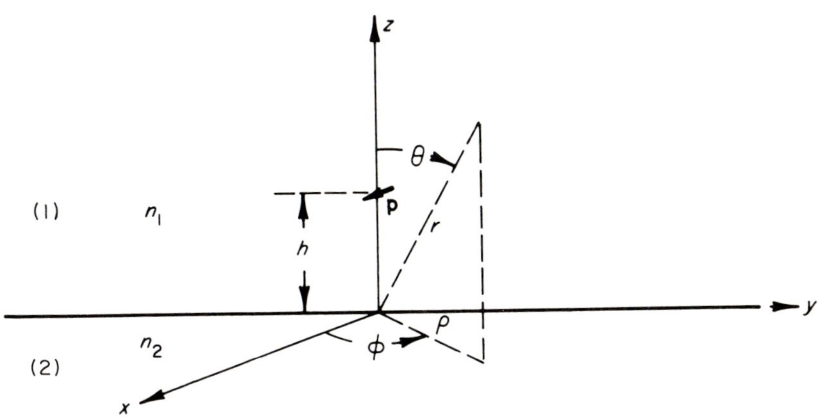

Fig. 6.19.

and determine all the singularities of the integrands in the λ plane.
(c) Transform to the β plane and show the location of the singularities when the upper medium is air ($n_1 = 1$) and the lower medium is ground, while the operating frequency is such that the displacement current can be neglected (vlf, lf).
(d) Find the radiation fields in the upper medium (air), neglecting the contribution of the singularities, and show that in the plane of the dipole ($\varphi = 0, \pi$) the electric field has only a vertical component and in the plane normal to the direction of the dipole the electric field has only a horizontal component.
(e) Explain how the polarization of the field would be changed if a horizontal magnetic dipole were substituted in place of the electric dipole considered.

6.7 The electromagnetic effects of a small wire loop carrying an alternating current $Ie^{-i\omega t}$ can be adequately represented by its equivalent magnetic dipole moment $\mathbf{m} = Ie^{-i\omega t}\mathbf{S}$, where $\mathbf{S}$ is the surface vector of area enclosed by the loop. Consider such a dipole located horizontally at a distance h above the interface of two media as shown in Fig. 6.20. Formulate the solution in terms of the Hertzian vector of the magnetic type and
(a) Show that, in general, one component of the Hertzian vector is not sufficient to satisfy the boundary conditions except in the case when the lower medium has infinite conductivity.

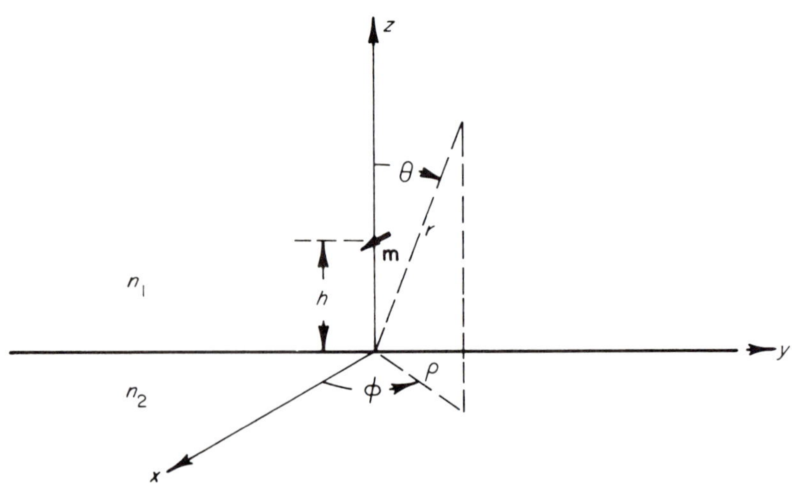

Fig. 6.20.

(b) Find the integral representation of the fields in the upper medium and determine all the singular points of the integrants in the λ plane.
(c) Transform to the β plane and show the location of the singularities when the upper medium is air ($n_1 = 1$) and the lower medium is ground, while the operating frequency is such that the displacement current can be neglected (vlf, lf).
(d) Find the radiation fields in the upper medium (air), neglecting the contribution of the singularities, and show that in the plane of the dipole ($\varphi = 0, \pi$) the electric field has only a vertical component E_θ, and in the plane normal to the direction of the dipole ($\varphi = \pm\pi/2$) the electric field has only a horizontal component E_φ.
(e) Explain how the polarization of the field would be changed if a horizontal electric dipole were substituted in the place of the magnetic dipole considered.

6.8 In the formulation of the Sommerfeld problem (vertical electric dipole above conducting earth) at frequencies low enough ($n_2 \to \infty$), we can represent the ground by a surface impedance $Z_s = Z_0/n_2$ and obtain the integral representation of the fields in the air in a more straightforward manner. Demonstrate this fact by actually formulating the problem this way and show that your result agrees with (6.38) when $n \to \infty$.

6.9 The three-layer problem formulated in Section 6.3 can be used with small changes to find the field of small annular slot on a perfectly conducting plane clad with a dielectric layer as shown in Fig. 6.21. Let the dielectric layer be a lossless plasma $n_1 = [1 - (\omega_p/\omega)^2]^{1/2}$.
(a) Find the integral expression for the fields in the air.

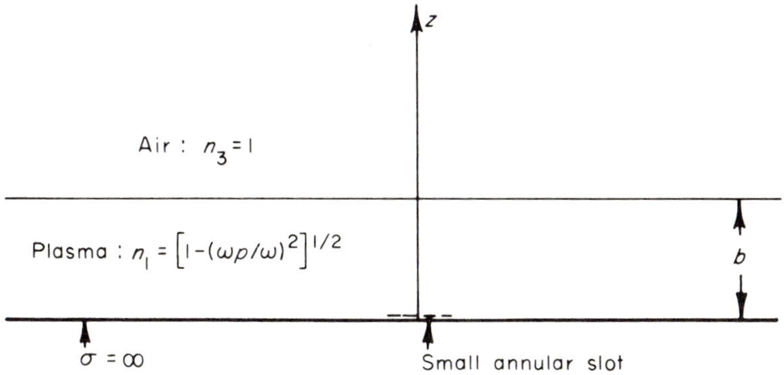

Fig. 6.21.

(b) Show how the solution in (a) is related to the solution of a centrally situated dipole in a slab of thickness $2b$ bounded by air on both sides.
(c) With due regard to the various singularities, evaluate the integral and find all of the electromagnetic field components in spherical coordinates when the distance from the slot is large.

6.10 (a) Formulate the solution to the problem of a vertical magnetic dipole in a three-layer medium as in Fig. 6.15 (find integral representation of the fields).
(b) When $n_3 = n_2 = \infty$, evaluate the field integral in the wave guide region [find expression equivalent to (6.129)].

References

Abramowitz, M., and Stegun, L. A., eds. (1964). "Handbook of Mathematical Functions." U.S. Govt. Printing Office, Washington, D.C. (*Natl. Bur. Std. U.S. Appl. Math. Ser.* **55**.)

Baños, A., Jr. (1966). "Dipole Radiation in the Presence of a Conducting Half-Space." Pergamon, Oxford.

Brekhovskikh, L. M. (1960). "Waves in Layered Media." Academic Press, New York.

Bremmer, H. (1958). Propagation of electromagnetic waves, *in* "Handbuch der Physik," Band XVI, "Elektrische Felder und Wellen" (S. Flugge, ed.). Springer, Berlin.

Budden, K. G. (1961). "The Wave-Guide Mode Theory of Wave Propagation." Prentice-Hall, Englewood Cliffs, New Jersey.

Karbowiak, A. E. (1959). Radiation and guided waves, *IRE Trans. Antennas Propagation Suppl.* **AP-7**, 191–200.

Sommerfeld, A. (1909). Uber die Ausbreitung der Wellen in der drahtlosen Telegraphie, *Ann. Physik.* **28**, 665–737.

Sommerfeld, A. (1949). "Partial Differential Equations in Physics." Academic Press, New York.

Stratton, J. A. (1941). "Electromagnetic Theory." McGraw-Hill, New York.

Wait, J. R. (1962). "Electromagnetic Waves in Stratified Media." McGraw-Hill, New York.

CHAPTER 7

RADIATION IN ANISOTROPIC PLASMA

In this chapter we consider the problem of radiation from discrete sources immersed in a plasma in the presence of a magnetostatic field. Since even a small magnetostatic field renders plasma anisotropic it can be recognized from the outset that the present problem is indeed a formidable one.

We shall begin with formulation of a solution to the problem of an arbitrarily oriented electric dipole in an unbounded magnetoplasma. Soon enough it will become apparent that simplifying assumptions concerning the orientation of the dipole with respect to the direction of the magnetostatic field must be made in order to express the solution in a manageable form. An asymptotic evaluation of the field integrals will be difficult to make due to the difficulties encountered in finding the saddle points.

Considerable simplification will result in the case of radiating line sources. It is found that in some of these cases the field integrals can be evaluated explicitly.

Finally, we shall consider the radiation from sources in the presence of a plane magnetoplasma–air boundary with particular emphasis on the fields in the air. Such a model is pertinent, for instance, when assessing the fields of an antenna situated in the ionosphere in the presence of the earth's magnetic field.

7.1 Electric Dipole in Magnetoplasma

Consider an arbitrarily oriented electric dipole $\mathbf{p}^{(e)}$ in an unbounded magnetoplasma. Let the dipole be situated on the z axis at $z = h$, and let the magnetostatic field be oriented in the positive z direction. The relative permeability of the magnetoplasma is unity, $\mu = 1$, and its electric permittivity in this case is

$$\boldsymbol{\varepsilon} = \begin{pmatrix} \varepsilon & -i\eta & 0 \\ i\eta & \varepsilon & 0 \\ 0 & 0 & \zeta \end{pmatrix}, \tag{7.1}$$

where the components of the matrix are given in (3.46) in terms of the wave frequency, plasma frequency, collision frequency, and magnetostatic field strength. For the time dependence of the form $\exp(-i\omega t)$ and the current distribution consisting of a single dipole, the vector wave equation (1.39) becomes

$$(-\nabla \times \nabla \times + k_0{}^2 \boldsymbol{\varepsilon})\mathbf{E} = -i\omega\mu_0 \mathbf{p}^{(e)}\,\delta(x)\,\delta(y)\,\delta(z-h) \tag{7.2}$$

and the magnetic fields can be found from

$$\mathbf{H} = (i\omega\mu_0)^{-1}\nabla \times \mathbf{E}. \tag{7.3}$$

The expansion of the double curl operator is given in (1.42). It will be convenient to reduce the vector wave equation (7.2) to a system of three algebraic equations by means of the triple Fourier transform pair given in (5.12). In doing this, it is expedient to note that the partial derivatives $\partial/\partial x$, $\partial/\partial y$, and $\partial/\partial z$ in the configuration space correspond to $i\alpha_1$, $i\alpha_2$, and $i\alpha_3$, respectively, in the transform space. Thus, we can write

$$\begin{pmatrix} k_0{}^2\varepsilon - \alpha_2{}^2 - \alpha_3{}^2 & \alpha_1\alpha_2 - ik_0{}^2\eta & \alpha_1\alpha_3 \\ \alpha_1\alpha_2 + ik_0{}^2\eta & k_0{}^2\varepsilon - \alpha_1{}^2 - \alpha_3{}^2 & \alpha_2\alpha_3 \\ \alpha_1\alpha_3 & \alpha_2\alpha_3 & k_0{}^2\zeta - \alpha_1{}^2 - \alpha_2{}^2 \end{pmatrix} \begin{pmatrix} \tilde{\tilde{E}}_x \\ \tilde{\tilde{E}}_y \\ \tilde{\tilde{E}}_z \end{pmatrix}$$

$$= -\frac{i\omega\mu_0 \exp(-i\alpha_3 h)}{(2\pi)^{3/2}} \begin{pmatrix} p_x^{(e)} \\ p_y^{(e)} \\ p_z^{(e)} \end{pmatrix}. \tag{7.4}$$

The system of simultaneous algebraic equations in (7.4) can be solved using Cramer's rule. For the determinant of the square matrix on the left we find

$$\Delta = k_0{}^2\zeta(\alpha_3{}^2 - \tau_1{}^2)(\alpha_3{}^2 - \tau_2{}^2), \tag{7.5a}$$

where

$$\tau_{1,2}^2 = k_0{}^2\varepsilon - \frac{\zeta + \varepsilon}{2\zeta}\lambda^2 \pm \left[k_0{}^4\eta^2 - \frac{k_0{}^2\eta^2}{\zeta}\lambda^2 + \left(\frac{\zeta - \varepsilon}{2\zeta}\right)^2 \lambda^4\right]^{1/2} \tag{7.5b}$$

and $\lambda^2 = \alpha_1{}^2 + \alpha_2{}^2$. Moreover, we find

$$\begin{aligned} \tilde{\tilde{E}}_x = -\frac{i\omega\mu_0 \exp(-i\alpha_3 h)}{(2\pi)^{3/2}\Delta} &\{p_x^{(e)}[(k_0{}^2\varepsilon - \alpha_1{}^2 - \alpha_3{}^2)(k_0{}^2\zeta - \lambda^2) - \alpha_2{}^2\alpha_3{}^2] \\ &- p_y^{(e)}[(k_0{}^2\zeta - \lambda^2)(\alpha_1\alpha_2 - ik_0{}^2\eta) - \alpha_1\alpha_2\alpha_3{}^2] \\ &- p_z^{(e)}\alpha_3[\alpha_1(k_0{}^2\varepsilon - \alpha_1{}^2 - \alpha_2{}^2 - \alpha_3{}^2) + ik_0{}^2\eta\alpha_2]\}, \end{aligned} \tag{7.6a}$$

$$\tilde{\tilde{E}}_y = -\frac{i\omega\mu_0 \exp(-i\alpha_3 h)}{(2\pi)^{3/2}\Delta}\{-p_x^{(e)}[(k_0{}^2\zeta - \lambda^2)(\alpha_1\alpha_2 + ik_0{}^2\eta) - \alpha_1\alpha_2\alpha_3{}^2]$$
$$+ p_y^{(e)}[(k_0{}^2\zeta - \lambda^2)(k_0{}^2\varepsilon - \alpha_2{}^2 - \alpha_3{}^2) - \alpha_1{}^2\alpha_3{}^2]$$
$$- p_z^{(e)}\alpha_3[\alpha_2(k_0{}^2\varepsilon - \alpha_1{}^2 - \alpha_2{}^2 - \alpha_3{}^2) - ik_0{}^2\eta\alpha_1]\}, \tag{7.6b}$$

$$\tilde{\tilde{E}}_z = -\frac{i\omega\mu_0 \exp(-i\alpha_3 h)}{(2\pi)^{3/2}\Delta}\{p_x^{(e)}\alpha_3[ik_0{}^2\eta\alpha_2 - \alpha_1(k_0{}^2\varepsilon - \alpha_1{}^2 - \alpha_2{}^2 - \alpha_3{}^2)]$$
$$- p_y^{(e)}\alpha_3[ik_0{}^2\eta\alpha_1 + \alpha_2(k_0{}^2\varepsilon - \alpha_1{}^2 - \alpha_2{}^2 - \alpha_3{}^2)]$$
$$+ p_z^{(e)}[(k_0{}^2\varepsilon - \alpha_2{}^2 - \alpha_3{}^2)(k_0{}^2\varepsilon - \alpha_1{}^2 - \alpha_3{}^2)$$
$$- (\alpha_1{}^2\alpha_2{}^2 + k_0{}^4\eta^2)]\}. \tag{7.6c}$$

The integral representation of the field components can be obtained by inversion according to (5.12b). Thus, any one of the field components can be written in the form

$$E = -\frac{i}{(2\pi)^3\omega\varepsilon_0\zeta}\iint_{-\infty}^{\infty} I(\alpha_1, \alpha_2)\exp[i(\alpha_1 x + \alpha_2 y)]\,d\alpha_1\,d\alpha_2, \tag{7.7a}$$

where

$$I(\alpha_1, \alpha_2) = \int_{-\infty}^{\infty}\frac{f(\alpha_1, \alpha_2, \alpha_3)}{(\alpha_3{}^2 - \tau_1{}^2)(\alpha_3{}^2 - \tau_2{}^2)}\exp[i\alpha_3(z-h)]\,d\alpha_3. \tag{7.7b}$$

The integral in (7.7b) can be evaluated by assuming α_3 to be complex and closing the contour with a large semicircle in the upper half plane for $z > h$ and in the lower half plane for $z < h$, as shown in Fig. 7.1. It is not difficult to show that the contribution from the semicircle will vanish in each case as its radius approaches infinity. The value of the integral is then equal to $2\pi i$ times the sum of residues at the enclosed poles. Thus, we obtain

$$I_{\gtrless} = \pi i\left[\frac{f(\alpha_1, \alpha_2, \pm\tau_1)}{\tau_1(\tau_1{}^2 - \tau_2{}^2)}\exp(i\tau_1|z-h|) + \frac{f(\alpha_1, \alpha_2, \pm\tau_2)}{\tau_2(\tau_2{}^2 - \tau_1{}^2)}\exp(i\tau_2|z-h|)\right], \tag{7.8}$$

which must be substituted into (7.7a) to obtain the integral representation of the various field components.

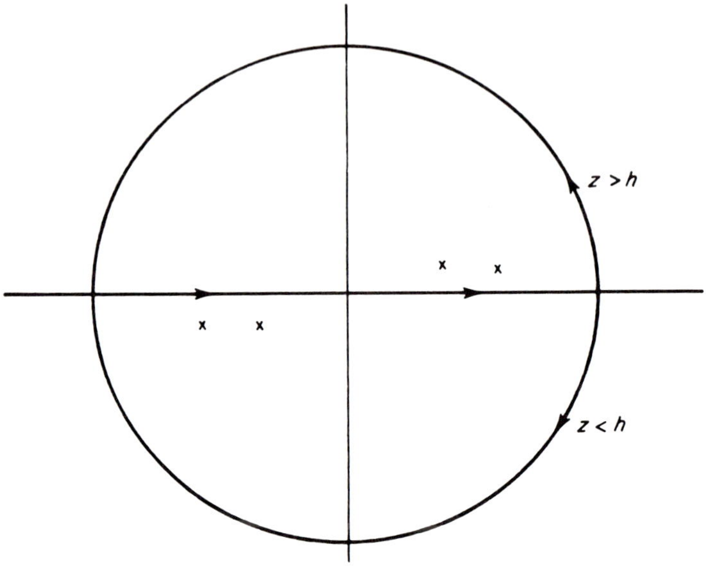

Fig. 7.1. The complex α_3 plane for evaluation of the integral in (7.7b).

7.1.1 Dipole along the Magnetostatic Field in an Unbounded Plasma

In the case when the dipole is oriented in the direction of the magnetostatic field $p_x^{(e)} = p_y^{(e)} = 0$, $p_z^{(e)} = p$ and the field components can be written

$$E_x = Q \frac{\partial}{\partial z}\left[\frac{\partial}{\partial x}(\nabla^2 + k_0{}^2\varepsilon) + ik_0{}^2\eta \frac{\partial}{\partial y}\right](I_1 - I_2), \tag{7.9a}$$

$$E_y = Q \frac{\partial}{\partial z}\left[\frac{\partial}{\partial y}(\nabla^2 + k_0{}^2\varepsilon) - ik_0{}^2\eta \frac{\partial}{\partial x}\right](I_1 - I_2), \tag{7.9b}$$

$$E_z = Q\left[\left(\frac{\partial^2}{\partial z^2} + k_0\varepsilon\right)(\nabla^2 + k_0{}^2\varepsilon) - k_0{}^4\eta^2\right](I_1 - I_2), \tag{7.9c}$$

where $Q = \omega\mu_0 p/8\pi^2 k_0{}^2 \zeta$ and

$$I_1 = \iint_{-\infty}^{\infty} \frac{\exp\{i[\alpha_1 x + \alpha_2 y + \tau_1|z - h|]\}}{\tau_1(\tau_1{}^2 - \tau_2{}^2)}\, d\alpha_1\, d\alpha_2\,, \tag{7.10a}$$

$$I_2 = \iint_{-\infty}^{\infty} \frac{\exp\{i[\alpha_1 x + \alpha_2 y + \tau_2\,|z - h|]\}}{\tau_2(\tau_1{}^2 - \tau_2{}^2)}\, d\alpha_1\, d\alpha_2\,. \tag{7.10b}$$

Moreover, in writing (7.9) we have availed ourselves of the following correspondences.

$$\alpha_1 \Leftrightarrow -i\frac{\partial}{\partial x}, \qquad \alpha_2 \Leftrightarrow -i\frac{\partial}{\partial y}, \qquad \alpha_3 \Leftrightarrow -i\frac{\partial}{\partial z}. \tag{7.11}$$

The magnetic field components can be found using (7.9) and (7.3). We obtain

$$H_x = -iQ\omega\varepsilon_0\,\varepsilon\left[\frac{\partial}{\partial y}(\nabla^2 + k_0{}^2\chi) + i\kappa\frac{\partial^3}{\partial x\,\partial z^2}\right](I_1 - I_2), \tag{7.12a}$$

$$H_y = iQ\omega\varepsilon_0\,\varepsilon\left[\frac{\partial}{\partial x}(\nabla^2 + k_0{}^2\chi) - i\kappa\frac{\partial^3}{\partial y\,\partial z^2}\right]I_1 - I_2), \tag{7.12b}$$

$$H_z = -Q\omega\varepsilon_0\,\eta\frac{\partial}{\partial z}\left(\frac{\partial^2}{\partial x^2} + \frac{\partial^2}{\partial y^2}\right)(I_1 - I_2). \tag{7.12c}$$

Employing the transformations

$$\begin{aligned} \alpha_1 &= \lambda\cos\gamma, \qquad & x &= \rho\cos\varphi, \\ \alpha_2 &= \lambda\sin\gamma, \qquad & y &= \rho\sin\varphi, \end{aligned} \tag{7.13}$$

the integrals in (7.10) can be recast as follows.

$$I_1 = \pi\int_{-\infty}^{\infty}\frac{H_0^{(1)}(\lambda\rho)\exp(i\tau_1|z-h|)}{\tau_1(\tau_1{}^2 - \tau_2{}^2)}\lambda\,d\lambda, \tag{7.14a}$$

$$I_2 = \pi\int_{-\infty}^{\infty}\frac{H_0^{(1)}(\lambda\rho)\exp(i\tau_2|z-h|)}{\tau_2(\tau_1{}^2 - \tau_2{}^2)}\lambda\,d\lambda, \tag{7.14b}$$

and it is worthwhile to note that the integrals I_1 and I_2 do not depend on φ, which could have been predicted from the symmetry of the problem.

The field components take on a somewhat simpler form in cylindrical polar coordinates. Using the transformations

$$\begin{pmatrix} E_\varphi \\ E_\rho \end{pmatrix} = \begin{pmatrix} -\sin\varphi & \cos\varphi \\ \cos\varphi & \sin\varphi \end{pmatrix}\begin{pmatrix} E_x \\ E_y \end{pmatrix} \tag{7.15a}$$

and

$$\begin{pmatrix} \partial_x \\ \partial_y \end{pmatrix} = \begin{pmatrix} \cos\varphi & -\sin\varphi \\ \sin\varphi & \cos\varphi \end{pmatrix}\begin{pmatrix} \partial_\rho \\ \rho^{-1}\partial_\varphi \end{pmatrix}, \tag{7.15b}$$

and since $\partial_\varphi = 0$ in this case, it readily follows that

$$E_\varphi = -iQk_0{}^2\eta\frac{\partial^2}{\partial\rho\,\partial z}(I_1 - I_2), \tag{7.16a}$$

$$E_\rho = Q\frac{\partial^2}{\partial\rho\,\partial z}(\nabla^2 + k_0{}^2\varepsilon)(I_1 - I_2), \tag{7.16b}$$

and E_z has the same form as before. The magnetic field components can be found by using (7.3). We obtain

$$H_\varphi = iQ\omega\varepsilon_0\,\varepsilon\frac{\partial}{\partial\rho}(\nabla^2 + k_0{}^2\chi)(I_1 - I_2), \tag{7.17a}$$

$$H_\rho = Q\omega\varepsilon_0\,\eta\frac{\partial^3}{\partial\rho\,\partial z^2}(I_1 - I_2). \tag{7.17b}$$

It is instructive to note the salient features of the solution of the present problem as compared with the solution to the vertical electric dipole in an isotropic medium of the previous chapter. Where in the isotropic case the only field components present were E_ρ, E_z, and H_φ, in the present case all of the field components are nonzero, which reveals that the field structure is considerably more complex.

Alternately, we can express the results in spherical coordinates, using the transformations

$$\begin{pmatrix} A_\theta \\ A_r \end{pmatrix} = \begin{pmatrix} \cos\theta & -\sin\theta \\ \sin\theta & \cos\theta \end{pmatrix}\begin{pmatrix} A_\rho \\ A_z \end{pmatrix}, \tag{7.18a}$$

$$\begin{pmatrix} \partial_\rho \\ \partial_z \end{pmatrix} = \begin{pmatrix} \cos\theta & \sin\theta \\ -\sin\theta & \cos\theta \end{pmatrix}\begin{pmatrix} r^{-1}\,\partial_\theta \\ \partial_r \end{pmatrix} \tag{7.18b}$$

on the various field components. Moreover, in I_1, we put

$$\begin{aligned} \lambda &= k_0 n_1(\beta)\sin\beta, \qquad & \rho &= r\sin\theta, \\ \tau_1 &= k_0 n_1(\beta)\cos\beta, \qquad & |z - h| &= r\cos\theta, \qquad -\pi/2 \le \theta \le \pi/2, \end{aligned} \tag{7.19}$$

and obtain

$$I_1 = \frac{\pi\zeta\exp(-i\pi/4)}{k_0}\int_{C_1}\frac{\sin\beta[n_1 + \tan\beta(dn_1/d\beta)]\check{H}_0^{(1)}(k_0\rho n_1\sin\beta)}{[(\zeta-\varepsilon)^2 n_1{}^2\sin^4\beta - 4\zeta\eta^2(n_1{}^2\sin^2\beta - \zeta)]^{1/2}}$$
$$\times\exp[ik_0 r n_1\cos(\beta - \theta)]\,d\beta, \tag{7.20a}$$

where

$$n_1{}^2 = \frac{\varepsilon\{2\zeta + (\chi - \zeta)\sin^2\beta + [(\chi-\zeta)^2\sin^4\beta + 4\zeta^2\kappa^2\cos^2\beta]^{1/2}\}}{2(\zeta\cos^2\beta + \varepsilon\sin^2\beta)} \tag{7.20b}$$

and the path C_1 is specified by (7.19). Making corresponding transformations in I_2, we obtain

$$I_2 = \frac{\pi\zeta\exp(-i\pi/4)}{k_0{}^2}\int_{C_2}\frac{\sin\beta[n_2 + \tan\beta(dn_2/d\beta)]\check{H}_0^{(1)}(k_0\rho n_2\sin\beta)}{[(\zeta-\varepsilon)^2 n_2{}^2\sin^4\beta - 4\eta^2\zeta(n_2{}^2\sin^2\beta - \zeta)]^{1/2}}$$
$$\times\exp[ik_0 r_2 n_2\cos(\beta - \theta)]\,d\beta, \tag{7.21a}$$

where

$$n_2{}^2 = \frac{\varepsilon\{2\zeta + (\chi - \zeta)\sin^2\beta - [(\chi - \zeta)^2 \sin^4\beta + 4\zeta^2\kappa^2\cos^2\beta]^{1/2}\}}{2(\zeta\cos^2\beta + \varepsilon\sin^2\beta)} \tag{7.21b}$$

and the path C_2 is obtained from the transformation $\lambda = k_0 n_2(\beta)\sin\beta$.

The further success in finding the radiation field is contingent upon being able to find the saddle points of the integrands in I_1 and I_2 given by

$$(d/d\beta)[n_{1,2}\cos(\beta - \theta)] = 0$$

or equivalently

$$\tan(\beta - \theta) = \frac{dn_{1,2}/d\beta}{n_{1,2}}, \tag{7.22}$$

which, in general, is impossible to solve. Graphical methods have been employed (Arbel and Felsen, 1963) to extract information from the integrals about the nature of the radiation field.

7.2 Magnetic Line Source in Magnetoplasma

The analysis of electromagnetic fields in magnetoplasma is somewhat simpler in the case of cylindrical waves such as those that are excited by a line source. In this section we shall formulate solutions to wave excitation by a magnetic current line source when the magnetostatic field is either perpendicular or parallel to the line source.

7.2.1 H_{DC} Perpendicular to the Line Source

Consider the geometry shown in Fig. 7.2. The appropriate form of

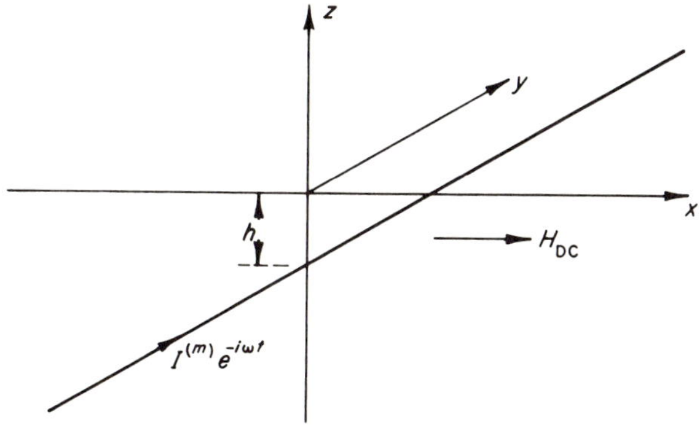

Fig. 7.2. Geometry of the problem of a magnetic current line source when the H_{DC} is perpendicular to the line source.

Maxwell's equations to be solved in this case is

$$\nabla \times \mathbf{E} = i\omega\mu_0 \mathbf{H} - I^{(m)}\,\delta(x)\,\delta(z+h)\mathbf{1}_y, \tag{7.23a}$$

$$\nabla \times \mathbf{H} = -i\omega\varepsilon_0 \boldsymbol{\varepsilon}\mathbf{E}, \tag{7.23b}$$

$$\nabla \cdot \boldsymbol{\varepsilon}\mathbf{E} = 0, \tag{7.23c}$$

$$\nabla \cdot \mathbf{H} = \rho_m/\mu_0, \tag{7.23d}$$

where ρ_m is a fictitious "magnetic charge" density related to the "magnetic current" density $\mathbf{J}^{(m)}$ by the continuity equation $\boldsymbol{\Delta} \cdot \mathbf{J}^{(m)} = i\omega\rho_m$ and $\boldsymbol{\varepsilon}$ in this case is

$$\boldsymbol{\varepsilon} = \begin{pmatrix} \zeta & 0 & 0 \\ 0 & \varepsilon & i\eta \\ 0 & -i\eta & \varepsilon \end{pmatrix}. \tag{7.24}$$

The vector wave equation for $\mathbf{H}$ can be obtained by eliminating $\mathbf{E}$ between (7.23a) and (7.23b) to give

$$-\nabla \times \boldsymbol{\varepsilon}^{-1}\nabla \times \mathbf{H} + k_0{}^2\mathbf{H} = -\,i\omega\varepsilon_0 I^{(m)}\,\delta(x)\,\delta(z+h)\mathbf{1}_y \tag{7.25}$$

where $\boldsymbol{\varepsilon}^{-1}$ can be found to be

$$\boldsymbol{\varepsilon}^{-1} = (\chi)^{-1}\begin{pmatrix} \chi/\zeta & 0 & 0 \\ 0 & 1 & -i\kappa \\ 0 & i\kappa & 1 \end{pmatrix}. \tag{7.26}$$

Noting that $\partial/\partial y = 0$, we can rewrite (7.25) in an expanded form to obtain

$$\begin{pmatrix} \chi k_0{}^2 + \partial_z{}^2 & -i\kappa\,\partial_x\,\partial_z & -\partial_x\,\partial_z \\ i\kappa\,\partial_x\,\partial_z & \chi k_0{}^2 + \partial_x{}^2 + \dfrac{\chi}{\zeta}\,\partial_z{}^2 & -i\kappa\,\partial_x{}^2 \\ -\partial_x\,\partial_z & i\kappa\,\partial_x{}^2 & \chi k_0{}^2 + \partial_x{}^2 \end{pmatrix}\begin{pmatrix} H_x \\ H_y \\ H_z \end{pmatrix} = -\,i\omega\varepsilon_0\,\chi I^{(m)}\delta(x)\delta(z+h)\begin{pmatrix} 0 \\ 1 \\ 0 \end{pmatrix}. \tag{7.27}$$

Moreover, using (7.23b), we can express the electric field components in terms of the magnetic field components as

$$\begin{pmatrix} E_x \\ E_y \\ E_z \end{pmatrix} = (i\omega\varepsilon_0\,\chi)^{-1}\begin{pmatrix} 0 & (\chi/\zeta)\partial_z & 0 \\ -\partial_z & i\kappa\,\partial_x & \partial_x \\ -i\kappa\,\partial_z & -\partial_x & i\kappa\,\partial_x \end{pmatrix}\begin{pmatrix} H_x \\ H_y \\ H_z \end{pmatrix}. \tag{7.28}$$

The system of three simultaneous differential equations in (7.27) can be reduced to a system of three simultaneous algebraic equations by using a two-dimensional Fourier transform on x and z. We obtain

$$\tilde{\tilde{H}}_x = -\frac{\omega\varepsilon_0\zeta\kappa I^{(m)}}{2\pi}\frac{\alpha_1\alpha_3}{(\alpha_3^2-\tau_1^2)(\alpha_3^2-\tau_2^2)}\exp(i\alpha_3 h), \tag{7.29a}$$

$$\tilde{\tilde{H}}_y = -\frac{i\omega\varepsilon_0\zeta I^{(m)}}{2\pi}\frac{k_0^2\chi-\alpha_1^2-\alpha_3^2}{(\alpha_3^2-\tau_1^2)(\alpha_3^2-\tau_2^2)}\exp(i\alpha_3 h), \tag{7.29b}$$

$$\tilde{\tilde{H}}_z = \frac{\omega\varepsilon_0\zeta\kappa I^{(m)}}{2\pi}\frac{\alpha_1^2}{(\alpha_3^2-\tau_1^2)(\alpha_3^2-\tau_2^2)}\exp(i\alpha_3 h), \tag{7.29c}$$

where in this case

$$\tau_{1,2}^2 = \tfrac{1}{2}\Bigg((\chi+\zeta)k_0^2 - \frac{\varepsilon+\zeta}{\varepsilon}\alpha_1^2 \pm \left\{\left[(\chi-\zeta)k_0^2 - \frac{\varepsilon-\zeta}{\varepsilon}\alpha_1^2\right]^2 + 4\kappa^2\zeta k_0^2\alpha_1^2\right\}^{1/2}\Bigg). \tag{7.30}$$

The inversion with respect to the α_3 transform variable can be performed immediately using the results of (7.8). Inverting also with respect to α_1, we obtain the following integral representation of the magnetic fields.

$$H_x = iQ\kappa\frac{\partial^2}{\partial x\,\partial z}(I_1 - I_2), \tag{7.31a}$$

$$H_y = Q\left(\frac{\partial^2}{\partial x^2} + \frac{\partial^2}{\partial z^2} + k_0^2\chi\right)(I_1 - I_2), \tag{7.31b}$$

$$H_z = -iQ\kappa\frac{\partial^2}{\partial x^2}(I_1 - I_2), \tag{7.31c}$$

where in this case

$$I_1 = \int_{-\infty}^{\infty}\frac{\exp(i\tau_1|z+h|)}{\tau_1(\tau_1^2-\tau_2^2)}\exp(i\alpha_1 x)\,d\alpha_1, \tag{7.32a}$$

$$I_2 = \int_{-\infty}^{\infty}\frac{\exp(i\tau_2|z+h|)}{\tau_2(\tau_1^2-\tau_2^2)}\exp(i\alpha_1 x)\,d\alpha_1, \tag{7.32b}$$

and $Q = \omega\varepsilon_0\zeta I^{(m)}/4\pi$. Alternately, using the transformations

$$\alpha_1 = k_0 n_1(\beta)\sin\beta, \qquad x = \rho\sin\theta,$$
$$\tau_1 = k_0 n_1(\beta)\cos\beta, \qquad |z+h| = \rho\cos\theta, \qquad -\pi/2 \le \theta \le \pi/2, \tag{7.33}$$

where

$$n_1{}^2 = \frac{\varepsilon\{2\zeta + (\chi - \zeta)\cos^2\beta + [(\chi - \zeta)^2 \cos^4\beta + 4\zeta^2\kappa^2 \sin^2\beta]^{1/2}\}}{2[\varepsilon\cos^2\beta + \zeta\ \mathrm{s\ n}^2\ \beta]}, \tag{7.34}$$

we obtain

$$I_1 = \frac{\varepsilon}{k_0{}^2}\int_{C_1} \frac{[1 + \tan\beta\, d/d\beta(\ln n_1)]\exp[ik_0 n_1 \rho\cos(\beta - \theta)]}{\{[\varepsilon(\chi - \zeta) - (\varepsilon - \zeta)n_1{}^2 \sin^2\beta]^2 + 4\eta^2\zeta\sin^2\beta\}^{1/2}}\, d\beta \tag{7.35}$$

where C_1 is a contour of integration in the β plane specified by the transformation $\alpha_1 = k_0 n_1 \sin\beta$.

Similarly, making the transformations $\alpha_1 = k_0 n_2 \sin\beta$, $\tau_2 = k_0 n_2 \cos\beta$, we obtain for the integral in (7.32b)

$$I_2 = \frac{\varepsilon}{k_0{}^2}\int_{C_2} \frac{[1 + \tan\beta\, d/d\beta(\ln n_2)]\exp[ik_0 n_2 \rho\cos(\beta - \theta)]}{\{[\varepsilon(\chi - \zeta) - (\varepsilon - \zeta)n_2{}^2 \sin^2\beta]^2 + 4\eta^2\zeta\sin^2\beta\}^{1/2}}\, d\beta, \tag{7.36}$$

where

$$n_2{}^2 = \frac{\varepsilon\{2\zeta + (\chi - \zeta)\cos^2\beta - [(\chi - \zeta)^2 \cos^4\beta + 4\zeta^2\kappa^2 \sin^2\beta]^{1/2}\}}{2[\varepsilon\cos^2\beta + \zeta\sin^2\beta]}. \tag{7.37}$$

The integrals in (7.35) and (7.36) are similar in all essential aspects to the integrals pertinent to the dipole solution in (7.20a) and (7.21a), which we found impossible to evaluate by using the standard method of saddle-point integration.

7.2.2 H_{DC} Parallel to the Line Source

If the magnetostatic field is parallel to the line source (in the positive y direction), then the permittivity tensor has the form

$$\boldsymbol{\varepsilon} = \begin{pmatrix} \varepsilon & 0 & i\eta \\ 0 & \zeta & 0 \\ -i\eta & 0 & \varepsilon \end{pmatrix} \tag{7.38a}$$

and its inverse is

$$\boldsymbol{\varepsilon}^{-1} = (\chi)^{-1}\begin{pmatrix} 1 & 0 & -i\kappa \\ 0 & \chi/\zeta & 0 \\ i\kappa & 0 & 1 \end{pmatrix}. \tag{7.38b}$$

Substituting (7.38b) into the wave equation (7.25), we obtain a system of simultaneous differential equations:

$$\begin{pmatrix} \partial_z^2 + k_0^2\zeta & 0 & -\partial_x\,\partial_z \\ 0 & \partial_x^2 + \partial_z^2 + k_0^2\chi & 0 \\ -\partial_x\,\partial_z & 0 & \partial_x^2 + k_0^2\zeta \end{pmatrix}\begin{pmatrix} H_x \\ H_y \\ H_z \end{pmatrix}$$

$$= -i\omega\varepsilon_0\,\chi I^{(m)}\,\delta(x)\,\delta(z+h)\begin{pmatrix} 0 \\ 1 \\ 0 \end{pmatrix}. \qquad (7.39)$$

Moreover, the electric field can be expressed in terms of the magnetic field as

$$\begin{pmatrix} E_x \\ E_y \\ E_z \end{pmatrix} = (i\omega\varepsilon_0\,\chi)^{-1}\begin{pmatrix} 0 & \partial_z + i\kappa\,\partial_x & 0 \\ -\dfrac{\chi}{\zeta}\,\partial_z & 0 & \dfrac{\chi}{\zeta}\,\partial_x \\ 0 & -\partial_x + i\kappa\,\partial_z & 0 \end{pmatrix}\begin{pmatrix} H_x \\ H_y \\ H_z \end{pmatrix}. \qquad (7.40)$$

From (7.39) it appears that a single component of the magnetic field is sufficient to solve the present problem. The corresponding differential equation is

$$\left(\frac{\partial^2}{\partial x^2} + \frac{\partial^2}{\partial z^2} + k_0^2\chi\right)H_y = -i\omega\varepsilon_0\,\chi I^{(m)}\,\delta(x)\,\delta(z+h), \qquad (7.41)$$

which is known to have the simple solution

$$H_y = -\frac{\omega\varepsilon_0\,\chi I^{(m)}}{4}\,H_0^{(1)}(k_0\sqrt{\chi}\rho), \qquad (7.42)$$

where $\rho = [x^2 + (z+h)^2]^{1/2}$.

7.3 Radiation from a Magnetic Current Line Source in the Presence of a Separation Boundary

In this section we shall consider problems of radiation from a magnetic current line source in the presence of a planar boundary separating the magnetoplasma and air. Such a model is pertinent to assessing, for instance, the fields of an antenna situated in the ionosphere at vlf.

7.3.1 H_{DC} Perpendicular to the Line Source

Consider the geometry of the problem as shown in Fig. 7.3. In the magnetoplasma region the fields must satisfy the differential equations

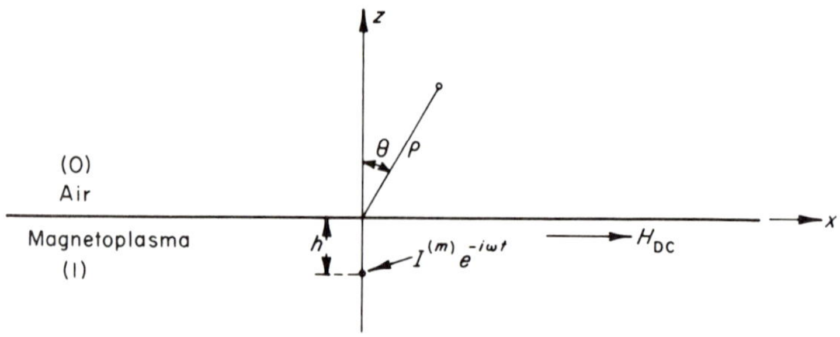

Fig. 7.3. Geometry of the problem of a magnetic line source, H_{DC} perpendicular, in the presence of a separation boundary.

$$\begin{aligned}\nabla \times \mathbf{E}_1 &= i\omega\mu_0\mathbf{H}_1 - I^{(m)}\,\delta(x)\,\delta(z+h)\mathbf{1}_y,\\ \nabla \times \mathbf{H}_1 &= -i\omega\varepsilon_0\boldsymbol{\varepsilon}\mathbf{E}_1,\\ \nabla\cdot\boldsymbol{\varepsilon}\mathbf{E} &= 0,\\ \nabla\cdot\mathbf{H} &= \rho_m/\mu_0,\end{aligned} \tag{7.43}$$

with $\boldsymbol{\varepsilon}$ given by (7.24). In the air region we must have

$$\begin{aligned}\nabla \times \mathbf{E}_0 &= i\omega\mu_0\mathbf{H}_0,\\ \nabla \times \mathbf{H}_0 &= -i\omega\varepsilon_0\mathbf{E}_0,\\ \nabla\cdot\mathbf{E}_0 &= 0,\\ \nabla\cdot\mathbf{H}_0 &= 0.\end{aligned} \tag{7.44}$$

In addition, at $z = 0$ the tangential components of the electric and magnetic fields must be continuous and as $\rho \to \infty$, all of the field components must be finite. We now seek a proper formulation of the solution that satisfies all of the stated requirements.

In the magnetoplasma region, in addition to the particular solution of (7.31) we must have a suitable complementary solution satisfying the homogeneous vector wave equation

$$-\nabla \times \boldsymbol{\varepsilon}^{-1}\nabla \times \mathbf{H}_1^{(c)} + k_0^2\mathbf{H}_1^{(c)} = 0 \tag{7.45a}$$

or in matrix form

$$\begin{pmatrix} \partial_z^2 + k_0^2\chi & -i\kappa\,\partial_x\,\partial_z & -\partial_x\,\partial_z \\ i\kappa\,\partial_x\,\partial_z & \partial_x^2 + \dfrac{\chi}{\zeta}\partial_z^2 + k_0^2\chi & -i\kappa\,\partial_x^2 \\ -\partial_x\,\partial_z & i\kappa\,\partial_x^2 & \partial_x^2 + k_0^2\chi \end{pmatrix}\begin{pmatrix} H_{1x}^{(c)} \\ H_{1y}^{(c)} \\ H_{1z}^{(c)} \end{pmatrix} = 0. \tag{7.45b}$$

Next we eliminate the x dependence via the Fourier transform and show that

$$\left(\frac{d^2}{dz^2} + \tau_1{}^2\right)\left(\frac{d^2}{dz^2} + \tau_2{}^2\right)\tilde{\mathbf{H}}_1^{(c)} = 0, \tag{7.46}$$

where $\tau_{1,2}$ is given by (7.30) and the tilde denotes the Fourier transform with respect to the x coordinate. We choose for the solution

$$\tilde{H}_{y1}^{(c)} = \frac{\omega\varepsilon_0\,\zeta I^{(m)}}{2(2\pi)^{1/2}}\,[A_1 \exp(-i\tau_1 z) + A_2 \exp(-i\tau_2 z)], \tag{7.47}$$

where $\mathrm{Im}(\tau_{1,2}) > 0$, as before. Since $z \leq 0$ in the magnetoplasma region, this choice of the solution ensures proper behavior of the fields as $|z| \to \infty$. The remaining transforms of the magnetic field components are found by substituting (7.47) back into (7.45) with $\partial_x \to -i\alpha_1$. Finally, inverting with respect to the α_1 transform variable, we find the desired integral representation of the complementary field

$$H_{x1}^{(c)} = \kappa Q \frac{\partial}{\partial x}\int_{-\infty}^{\infty}\left[\frac{\tau_1 A_1 \exp(-i\tau_1 z)}{k_0{}^2\chi - \alpha_1{}^2 - \tau_1{}^2} + \frac{\tau_2 A_2 \exp(-i\tau_2 z)}{k_0{}^2\chi - \alpha_1{}^2 - \tau_2{}^2}\right]\exp(i\alpha_1 x)\,d\alpha_1, \tag{7.48a}$$

$$H_{y1}^{(c)} = Q\int_{-\infty}^{\infty}[A_1 \exp(-i\tau_1 z) + A_2 \exp(-i\tau_2 z)]\exp(i\alpha_1 x)\,d\alpha_1, \tag{7.48b}$$

$$H_{z1}^{(c)} = -i\kappa Q \frac{\partial^2}{\partial x^2}\int_{-\infty}^{\infty}\left[\frac{A_1 \exp(-i\tau_1 z)}{k_0{}^2\chi - \alpha_1{}^2 - \tau_1{}^2} + \frac{A_2 \exp(-i\tau_2 z)}{k_0{}^2\chi - \alpha_1{}^2 - \tau_2{}^2}\right]\exp(i\alpha_1 x)\,d\alpha_1, \tag{7.48c}$$

where $Q = \omega\varepsilon_0\,\zeta I^{(m)}/4\pi$ as before. The complementary fields in (7.48) together with the primary fields in (7.31) constitute the total magnetic fields in the magnetoplasma region. The corresponding electric fields are determined by the relations of (7.28).

In the air region the magnetic field components must satisfy the wave equation

$$(\nabla_{xz}^2 + k_0{}^2)\mathbf{H}_0 = 0. \tag{7.49}$$

The appropriate solution to (7.49) can be written at once as follows.

$$H_{x0} = -\kappa Q \frac{\partial}{\partial x}\int_{-\infty}^{\infty} B_2 \exp[i(\alpha_1 x + \tau_0 z)]\,d\alpha_1, \tag{7.50a}$$

$$H_{y0} = Q\int_{-\infty}^{\infty} B_1 \exp[i(\alpha_1 x + \tau_0 z)\,d\alpha_1, \tag{7.50b}$$

$$H_{z0} = -i\kappa Q \frac{\partial^2}{\partial x^2}\int_{-\infty}^{\infty}\frac{B_2}{\tau_0}\exp[i(\alpha_1 x + \tau_0 z)]\,d\alpha_1, \tag{7.50c}$$

where $\tau_0 = (k_0{}^2 - \alpha_1{}^2)^{1/2}$, $\text{Im}(\tau_0) > 0$. We can readily show that the magnetic field components in (7.50) also satisfy the divergence equation $\nabla \cdot \mathbf{H}_0 = 0$. The corresponding electric field components are determined by $\mathbf{E}_0 = (i/\omega\varepsilon_0)\nabla \times \mathbf{H}_0$.

The boundary conditions to be satisfied by the components of the field vectors require continuity of the tangential components of the electric and magnetic fields at the interface $z = 0$. In terms of the magnetic field only, these can be written

$$H_{x0} = H_{x1}, \tag{7.51a}$$

$$H_{y0} = H_{y1}, \tag{7.51b}$$

$$\zeta \frac{\partial H_{y0}}{\partial z} = \frac{\partial H_{y1}}{\partial z}, \tag{7.51c}$$

$$\chi\left(\frac{\partial H_{x0}}{\partial z} - \frac{\partial H_{z0}}{\partial x}\right) = \frac{\partial H_{x1}}{\partial z} - i\kappa\frac{\partial H_{y1}}{\partial x} - \frac{\partial H_{z1}}{\partial x}, \tag{7.51d}$$

which supplies four simultaneous algebraic equations to determine the four unknown coefficients A_1, A_2, B_1, and B_2. In particular, the coefficients B_1 and B_2 pertinent to the fields in the air region are found to be

$$B_1 = -2\frac{(\tau_2 - \tau_0)P_H(\tau_1)\exp(i\tau_1 h) - (\tau_1 + \tau_0)P_H(\tau_2)\exp(i\tau_2 h)}{N_H}, \tag{7.52a}$$

$$B_2 = -2\tau_0\frac{(\tau_2 + \zeta\tau_0)\exp(i\tau_1 h) - (\tau_1 + \zeta\tau_0)\exp(i\tau_2 h)}{N_H}, \tag{7.52b}$$

where

$$P_H(\tau_{1,2}) = \chi k_0{}^2 - \alpha_1{}^2 - \tau_{1,2}^2 \tag{7.53}$$

and

$$N_H = (\tau_1 + \zeta\tau_0)(\tau_2 + \tau_0)P_H(\tau_1) - (\tau_2 + \zeta\tau_0)(\tau_1 + \tau_0)P_H(\tau_2). \tag{7.54}$$

In what follows it will be more convenient to express the field components in cylindrical coordinates, using the transformations

$$\alpha_1 = k_0 \sin\beta, \qquad x = \rho\sin\theta, \qquad z = \rho\cos\theta, \tag{7.55}$$

to obtain

$$H_{y0} = \frac{\omega\varepsilon_0\zeta I^{(m)}}{2\pi}\int_{\Gamma_1} G_{y0}(\beta)\cos\beta\exp[ik_0\rho\cos(\beta - \theta)]\,d\beta, \tag{7.56a}$$

$$H_{\theta 0} = \frac{\omega\varepsilon_0 \zeta\eta I^{(m)}}{2\pi k_0} \frac{\partial}{\partial x} \int_{\Gamma_1} G_{\theta 0}(\beta) \cos\beta \cos(\beta - \theta) \exp[ik_0 \rho \cos(\beta - \theta)] \, d\beta, \tag{7.56b}$$

$$H_{\rho 0} = \frac{-\omega\varepsilon_0 \zeta\eta I^{(m)}}{2\pi k_0} \frac{\partial}{\partial x} \int_{\Gamma_1} G_{\theta 0}(\beta) \cos\beta \sin(\beta - \theta) \exp[ik_0 \rho \cos(\beta - \theta)] \, d\beta, \tag{7.56c}$$

where

$$G_{y0}(\beta) = \frac{(\sigma_2 + \cos\beta)P_2(\beta)\exp(ik_0 h\sigma_1) - (\sigma_1 + \cos\beta)P_1(\beta)\exp(ik_0 h\sigma_2)}{\mathfrak{N}(\beta)}, \tag{7.57}$$

$$G_{\theta 0}(\beta) = \frac{(\sigma_2 + \zeta\cos\beta)\exp(ik_0 h\sigma_1) - (\sigma_1 + \zeta\cos\beta)\exp(ik_0 h\sigma_2)}{\mathfrak{N}(\beta)}, \tag{7.58}$$

$$\mathfrak{N}(\beta) = (\sigma_1 + \cos\beta)(\sigma_2 + \zeta\cos\beta)P_1(\beta) - (\sigma_2 + \cos\beta)(\sigma_1 + \zeta\cos\beta)P_2(\beta), \tag{7.59}$$

$$P_{1,2}(\beta) = R(\beta) \mp (R^2(\beta) + \eta^2\zeta\sin^2\beta)^{1/2}, \tag{7.60}$$

$$\sigma_{1,2} = \left[\frac{\zeta(\varepsilon - \sin^2\beta) - P_{1,2}(\beta)}{\varepsilon}\right]^{1/2} \tag{7.61}$$

$$R(\beta) = \tfrac{1}{2}[\varepsilon\zeta + (\varepsilon - \zeta)\sin^2\beta - (\varepsilon^2 - \eta^2)], \tag{7.62}$$

and Γ_1 is the appropriate path in the complex β plane.

In what follows we shall evaluate the integrals in (7.56) by the method of steepest descents. To this end we must investigate the singularities of the integrands that may be found between the original path of integration Γ_1 and the path of steepest descents Γ.

Any possible poles of the integrands are the zeros of the denominator given by $\mathfrak{N}(\beta_P) = 0$. Unfortunately, an analytical solution to this equation is extremely difficult to find. By analogy with a corresponding isotropic case, however, we would not expect the poles to play an important role, particularly in the evaluation of the far fields.

The integrands under consideration also contain radicals associated with σ_1 and σ_2 as a result of which the points $\beta = \theta_{B1}$ and $\beta = \theta_{B2}$, at which σ_1 and σ_2, respectively, vanish, will be branch points. At each point β these integrands can take on four different values, depending on which sign we choose for the radicals. It will be convenient, in this case, to think of four sheets of the β plane (formed by a four-sheeted Riemann surface) on which each integrand is single valued. We shall call the "proper sheet" one on which $\text{Im}(\sigma_{1,2}) > 0$.

Now it can be readily found that the branch points are located at

$$\pm\theta_{B1,2} = \arcsin(\varepsilon \pm \eta)^{1/2}. \tag{7.63}$$

It will be recalled that $(\varepsilon + \eta)^{1/2}$ and $(\varepsilon - \eta)^{1/2}$ are, respectively, the indices of refraction of the right- and left-hand circularly polarized plane waves propagating in the direction of the steady magnetic field in the plasma.

To evaluate the field integrals by the method of steepest descents, we note first that each of the integrals is of the form

$$W = \int_{\Gamma_1} F(\beta) \exp[ik_0 \rho \cos(\beta - \theta)]\, d\beta \tag{7.64}$$

and we wish to evaluate these integrals approximately when the radial distance from the source ρ is large. We note that the saddle point of the integrands occurs when $\beta = \theta$ and the path of steepest descents intersects the real axis at an angle of $-\pi/4$ radians.

Now we focus our attention on the following. As remarked earlier, the integrals under consideration are four valued since they contain the radicals of σ_1 and σ_2. The original path of integration passes over the proper sheet of the four-sheeted Riemann surface and it can be deformed into the path of steepest descents only when at least the beginning and end of it lie on this sheet. In the case when the angle θ does not exceed one of the criticl angles θ_{B1} or θ_{B2}, the transition from the path Γ_1 to the path Γ is accomplished without complications and the only contribution to the field integrals will be from the saddle point.

The situation is quite different when the angle θ exceeds one of the critical angles. A more complicated path of integration must be employed, supplementing the path of steepest descents Γ by a contour encompassing each branch cut in such a way that the beginning and end of the more complicated path will lie on the proper sheet. The new contour, shown in Fig. 7.4, is similar to that of Fig. 6.12 pertinent to isotropic plasma with the only difference that now two branch cuts are involved in integration.

As a consequence, the complete expression for any of the field integrals will consist of three parts:

$$W = W_S + W_{B1} u(\theta - \theta_{B1}) + W_{B2}\, u(\theta - \theta_{B2}), \tag{7.65}$$

where W_S denotes the contribution from the saddle point and W_{B1} and W_{B2} denote contributions from the borders of the branch cuts associated with the branch points β_{B1} and β_{B2}, respectively. The symbol $u(\theta - \theta_B)$ is used to denote that the contribution W_B is not present unless $\theta \geq \theta_B$.

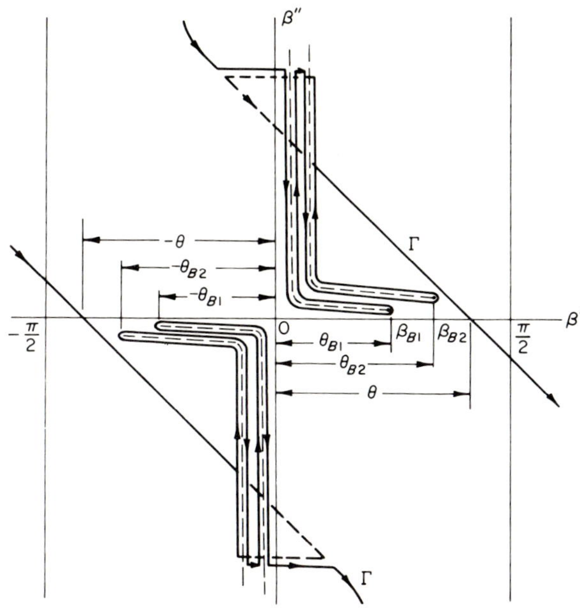

Fig. 7.4. A Modified path of integration to account for the presence of two branch cuts.

While the saddle-point contribution W_S represents the radiation field, the branch cut contributions W_{B1} and W_{B2} are associated with the phenomenon of lateral waves, which for the case of isotropic interfaces was discussed in the preceding chapter. If the losses in the plasma are negligible, the lateral field may contribute significantly to the total field, especially near the interface. If the plasma is sufficiently lossy (high collision frequency), the critical angles θ_{B1} and θ_{B2} will be complex and a high attenuation of the lateral field will result.

The radiation field, given by the leading term of saddle-point contribution, can be readily found by using the formula of (5.94). We find

$$H_{y0}^{(R)} = \frac{\omega\varepsilon_0 \zeta I^{(m)}}{(2\pi)^{1/2}} G_{y0}(\theta) \cos\theta \, \frac{\exp[i(k_0\rho - \pi/4)]}{(k_0\rho)^{1/2}}, \tag{7.66a}$$

$$H_{\theta 0}^{(R)} = \frac{i\omega\varepsilon_0 \zeta\eta I^{(m)}}{(2\pi)^{1/2}} G_{\theta 0}(\theta) \sin\theta \cos\theta \, \frac{\exp[i(k_0\rho - \pi/4)]}{(k_0\rho)^{1/2}}, \tag{7.66b}$$

$$H_{\rho 0}^{(R)} = 0. \tag{7.66c}$$

Note that the magnetic field is purely transverse to the radial direction but in general it is not linearly polarized. The time-averaged Poynting vector has a single component in the radial direction, given by

$$S_{\rho 0}^{(R)} = \frac{\omega\varepsilon_0(\zeta I^{(m)} \cos\theta)^2}{4\pi\rho} [|G_{y0}(\theta)|^2 + |\eta \sin\theta G_{\theta 0}(\theta)|^2]. \tag{7.67}$$

As remarked earlier, the integration along the borders of the branch cuts is associated with the phenomenon of lateral waves. To evaluate these contributions, we integrate formally on both sides of the cuts and obtain

$$\begin{aligned} W_B &= \int_{i\infty}^{\beta_B} F_+(\beta) \exp[ik_0 \rho \cos(\beta - \theta)]\, d\beta + \int_{\beta_B}^{i\infty} F_-(\beta) \exp[ik_0 \rho \cos(\beta - \theta)]\, d\beta \\ &= \int_{i\infty}^{\beta_B} [F_+(\beta) - F_-(\beta)] \exp[ik_0 \rho \cos(\beta - \theta)]\, d\beta, \end{aligned} \tag{7.68}$$

where $F_+(\beta)$ and $F_-(\beta)$, respectively, denote the values of the function $F(\beta)$ on the right and left sides of the branch cut $\operatorname{Im}(\tau_{1,2}) = 0$ facing the corresponding branch point. On their respective branch cuts, the imaginary part of τ_1 and τ_2 is zero, and the real part is positive on the right side of the branch cut facing the branch point and negative on the opposite side.Hence in $F_+(\beta)$. we take $+\tau_{1,2}$, whereas in $F_-(\beta)$ we must take $-\tau_{1,2}$.

Next we deform the path of integration in such a way that it begins at the branch point and proceeds along a line on which the exponent decreases most rapidly. This is equivalent to changing the variable of integration as follows.

$$\beta = \pm\,\theta_{B1,2} + is^2, \tag{7.69a}$$

where the upper (positive) sign in front of $\theta_{B1,2}$ is appropriate to $\theta > 0$ and the lower (negative) sign is appropriate to $\theta < 0$. Substitution of (7.69a) into (7.68) gives

$$\begin{aligned} W_B = {} & 2i \exp[ik_0 \rho \cos(|\theta| - \theta_B)] \\ & \times \int_0^{\infty} s[F_-(\beta) - F_+(\beta)] \exp[-k_0 \rho \sin(|\theta| - \theta_B) s^2]\, ds. \end{aligned} \tag{7.69b}$$

Evaluation of the field integrals in (7.56) using the formula (7.69b) can now be carried out. When the elevation of the source h is moderate, we obtain for the leading terms

$$H_{y0}^{(L_{1,2})} = \frac{i\omega\varepsilon_0\, \zeta I^{(m)}}{(2\pi)^{1/2}} M_{y0}^{(1,2)} \frac{\exp\{i[k_0 \rho \cos(|\theta| - \theta_{B1,2}) - \pi/4]\}}{[k_0 \rho \sin(|\theta| - \theta_{B1,2})]^{3/2}} \tag{7.70a}$$

$$\begin{aligned} H_{\theta 0}^{(L_{1,2})} = {} & \pm \operatorname{sgn}(\theta) \frac{\omega\varepsilon_0\, \zeta I^{(m)}}{(2\pi)^{1/2}} M_{\theta 0}^{(1,2)} \cos(|\theta| - \theta_{B1,2}) \\ & \times \frac{\exp[i(k_0 \rho \cos(|\theta| - \theta_{B1,2}) - \pi/4)]}{[k_0 \rho \sin(|\theta| - \theta_{B1,2})]^{3/2}}, \end{aligned} \tag{7.70b}$$

$$H_{\rho 0}^{(L_1,2)} = \operatorname{sgn}(\theta) H_{\theta 0}^{(L_1,2)} \tan(|\theta| - \theta_{B1,\,2}), \tag{7.70c}$$

where

$$\begin{aligned} M_{y0}^{(1,\,2)} = {} & \frac{(\varepsilon \pm \eta)\{2\zeta(1-(\varepsilon \pm \eta))[(\varepsilon \pm \eta)(1-(\varepsilon \pm \eta))]\}^{1/2}}{b_{H1,\,2}^2[\zeta + (\varepsilon \pm \eta)]^{1/2}} \\ & \times \{[1-(\varepsilon \pm \eta)]^{1/2} + [\mp\kappa(\zeta + \varepsilon \pm \eta)]^{1/2}\} \\ & \times [ik_0 h b_{H1,\,2} - a_{H1,\,2} + (1-\zeta^2)\,[1-(\varepsilon \pm \eta)]^{1/2} \\ & \times \exp\{ik_0 h[\mp\kappa(\zeta + \varepsilon \pm \eta)]^{1/2}\}], \end{aligned} \tag{7.71}$$

$$M_{\theta 0}^{(1,\,2)} = \frac{\zeta[1-(\varepsilon \pm \eta)]^{1/2} + [\mp\kappa(\zeta + \varepsilon \pm \eta)]^{1/2}}{(\varepsilon \pm \eta)^{1/2}\{[1-(\varepsilon \pm \eta)]^{1/2} + [\mp\kappa(\zeta + \varepsilon \pm \eta)]^{1/2}\}} M_{y0}^{(1,\,2)}, \tag{7.72}$$

$$\begin{aligned} a_{H1,\,2} = {} & [\mp\kappa(\zeta + \varepsilon \pm \eta)]^{1/2}[1-(\zeta + \varepsilon \pm \eta)] \\ & - [1-(\varepsilon \pm \eta)]^{1/2}[\zeta^2 + (\varepsilon \pm \eta - \zeta)], \end{aligned} \tag{7.73}$$

$$\begin{aligned} b_{H1,\,2} = {} & [1-(\varepsilon \pm \eta)]^{1/2}\{[1-\zeta(1+\varepsilon \pm \eta)][\mp\kappa(\zeta + \varepsilon \pm \eta)]^{1/2} \\ & + \zeta[1-\zeta-(\varepsilon \pm \eta)][1-(\varepsilon \pm \eta)]^{1/2}\}. \end{aligned} \tag{7.74}$$

It will be observed that the leading terms of the lateral field components lose their meaning when the plasma is lossless, so that the critical angles $\theta_{B1,\,2}$ are real and $\theta = \theta_{B1,\,2}$. In this case the field expressions in (7.70) are not valid since in their derivations we used the method of steepest decents, which assumed that the function $F(\beta)$ in (7.64) was slowly varying. This, is, evidently, not true in the vicinity of $\beta \to \theta_{B1,\,2}$. The way of dealing with such a situation was presented in the preceding chapter, but we shall not follow it here because of the unavoidable algebraic complexity. The point we want to stress is that the lateral field is not necessarily singular at $\theta = \theta_{B1,\,2}$ and the singularities in the field expressions in (7.70) are artificial ones induced by the approximate evaluation of the field integrals.

As in the corresponding isotropic case of the previous chapter, the geometrical interpretation of the lateral field becomes more apparent if we make use of the transformations

$$x = d_{1,\,2} + g_{1,\,2} \sin\theta_{B1,\,2} \tag{7.75}$$

as a consequence of which we obtain

$$\begin{aligned} k_0\,\rho\cos(\theta - \theta_{B1,\,2}) &= k_0(\varepsilon \pm \eta)^{1/2}\,d_{1,\,2} + k_0\,g_{1,\,2}, \\ \rho\sin(\theta - \theta_{B1,\,2}) &= [1-(\varepsilon \pm \eta)]^{1/2}\,d_{1,\,2}. \end{aligned} \tag{7.76}$$

Using these results, we can recast the components of the lateral field as follows.

$$H_0^{(L1,2)} \sim d_{1,2}^{-3/2} \exp\{i[k_0(\varepsilon \pm \eta)^{1/2} d_{1,2} + k_0 g_{1,2}]\}. \tag{7.77}$$

In (7.77) we recognize $k_0(\varepsilon \pm \eta)^{1/2}$ as the propagation constants of the plane extraordinary and ordinary rays, respectively, in a magnetoplasma when the direction of propagation coincides with the direction of the steady magnetic field. The total phase of each one of the lateral waves consists of two parts. The first part, $k_0(\varepsilon \pm \eta)^{1/2}$, tells us that the wave travels along the interface in the magnetoplasma through a distance $d_{1,2}$ with the phase velocity $c(\varepsilon \pm \eta)^{-1/2}$ and the second part, $k_0 g_{1,2}$, stands for the phase change due to the travel in free space with phase velocity c through a distance $g_{1,2}$.

The geometry of both lateral waves is shown in Fig. 7.5, where, for convenience, we assumed that the source is situated at the interface. The ray OB

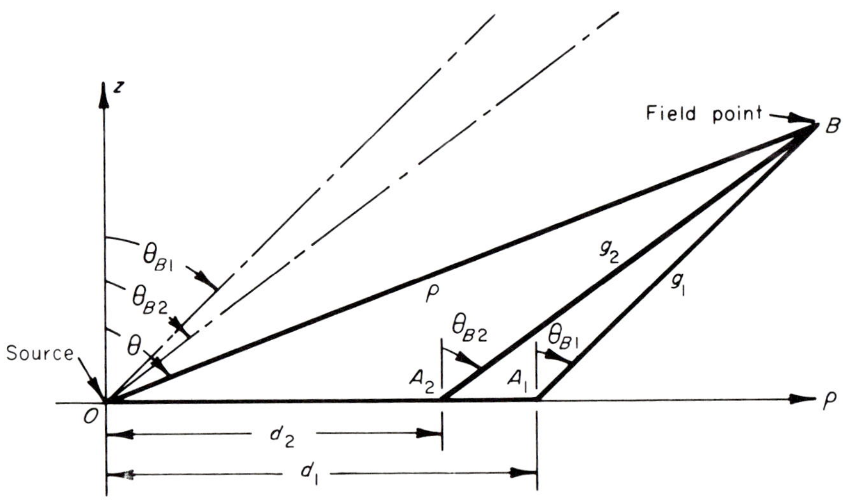

Fig. 7.5. Geometry of lateral waves on an anisotropic plasma interface.

is a direct ray from the source representing the radiation field as found from the saddle-point contribution. The lateral waves represented by the rays OA_1 and OA_2 propagate along the interface with the phase velocity, respectively, of the extraordinary and ordinary rays in magnetoplasma. Since these velocities are different from those "allowed" by free space, there is a disturbance created along the interface in the air as a result of which new waves are produced, represented by the rays $A_1 B$ and $A_2 B$. Since the spatial period of the disturbance along the interface due to, say, the extraordinary wave is equal to λ_1, the wavelength in the air can "fit" this period only if the angle θ_{B1} between the normal to the boundary and the direction of propagation is such that the wavelength in the air $\lambda_0 = \lambda_1 \sin \theta_{B1}$. This is just the direction of the lateral wave. The same argument holds for the second lateral wave, represented by the ray OA_2B. The resultant field at the point B is a superposition of the three waves, as concluded earlier in (7.65).

Figure 7.4 depicts the situation when the magnetoplasma is lossless and $\omega_p^2 < \omega(\omega \pm \omega_H)$. If this inequality is not satisfied, one or both lateral waves will either be highly attenuated or disappear completely.

7.3.2 H_{DC} Parallel to Line Source

In this case the magnetic current line source is parallel to the y axis as well as the magnetostatic field H_{DC}. The situation is depicted in Fig. 7.6.

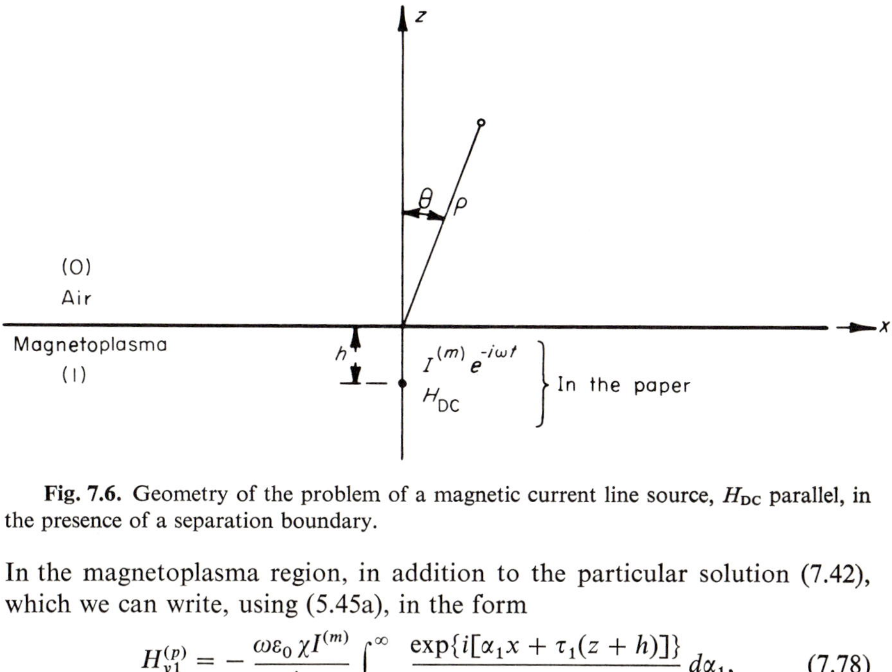

Fig. 7.6. Geometry of the problem of a magnetic current line source, H_{DC} parallel, in the presence of a separation boundary.

In the magnetoplasma region, in addition to the particular solution (7.42), which we can write, using (5.45a), in the form

$$H_{y1}^{(p)} = -\frac{\omega\varepsilon_0 \chi I^{(m)}}{4\pi}\int_{-\infty}^{\infty}\frac{\exp\{i[\alpha_1 x + \tau_1(z+h)]\}}{\tau_1}\,d\alpha_1, \tag{7.78}$$

where $\tau_1 = (k_0{}^2\chi - \alpha_1{}^2)^{1/2}$ in this case, we must have a suitable solution to the homogeneous equation

$$\left(\frac{\partial^2}{\partial x^2} + \frac{\partial^2}{\partial z^2} + k_0{}^2\chi\right)H_{y1}^{(c)} = 0. \tag{7.79}$$

The latter can be written at once as follows.

$$H_{y1}^{(c)} = -\frac{\omega\varepsilon_0 \chi I^{(m)}}{4\pi}\int_{-\infty}^{\infty} A\exp[i(\alpha_1 x - \tau_1 z)]\,d\alpha_1 \tag{7.80}$$

In the air region it is clear that we must have

$$H_{y0} = -\frac{\omega\varepsilon_0\chi I^{(m)}}{4\pi}\int_{-\infty}^{\infty} B\exp[i(\alpha_1 x + \tau_0 z)]\,d\alpha_1, \tag{7.81}$$

where $\tau_0 = (k_0{}^2 - \alpha_1{}^2)^{1/2}$.

The boundary conditions to be satisfied at the interface $z = 0$ require the continuity of the tangential magnetic and electric field components. Hence at $z = 0$

$$H_{y0} = H_{y1}, \tag{7.82a}$$

$$\chi\frac{\partial H_{y0}}{\partial z} = \left(\frac{\partial}{\partial z} + i\kappa\frac{\partial}{\partial x}\right)H_{y1}. \tag{7.82b}$$

Solving (7.82) simultaneously, we determine the unknown coefficients A and B, and consequently the field components become

$$H_{y1} = -\frac{\omega\varepsilon_0\chi I^{(m)}}{4\pi}\int_{-\infty}^{\infty}\left(\frac{\exp[i(\alpha_1 x + \tau_1|z+h|)]}{\tau_1} - \frac{\exp\{i[\alpha_1 x - \tau_1(z-h)]\}}{\tau_1} + \frac{2\exp\{i[\alpha_1 x - \tau_1(z-h)]\}}{\tau_1 + \chi\tau_0 - i\kappa\alpha_1}\right)d\alpha_1 \tag{7.83}$$

$$H_{y0} = -\frac{\omega\varepsilon_0\chi I^{(m)}}{2\pi}\int_{-\infty}^{\infty}\frac{\exp(i\tau_1 h)}{\tau_1 + \chi\tau_0 - i\kappa\alpha_1}\exp[i(\alpha_1 x + \tau_0 z)]\,d\alpha_1. \tag{7.84}$$

In the air region we employ the transformations of (7.55) and obtain in cylindrical coordinates

$$H_{y0} = -\frac{\omega\varepsilon_0\chi I^{(m)}}{2\pi}\int_{\Gamma_1}\frac{\cos\beta\exp[ik_0 h(\chi - \sin^2\beta)^{1/2}]\exp[ik_0\rho\cos(\beta-\theta)]}{(\chi-\sin^2\beta)^{1/2} + \chi\cos\beta - i\kappa\sin\beta}\,d\beta, \tag{7.85}$$

where Γ_1 is the appropriate integration path in the complex β plane. The integrand contains the radical $(\chi - \sin^2\beta)^{1/2}$, as a result of which the point $\beta = \theta_B$ where $\theta_B = \pm\arcsin\sqrt{\chi}$ will be a branch point. As a consequence, the the field component H_{y0} will consist of radiation field and lateral field as before. The leading term of each of these fields is given by

$$H_{y0}^{(R)} = -\frac{\omega\varepsilon_0\chi I^{(m)}}{(2\pi)^{1/2}}\cdot\frac{\cos\theta\exp(ik_0 h)}{(\chi-\sin^2\theta)^{1/2} + \chi\cos\theta - i\kappa\sin\theta}\,\frac{\exp[i(k_0\rho - \pi/4)]}{(k_0\rho)^{1/2}}, \tag{7.86a}$$

$$\underset{\theta \gtrless 0}{H_{y0}^{(L)}} = \frac{\omega \varepsilon_0 I^{(m)} [\chi(1-\chi)^3]^{1/4} (1 - ik_0 h(\chi)^{1/2} \{[\chi(1-\chi)]^{1/2} \mp i\kappa\})}{(2\pi)^{1/2} \{[\chi(1-\chi)]^{1/2} \mp i\kappa\}^2 [k_0 \rho \sin(|\theta| - \theta_B)]^{3/2}}$$
$$\times \exp\{i[k_0 \rho \cos(|\theta| - \theta_B) - \pi/4]\}. \tag{7.86b}$$

Note that the radiation field and lateral field in the air region will not, in general, be symmetrical about the vertical axis $\theta = 0$. Moreover, it will be observed that the reversal of the orientation of the magnetostatic field has the same effect as the reversal of the sign of θ.

The electric field components are found from the appropriate Maxwell equation. The only nonvanishing component is $E_{\theta 0}$ and is given by $E_{\theta 0} = Z_0 H_{y0}$ to the first order.

The Poynting vector has only a radial component, which in the case of the radiation field only is given by

$$S_{\rho 0}^{(R)} = \frac{\omega \varepsilon_0 (\chi I^{(m)})^2}{4\pi\rho} \left| \frac{\cos\theta \exp[ik_0 h(\chi - \sin^2\theta)^{1/2}]}{(\chi - \sin^2\theta)^{1/2} + \chi\cos\theta - i\kappa\sin\theta} \right|^2. \tag{7.87}$$

Examination of (7.87) reveals some interesting possibilities. For when $\chi > \sin^2\theta$, we can write, for a lossless magnetoplasma,

$$S_{\rho 0}^{(R)} = \frac{\omega \varepsilon_0 (\chi I^{(m)})^2}{4\pi\rho} \frac{\cos^2\theta}{((\chi - \sin^2\theta)^{1/2} + \chi\cos\theta)^2 + \kappa^2 \sin^2\theta}, \tag{7.88a}$$

and when $\chi < \sin^2\theta$,

$$S_{\rho 0}^{(R)} = \frac{\omega \varepsilon_0 (\chi I^{(m)})^2}{4\pi\rho} \cdot \frac{\cos^2\theta \exp[-2k_0 h(\sin^2\theta - \chi)^{1/2}]}{\chi^2\cos^2\theta + ((\sin^2\theta - \chi)^{1/2} - \kappa\sin\theta)^2}. \tag{7.88b}$$

Then it follows that for $\chi > \sin^2\theta$ the power pattern is symmetrical about the axis $\theta = 0$ and unaffected by the depth of the source burial h. When $\chi < \sin^2\theta$, the opposite is true; the pattern, in general, is not symmetrical and its magnitude decreases with the increasing depth of source burial h. Moreover, we note that the reversal of the orientation of the magnetostatic field, which changes the sign of κ, has the same effect as the reversal of the sign of the angle θ. Thus, the pattern corresponding to $-H_{DC}$ has a mirror symmetry with respect to the one corresponding to $+H_{DC}$.

The time-averaged Poynting vector found in (7.87) is plotted in Fig. 7.7 for several frequencies and a situation pertinent to a source situated in the lower region of the earth's ionosphere. For the purpose of this example, the inosophere was assumed to be homogeneous and sharply bounded, having an electron density $N = 750$ electrons per cubic centimeter and the earth's magnetic field $H_{DC} = 0.4$ gauss. The corresponding patterns for $H_{DC} = 0$ are

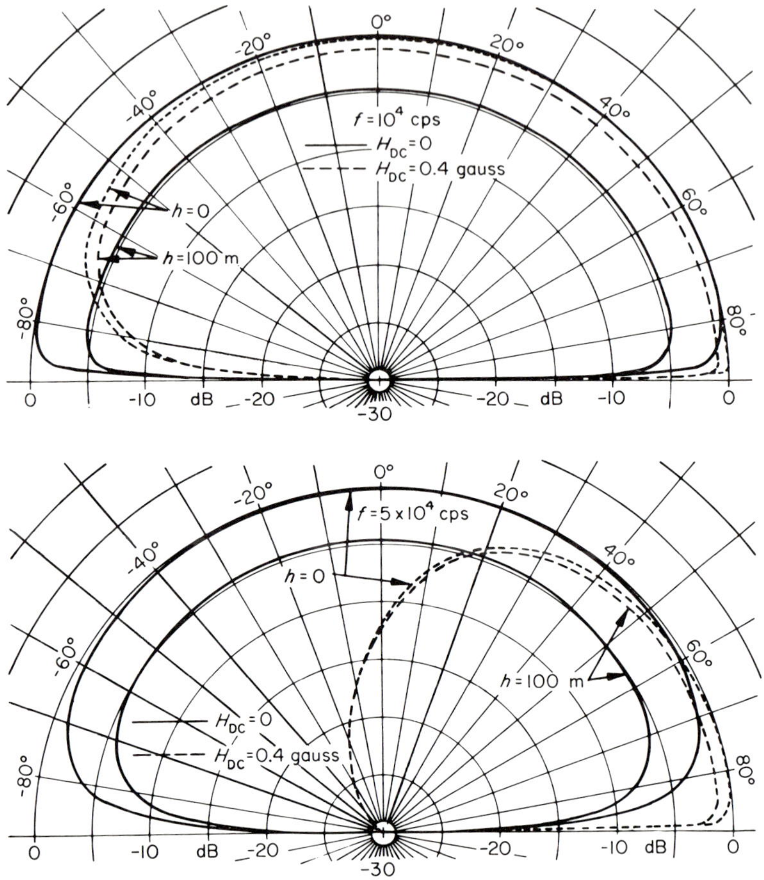

Fig. 7.7. Power pattern in air due to a magnetic current line source in magnetoplasma; H_{DC} parallel to the line source, $N = 750$ electrons per cubic centimeter.

superposed for comparison and all patterns are normalized so that $S_{\rho 0}^{(R)} = 1$ when H_{DC} and θ are zero. As concluded earlier, the radiation patterns are not symmetrical about the vertical axis and the degree of the nonsymmetry varies with the wave frequency.

Problems

7.1 Derive (7.26) from (7.24) and (7.38b) from (7.38a), and hence verify (7.28) and (7.40).

7.2 Solve (7.41) using a two-dimensional transform on x and z coordinates and verify the solution in (7.42).

7.3 If in Fig. 7.2 the magnetic current line source is replaced by an electric current line source, show that the radiation field in the air becomes

$$E_{y0}^{(R)} = -\frac{\omega\mu_0 I^{(e)}}{(2\pi)^{1/2}} F_{y0}(\theta)\cos\theta \frac{\exp[i(k_0\rho - \pi/4)]}{(k_0\rho)^{1/2}},$$

$$E_{\theta 0}^{(R)} = \frac{i\kappa\zeta\omega\mu_0 I^{(e)}}{(2\pi)^{1/2}} F_{\theta 0}(\theta)\sin\theta\cos\theta \frac{\exp[i(k_0\rho - \pi/4)]}{(k_0\rho)^{1/2}},$$

$$E_{\rho 0}^{(R)} = 0,$$

where

$$F_{y0} = \frac{(\sigma_2 + \zeta\cos\theta)P_1(\theta)\exp(ik_0 h\sigma_1) - (\sigma_1 + \zeta\cos\theta)P_2(\theta)\exp(ik_0 h\sigma_2)}{\mathfrak{N}(\theta)},$$

$$F_{\theta 0} = \frac{(\sigma_2 + \cos\theta)\exp(ik_0 h\sigma_1) - (\sigma_1 + \cos\theta)\exp(ik_0 h\sigma_2)}{\mathfrak{N}(\theta)}.$$

7.4 If in Fig. 7.5 the magnetic current line source is replaced by an electric current line source, show that the radiation field in the air becomes

$$E_{y0}^{(R)} = -\frac{\omega\mu_0 I^{(e)}}{(2\pi)^{1/2}} \frac{\cos\theta\exp[ik_0 h(\zeta - \sin^2\theta)^{1/2}]}{\cos\theta + (\zeta - \sin^2\theta)^{1/2}} \frac{\exp[i(k_0\rho - \pi/4)]}{(k_0\rho)^{1/2}}$$

and hence independent of the presence of the magnetostatic field. Can you give a simple explanation for this?

7.5 Consider the problem whose geometry is shown in Fig. 7.8.
(a) Find an integral representation for the fields in the magnetoplasma and air.

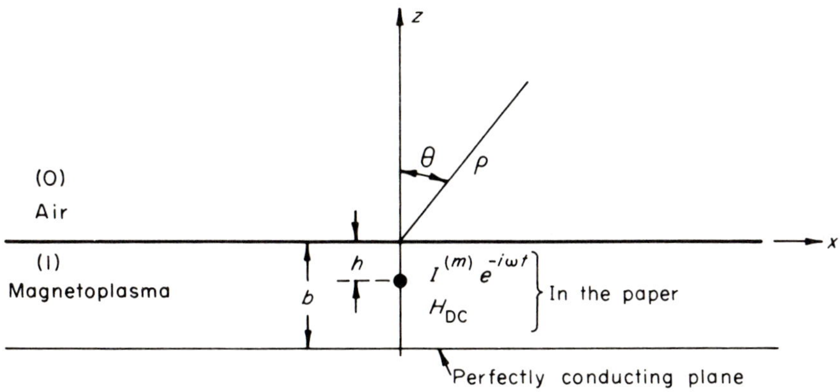

Fig. 7.8.

(b) If $h = b$, find the radiation fields in the air and show that, in general, the radiation pattern is not symmetrical.
(With proper interpretation of the magnetic current $I^{(m)}$, this problem is identical to the problem of a thin slot on a ground plane clad with magnetoplasma.)

References

Arbel, E., and Felsen, L. B. (1963). Theory of radiation from sources in anisotropic media, *in* "Electromagnetic Theory and Antennas" (E. C. Jordan, ed.), Part I, pp. 391–459. Pergamon, Oxford.

Barsukov, K. A. (1959). Radiation of electromagnetic waves in a gyrotropic medium with separation boundary, *Radio Eng. Electronics* **4**, 1–9.

Kuehl, H. (1960). "Radiation from an Electric Dipole in an Anisotropic Cold Plasma." Antenna Lab. Rept. 24, Calif. Inst. Technol. (October).

Mittra, R., and Deschamps, G. A. (1963). Field solution for a dipole in an anisotropic medium, *in* "Electromagnetic Theory and Antennas" (E. C. Jordan, ed.), Part I, pp. 495–512. Pergamon, Oxford.

Papas, C. H. (1965). "Theory of Electromagnetic Wave Propagation." McGraw-Hill, New York.

Tyras, G., Ishimaru, A., and Swarm, H. M. (1963). Lateral waves on air-magnetoplasma interfaces, *in* "Electromagnetic Theory and Antennas" (E. C. Jordan, ed.), Part I, pp. 517–535. Pergamon, Oxford.

CHAPTER 8

GREEN'S FUNCTION METHOD OF SOLUTION

In this chapter we outline a systematic procedure for the formulation of solutions to certain electromagnetic boundary value problems with the aid of the Green or influence function. This procedure is extremely useful for problems involving regions bounded by either perfectly conducting or impedance boundaries.

We begin with the study of properties of the self-adjoint Liouville differential equation with homogeneous boundary conditions that together define the *self-adjoint boundary value problem* or the *Sturm–Liouville problem.* Green's function is introduced here as a natural consequence of solving the problem of source-excited regions by means of eigenfunction expansion.

Finally, we shall study the properties of Green's function per se and outline a systematic procedure for its construction from the solutions of the associated homogeneous differential equation.

8.1 Self-Adjoint Boundary Value Problems

In many cases the scalar wave equation

$$(\nabla^2 + k^2)\varphi(\mathbf{r}) = 0 \tag{8.1}$$

can be either separated or transformed so as to reduce the problem to the solution of the ordinary differential equation

$$\frac{d}{dx}\left[P(x)\frac{d\psi}{dx}\right] + [Q(x) + \lambda R(x)]\psi = 0, \tag{8.2}$$

called the Liouville equation. The parameter λ is either one or more separation constants or a transform parameter, and the coefficients $P(x)$, $Q(x)$, and $R(x)$ are characteristic of the coordinates used. As far as the boundary value problems of electromagnetic theory are concerned, the functions $P(x)$, $Q(x)$, and $R(x)$ can be assumed to be nonsingular and positive in the entire range of x except possibly on the boundaries.

8.1.1 Eigenvalues and Eigenfunctions

It is clear that a solution to (8.2) subject to some specified boundary conditions will be possible only for certain specific values of the parameter $\lambda = \lambda_n$, called the eigenvalue, and consequently we must associate with each λ_n the corresponding solution ψ_n, the eigenfunction. The equation (8.2) written for two different eigenfunctions ψ_m and ψ_n is

$$\frac{d}{dx}\left[P(x)\frac{d\psi_m}{dx}\right] + [Q(x) + \lambda_m R(x)]\psi_m = 0 \tag{8.3a}$$

and

$$\frac{d}{dx}\left[P(x)\frac{d\psi_n}{dx}\right] + [Q(x) + \lambda_n R(x)]\psi_n = 0, \tag{8.3b}$$

respectively. Multiplying (8.3a) by ψ_n and (8.3b) by ψ_m and subtracting gives

$$\frac{d}{dx}\left[P\left(\psi_n\frac{d\psi_m}{dx} - \psi_m\frac{d\psi_n}{dx}\right)\right] = (\lambda_n - \lambda_m)R\psi_m\psi_n. \tag{8.4}$$

The property that enables us to arrive at the results of (8.4) is that the Liouville equation (8.2) is self-adjoint. The necessary and sufficient condition for the general second-order equation $L(u) = fu'' + gu' + hu$ to be self-adjoint is simply $g = f'$.

Assuming that the range of x is $x_1 \leq x \leq x_2$, we get from the integration of (8.4)

$$(\lambda_n - \lambda_m)\int_{x_1}^{x_2} R\psi_m\psi_n\,dx = \left[P\left(\psi_n\frac{d\psi_m}{dx} - \psi_m\frac{d\psi_n}{dx}\right)\right]_{x_1}^{x_2}. \tag{8.5}$$

If the coefficient $P(x)$ vanishes at the end points x_1 and x_2, the right side of (8.5) vanishes as long as ψ_m, ψ_n, ψ_m', and ψ_n' are finite therein. If x_1 and x_2 are not the zeros of $P(x)$ but if the boundary conditions are homogeneous, i.e.,

$$\begin{aligned} a_{11}\psi(x_1) + a_{12}\psi'(x_1) &= 0, \\ a_{21}\psi(x_2) + a_{22}\psi'(x_2) &= 0, \end{aligned} \tag{8.6}$$

where a_{ij}'s are constants independent of λ, then the right side of (8.5) also vanishes. (It is easy to show that homogeneous boundary conditions (8.6) are equivalent in electromagnetic theory to the surface impedance concept discussed in Section 2.7.) Thus, in either one of the two cases,

$$(\lambda_n - \lambda_m)\int_{x_1}^{x_2} R\psi_m\psi_n\,dx = 0, \tag{8.7}$$

showing that if the eigenfunction ψ_m and ψ_n belong to two different eigenvalues $\lambda_m \neq \lambda_n$, then these eigenfunctions must be orthogonal. This is the basic property defining the Sturm–Liouville or self-adjoint boundary value problem. Note that in the case $\lambda_m = \lambda_n$ the integral

$$\int_{x_1}^{x_2} R\psi_m^2\,dx \neq 0, \tag{8.8}$$

since the integrand is positive over the entire range of x.

Without loss of generality it can be assumed that the eigenfunctions have been normalized, so that the orthogonality condition can be put in the form

$$\int_{x_1}^{x_2} R\psi_m\psi_n\,dx = \delta_{mn}, \tag{8.9}$$

where δ_{mn} is the *Kronecker delta* defined by

$$\delta_{mn} = \begin{cases} 1, & m = n, \\ 0, & m \neq n. \end{cases} \tag{8.10}$$

Thus far it has been tacitly assumed that the eigenfunctions ψ_m were real. Should they happen to be complex while the coefficients $P(x)$, $Q(x)$, and $R(x)$ are still real in the interval $x_1 \leq x \leq x_2$, then it can be shown (Problem 8.1) that the orthogonality relation is expressed in the form

$$\int_{x_1}^{x_2} R\psi_m\psi_n^*\,dx = \delta_{mn} \tag{8.11}$$

with the understanding that the eigenfunctions have been normalized, i.e.,

$$\int_{x_1}^{x_2} R\,|\psi_n|^2\,dx = 1. \tag{8.12}$$

It will follow that even in this case the eigenvalues are still real (Problem 8.1).

Since the eigenfunctions associated with the Sturm–Louiville problem are orthogonal we may attempt to expand any piecewise continuous function $F(x)$ in a Fourier series

$$F(x) = \sum_0^\infty A_m\psi_m(x) \tag{8.13}$$

in the range $x_1 \leq x \leq x_2$. Multiplying both sides by $R\psi_n^*\,dx$ and integrating with due regard to (8.11), we obtain for the Fourier coefficients

$$A_m = \int_{x_1}^{x_2} R(x')F(x')\psi_m^*(x')\,dx', \tag{8.14}$$

providing that the integral exists. The formula for finding the coefficients

A_m can be obtained by substituting back into (8.13) to give

$$F(x) = \int_{x_1}^{x_2} F(x') \left[\sum_0^\infty \psi_m(x)\psi_m(x')R(x') \right] dx'. \tag{8.15}$$

The quantity in the square brackets in (8.15) can be looked upon as an integrating instrument (delta function) that turns $F(x')$ into $F(x)$. Thus, we write

$$\delta(x - x') = R(x') \sum_0^\infty \psi_m(x)\psi_m{}^*(x') \tag{8.16}$$

and also deduce that

$$\frac{\delta(x - x')}{R(x')} = \frac{\delta^*(x' - x)}{R(x)}. \tag{8.17}$$

Since the series in (8.16) may not be absolutely convergent, it may be used only in cases where the integrated result converges.

Equation (8.13) is a statement of the completeness of the eigenfunctions. It expresses the fact that if it is possible to expand an arbitrary piecewise continuous function in terms of the eigenfunction, the set $\{\psi_n\}$ is complete. Equation (8.16) expresses the closure property of the orthonormal eigenfunctions as it follows directly from the completeness property.

8.1.2 Nonhomogeneous Problem

Let us now consider the nonhomogeneous Liouville equation with the delta function as the forcing function

$$\frac{d}{dx}\left(P\frac{dG}{dx}\right) + (Q + \lambda R)G = -\delta(x - x') \tag{8.18a}$$

and the associated homogeneous equation

$$\frac{d}{dx}\left(P\frac{d\psi_m}{dx}\right) + (Q + \lambda_m R)\psi_m = 0, \tag{8.18b}$$

where each eigenfunction $\psi_m(x)$ is assumed to satisfy the boundary conditions imposed on $G(x)$. If the boundary conditions are homogeneous, then the eigenfunctions $\psi_m(x)$ are orthogonal and we can attempt to expand the function $G(x)$ in a series of eigenfunctions:

$$G(x) = \sum_0^\infty B_m \psi_m(x). \tag{8.19}$$

Substitution of (8.19) into (8.18a) and using (8.18b) gives

$$\sum_0^\infty B_m(\lambda_m - \lambda)R(x)\psi_m(x) = \delta(x - x'). \tag{8.20}$$

Equation (8.20) is now multiplied by $\psi_n^*(x)\,dx$ and integrated over the range $x_1 \leq x \leq x_2$ to give the solution

$$G(x, x') = \sum_0^\infty \frac{\psi_m(x)\psi_m^*(x')}{\lambda_m - \lambda}. \tag{8.21}$$

The result (8.21) is evidently valid as long as λ in (8.18a) is not one of the eigenvalues of the associated homogeneous problem.

Now let us consider a more general nonhomogeneous problem

$$\frac{d}{dx}\left(P\frac{dF}{dx}\right) + (Q + \lambda R)F = -f(x) \tag{8.22}$$

and its associated homogeneous equation in (8.18b). If the boundary conditions are homogeneous, we may attempt to generate a solution in the form

$$F(x) = \sum_0^\infty C_m \psi_m(x), \tag{8.23}$$

which upon substitution into (8.22) enables us to determine the coefficients C_m as follows.

$$C_m = (\lambda_m - \lambda)^{-1} \int_{x_1}^{x_2} f(x')\psi_m^*(x')\,dx' \tag{8.24}$$

so that

$$F(x) = \int_{x_1}^{x_2} f(x')\left[\sum_0^\infty \frac{\psi_m(x)\psi_m^*(x')}{\lambda_m - \lambda}\right] dx' \tag{8.25}$$

The quantity in square brackets in (8.25) is identical to that found in (8.21). Hence, we write

$$F(x) = \int_{x_1}^{x_2} f(x')G(x, x')\,dx'. \tag{8.26}$$

What we have stumbled on in the foregoing analysis is the concept of the Green function $G(x, x')$, about which we still have a great deal to say in this chapter. The usefulness of this concept is apparent, however, even now. Once the Green function for a given boundary value problem is known, the response to an arbitrary excitation can be determined by using (8.26).

Before leaving this section it will be instructive to inquire about the solution to a problem that is analogous to the one just considered. Suppose that the governing differential equation is homogeneous, i.e.,

$$\frac{d}{dx}\left(P\frac{d\psi}{dx}\right) + (Q + \lambda R)\psi = 0 \tag{8.27}$$

everywhere inside the region not including the boundaries $x_1 < x < x_2$ and the boundary conditions are nonhomogeneous, i.e.,

$$\begin{aligned} a_{11}\psi(x_1) + a_{12}\psi'(x_1) &= u(x_1), \\ a_{21}\psi(x_2) + a_{22}\psi'(x_2) &= u(x_2). \end{aligned} \tag{8.28}$$

Suppose there exists a twice-differentiable function $\varphi(x)$ defined everywhere in the region, including the boundaries $x_1 \leq x \leq x_2$, such that

$$\begin{aligned} a_{11}\varphi(x_1) + a_{12}\varphi'(x_1) &= u(x_1), \\ a_{21}\varphi(x_2) + a_{22}\varphi'(x_2) &= u(x_2); \end{aligned} \tag{8.29}$$

then let us form

$$\Phi(x) = \psi(x) - \varphi(x). \tag{8.30}$$

It is clear that the function $\Phi(x)$ satisfies the homogeneous boundary conditions (8.6). Moreover, it is easy to show that $\Phi(x)$ satisfies the nonhomogeneous differential equation

$$\frac{d}{dx}\left(P\frac{d\Phi}{dx}\right) + (Q + \lambda R)\Phi = -f, \tag{8.31a}$$

where

$$f = \frac{d}{dx}\left(P\frac{d\varphi}{dx}\right) + (Q + \lambda R)\varphi. \tag{8.31b}$$

Thus, a problem described by a homogeneous differential equation and nonhomogeneous boundary condition is equivalent to one described by a nonhomogeneous differential equation and homogeneous boundary conditions that we know how to solve. The success of this method is, of course, dependent on the knowledge of the auxiliary function $\varphi(x)$, which, in general, may not be easy to find.

8.1.3 Nonhomogeneous Problem in Three Dimensions

The results of the preceding section can be readily extended to two or three dimensions. Consider, for instance, the Green function satisfying the scalar Helmholtz equation

$$(\nabla^2 + \lambda)G(\mathbf{r}, \mathbf{r}') = -\delta(\mathbf{r} - \mathbf{r}') \tag{8.32a}$$

and the associated homogeneous equation

$$(\nabla^2 + \lambda_m)\psi_m(\mathbf{r}) = 0, \tag{8.32b}$$

where G considered as a function of $\mathbf{r}$ is assumed to satisfy the same homogeneous boundary conditions on the bounding surface S as the eigenfunctions ψ_m.

First we show that the eigenfunctions are still orthogonal if the boundary conditions are homogeneous. To this end we write

$$\int_V (\psi_n^* \nabla^2 \psi_m - \psi_m \nabla^2 \psi_n^*)\, dv = (\lambda_n - \lambda_m) \int_V \psi_m \psi_n^*\, dv. \tag{8.33}$$

If V is a closed region of space bounded by a regular surface S, and the eigenfunctions ψ_m and ψ_n together with their first and second derivatives are continuous throughout V and on the surface S, then by using Green's theorem we get

$$\int_V (\psi_n^* \nabla^2 \psi_m - \psi_m \nabla^2 \psi_n^*)\, dv = \int_S \left(\psi_n^* \frac{\partial \psi_m}{\partial n} - \psi_m \frac{\partial \psi_n^*}{\partial n}\right) ds \tag{8.34}$$

where $\partial/\partial n$ is the normal derivative in the direction of the outward normal. Now if each one of the eigenfunctions ψ_i satisfies the homogeneous boundary conditions on the surface S

$$a\psi_i(S) + b\,\frac{\partial \psi_i(S)}{\partial n} = 0, \tag{8.35}$$

then the integral over the surface S will vanish and consequently

$$(\lambda_n - \lambda_m) \int_V \psi_m \psi_n^*\, dv = 0, \tag{8.36}$$

which shows that if λ_m and λ_n are two different eigenvalues, $\lambda_m \neq \lambda_n$, then the eigenfunctions ψ_m and ψ_n must be orthogonal.

Assuming that the eigenfunctions have been normalized so that

$$\int_V |\psi_m|^2\, dv = 1, \tag{8.37a}$$

then the orthogonality condition becomes

$$\int_V \psi_m \psi_n^*\, dv = \delta_{nm}. \tag{8.37b}$$

Coming back to the problem of finding the Green function $G(\mathbf{r}, \mathbf{r}')$ satisfying the differential equation (8.32a) and homogeneous boundary conditions in the form of (8.35), we attempt to find the solution in the form

$$G = \sum_0^\infty A_m \psi_m(\mathbf{r}). \tag{8.38}$$

Since the eigenfunctions are orthogonal the coefficients A_m can be easily found. Hence, we obtain

$$G(\mathbf{r}, \mathbf{r}') = \sum_0^\infty \frac{\psi_m(\mathbf{r})\psi_m{}^*(\mathbf{r}')}{\lambda_m - \lambda} \tag{8.39}$$

and

$$\delta(\mathbf{r} - \mathbf{r}') = \sum_0^\infty \psi_m(\mathbf{r})\psi_m{}^*(\mathbf{r}'). \tag{8.40}$$

Moreover, the solution to the three-dimensional nonhomogeneous Helmholtz equation

$$(\nabla^2 + k^2)F(\mathbf{r}) = -f(\mathbf{r}) \tag{8.41}$$

can be written

$$F(\mathbf{r}) = \int_{V'} f(\mathbf{r}')G(\mathbf{r}, \mathbf{r}')\, dv' \tag{8.42}$$

with $G(\mathbf{r}, \mathbf{r}')$ given by (8.39).

If the differential equation is homogeneous,

$$(\nabla^2 + \lambda)\psi(\mathbf{r}) = 0 \tag{8.43}$$

but the boundary conditions are nonhomogeneous

$$a\psi(S) + b\,\frac{\partial\psi(S)}{\partial n} = u(S), \tag{8.44}$$

then we can show as before that the problem can be transformed into one described by a nonhomogeneous differential equation and homogeneous boundary conditions. Suppose that there exists a twice-differentiable function $\varphi(\mathbf{r})$ defined in the volume V and on the surface S such that

$$a\varphi(S) + b\,\frac{\partial\varphi(S)}{\partial n} = u(S), \tag{8.45}$$

then the difference $\Phi(\mathbf{r}) = \psi(\mathbf{r}) - \varphi(\mathbf{r})$ satisfies the homogeneous boundary conditions and moreover

$$(\nabla^2 + \lambda)\Phi(\mathbf{r}) = -f(\mathbf{r}), \tag{8.46}$$

where $f(\mathbf{r}) = (\nabla^2 + \lambda)\varphi(\mathbf{r})$.

8.2 Examples of Eigenfunction Expansion

In this section we shall formulate solutions to problems of current-excited cylindrical regions of various configurations. To avoid confusion we shall denote by $\psi_{mn}^{(e)}$ the eigenfunctions of the electric type, i.e., those associated with the axial component of the electric field. Hence, $\psi_{mn}^{(e)} = 0$ on the walls. The eigenfunctions of the magnetic type will be denoted by $\psi_{mn}^{(m)}$ and will be associated with the axial component of the magnetic field. Hence, $\partial\psi_{mn}^{(m)}/\partial n = 0$ on the walls.

8.2.1 Finite Rectangular Region

Consider a rectangular region free of currents and bounded by perfectly conducting walls as shown in Fig. 8.1. The axial component of the electric

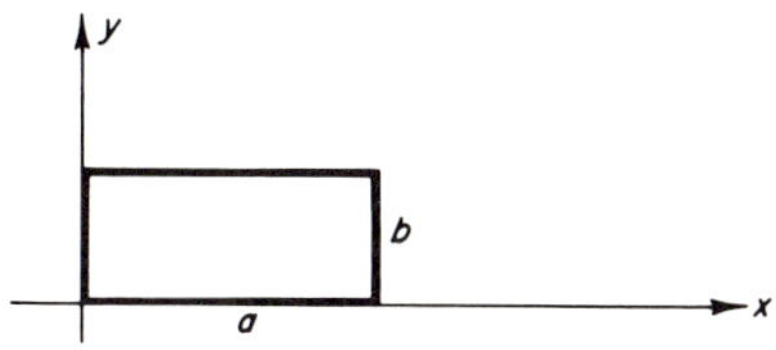

Fig. 8.1. Finite rectangular region.

or magnetic field must satisfy the differential equation

$$(\nabla_{xy}^2 + k^2)\psi(\boldsymbol{\rho}) = 0. \tag{8.47}$$

The proper eigenfunctions of the electric type must be as follows.

$$\psi_{mn}^{(e)}(\boldsymbol{\rho}) = \frac{2}{(ab)^{1/2}} \sin\frac{m\pi x}{a} \sin\frac{n\pi y}{b}, \tag{8.48a}$$

$$k_{mn}^2 = \left(\frac{m\pi}{a}\right)^2 + \left(\frac{n\pi}{b}\right)^2, \tag{8.48b}$$

since $\psi_{mn}^{(e)}$ must be zero on the walls. The eigenfunctions are normalized so that

$$\int_0^a \int_0^b |\psi_{mn}^{(e)}|^2\, dx\, dy = 1. \tag{8.49}$$

The eigenfunctions are orthogonal since

$$\frac{2}{a}\int_0^a \sin\frac{m\pi x}{a} \sin\frac{l\pi x}{a}\, dx = \delta_{lm}. \tag{8.50}$$

Using (8.16), the delta functions are expanded as follows.

$$\delta(x - x') = \frac{2}{a}\sum_1^\infty \sin\frac{m\pi x}{a} \sin\frac{m\pi x'}{a}, \tag{8.51a}$$

$$\delta(y - y') = \frac{2}{b}\sum_1^\infty \sin\frac{n\pi y}{b} \sin\frac{n\pi y'}{b}, \tag{8.51b}$$

so that the two-dimensional delta function $\delta(\boldsymbol{\rho}-\boldsymbol{\rho}')=\delta(x-x')\,\delta(y-y')$ becomes

$$\delta(\boldsymbol{\rho}-\boldsymbol{\rho}')=\frac{4}{ab}\sum_{1}^{\infty}\sum_{1}^{\infty}\sin\frac{m\pi x}{a}\sin\frac{n\pi y}{b}\sin\frac{m\pi x'}{a}\sin\frac{n\pi y'}{b}. \tag{8.52}$$

It is now a straightforward matter to solve the problem of an axial electric current distribution inside a rectangular pipe with perfectly conducting walls as shown in Fig. 8.2. The only possible component of the electric field is the

Fig. 8.2. Source-excited rectangular region.

axial one and it must satisfy the differential equation

$$(\nabla_{xy}^2+k^2)E_z=-i\omega\mu_0\,\mu J_z^{(e)}(x,y). \tag{8.53}$$

From what has been said in the preceding section, we know that the solution can be written in the form

$$E_z(x,y)=i\omega\mu_0\,\mu\int_0^a\int_0^b J_z^{(e)}(x',y')G^{(e)}(x,y:x',y')\,dx'\,dy' \tag{8.54}$$

so that the problem amounts to finding the Green function satisfying the differential equation

$$(\nabla_{xy}^2+k^2)G^{(e)}(\boldsymbol{\rho},\boldsymbol{\rho}')=-\delta(\boldsymbol{\rho}-\boldsymbol{\rho}') \tag{8.55}$$

and the boundary condition $G^{(e)}=0$ at $x=0$, a and $y=0$, b.

The task of finding the Green function satisfying (8.55) and the stated boundary conditions is simple inasmuch as the appropriate expansion of the delta function $\delta(\boldsymbol{\rho}-\boldsymbol{\rho}')$ has already been determined in (8.52). Consequently we assumed the expansion

$$G^{(e)}(\boldsymbol{\rho},\boldsymbol{\rho}')=\frac{4}{ab}\sum_{1}^{\infty}\sum_{1}^{\infty}A_{mn}\sin\frac{m\pi x}{a}\sin\frac{n\pi y}{b}\sin\frac{m\pi x'}{a}\sin\frac{n\pi y'}{b}. \tag{8.56}$$

Substitution of (8.56) and (8.52) into (8.55) permits determination of the unknown coefficient A_{mn}:

$$A_{mn}=(k_{mn}^2-k^2)^{-1}, \tag{8.57}$$

and hence

$$G^{(e)}(\boldsymbol{\rho}, \boldsymbol{\rho}') = \frac{4}{ab} \sum_1^\infty \sum_1^\infty \frac{\sin(m\pi x/a) \sin(n\pi y/b) \sin(m\pi x'/a) \sin(n\pi y'/b)}{k_{mn}^2 - k^2}. \quad (8.58)$$

The axial component of the electric field E_z can now be found for any prescribed axial electric current distribution by performing the double integration indicated in (8.54).

In the corresponding case of axial, magnetic current excitation we find

$$\delta(\boldsymbol{\rho} - \boldsymbol{\rho}') = \frac{1}{ab} \sum_0^\infty \sum_0^\infty \varepsilon_m \varepsilon_n \cos\frac{m\pi x}{a} \cos\frac{n\pi y}{b} \cos\frac{m\pi x'}{a} \cos\frac{n\pi y'}{b}, \quad (8.59\text{a})$$

$$G^{(m)}(\boldsymbol{\rho}, \boldsymbol{\rho}') = \frac{1}{ab} \sum_0^\infty \sum_0^\infty \frac{\varepsilon_m \varepsilon_n \cos(m\pi x/a) \cos(n\pi y/b) \cos(m\pi x'/a) \cos(n\pi y'/b)}{k_{mn}^2 - k^2}, \quad (8.59\text{b})$$

and

$$H_z = i\omega\varepsilon_0 \varepsilon \int_0^a \int_0^b J_z^{(m)}(x', y') G^{(m)}(x, y; x', y')\, dx'\, dy'. \quad (8.59\text{c})$$

8.2.2 Semi-Infinite Rectangular Region

If the dimension a in Fig. 8.1 is allowed to become infinite in the positive x direction, the finite rectangular region goes over into a semi-infinite rectangular region as shown in Fig. 8.3. The delta function expansion pertinent

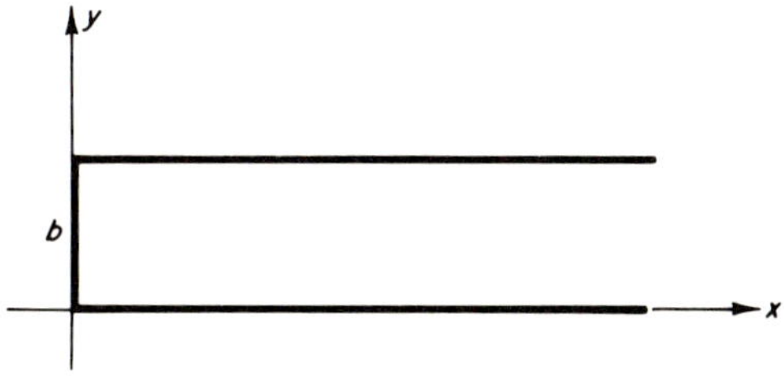

Fig. 8.3. Semi-infinite rectangular region.

to the axial electric field can be obtained from (8.51) and (8.52) by the following limiting process. We put

$$\alpha_m = m\pi/a, \qquad \Delta\alpha_m = \alpha_{m+1} - \alpha_m = \pi/a. \quad (8.60)$$

As $a \to \infty$, the eigenvalues $m\pi/a$, $m = 1, 2, 3 \ldots$, fall closer and closer together until they coalesce into a continuous spectrum. Thus,

$$\delta(x - x') = \frac{2}{\pi} \sum_1^\infty \sin \alpha_m x \sin \alpha_m x' \, \Delta\alpha_m ,$$

$$= \frac{2}{\pi} \int_0^\infty \sin \alpha x \sin \alpha x' \, d\alpha. \tag{8.61}$$

Since $\delta(y - y')$ is unaffected in this case, we obtain

$$\delta(\boldsymbol{\rho} - \boldsymbol{\rho}') = \frac{4}{\pi b} \int_0^\infty d\alpha \sin \alpha x \sin \alpha x' \sum_1^\infty \sin \frac{n\pi y}{b} \sin \frac{n\pi y'}{b}. \tag{8.62}$$

The Green function related to the axial component of the electric field and satisfying the differential equation (8.55) can be found by assuming its expansion in the form given by the delta function (8.62) and finding its appropriate coefficients. Alternately, the Green function can be found from (8.56) by going over to the limit as $a \to \infty$. In either case we obtain

$$G^{(e)}(\boldsymbol{\rho}, \boldsymbol{\rho}') = \frac{4}{\pi b} \sum_1^\infty \sin \frac{n\pi y}{b} \sin \frac{n\pi y'}{b} \int_0^\infty \frac{\sin \alpha x \sin \alpha x'}{\alpha^2 + (n\pi/b)^2 - k^2} \, d\alpha. \tag{8.63}$$

It is easy to show that in the case corresponding to the axial magnetic field we find

$$\delta(\boldsymbol{\rho} - \boldsymbol{\rho}') = \frac{2}{\pi b} \int_0^\infty d\alpha \cos \alpha x \cos \alpha x' \sum_0^\infty \varepsilon_n \cos \frac{n\pi y}{b} \cos \frac{n\pi y'}{b}, \tag{8.64a}$$

$$G^{(m)}(\boldsymbol{\rho}, \boldsymbol{\rho}') = \frac{2}{\pi b} \sum_0^\infty \varepsilon_n \cos \frac{n\pi y}{b} \cos \frac{n\pi y'}{b} \int_0^\infty \frac{\cos \alpha x \cos \alpha x'}{\alpha^2 + (n\pi/b)^2 - k^2} \, d\alpha. \tag{8.64b}$$

8.2.3 Quarter-Space Region

Consider a quarter-space region as shown in Fig. 8.4. To arrive at the desired expansions pertinent to the electric case, we let $b \to \infty$ in (8.62) and (8.63) and obtain

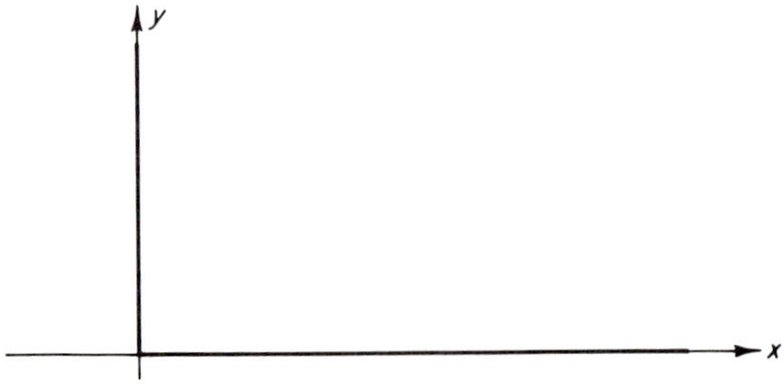

Fig. 8.4. Quarter-space region.

$$\delta(\boldsymbol{\rho} - \boldsymbol{\rho}') = \frac{4}{\pi^2} \iint\limits_0^\infty \sin \alpha_1 x \sin \alpha_1 x' \sin \alpha_2 y \sin \alpha_2 y' \, d\alpha_1 \, d\alpha_2 , \tag{8.65a}$$

$$G^{(e)}(\boldsymbol{\rho}, \boldsymbol{\rho}') = \frac{4}{\pi^2} \iint\limits_0^\infty \frac{\sin \alpha_1 x \sin \alpha_1 x' \sin \alpha_2 y \sin \alpha_2 y'}{\alpha_1{}^2 + \alpha_2{}^2 - k^2} \, d\alpha_1 \, d\alpha_2 . \tag{8.65b}$$

It can be shown that in the case corresponding to the axial magnetic field we find

$$\delta(\boldsymbol{\rho} - \boldsymbol{\rho}') = \frac{4}{\pi^2} \iint\limits_0^\infty \cos \alpha_1 x \cos \alpha_1 x' \cos \alpha_2 y \cos \alpha_2 y' \, d\alpha_1 \, d\alpha_2 , \tag{8.66a}$$

$$G^{(m)}(\boldsymbol{\rho}, \boldsymbol{\rho}') = \frac{4}{\pi^2} \iint\limits_0^\infty \frac{\cos \alpha_1 x \cos \alpha_1 x' \cos \alpha_2 y \cos \alpha_2 y'}{\alpha_1{}^2 + \alpha_2{}^2 - k^2} \, d\alpha_1 \, d\alpha_2 . \tag{8.66b}$$

8.2.4 Parallel Plate Region

Before considering the parallel plate region it will be helpful to shift the origin of the rectangular region of Fig. 8.1 to the mid-position, as shown in Fig. 8.5. Substituting $x + a/2$ and $x' + a/2$ for x and x', respectively, in the expression for the delta function representation of (8.51a), we write for the electric case

$$\begin{aligned}\delta(x - x') &= \frac{2}{a} \sum_1^\infty \sin m\pi\left(\frac{1}{2} + \frac{x}{a}\right) \sin m\pi\left(\frac{1}{2} + \frac{x'}{a}\right) \\ &= \frac{2}{a}\left(\sum_{1,3,5,\ldots} \cos \frac{m\pi x}{a} \cos \frac{m\pi x'}{a} + \sum_{2,4,6,\ldots} \sin \frac{m\pi x}{a} \sin \frac{m\pi x'}{a} \right).\end{aligned} \tag{8.67a}$$

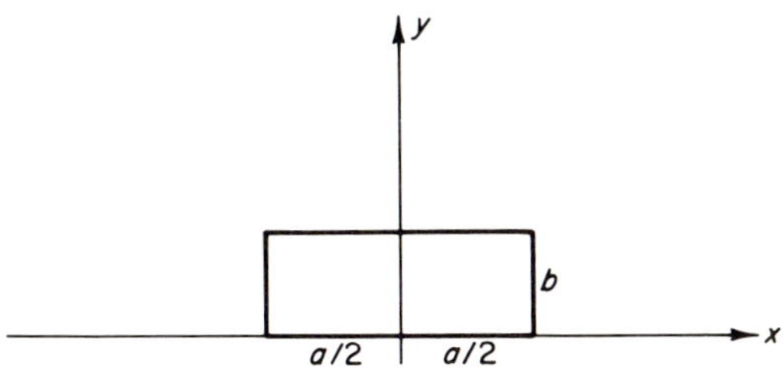

Fig. 8.5. Rectangular region, alternate coordinate system.

In the first series we put $m = 2n + 1$ and in the second series $m = 2n$, $0 \le n \le \infty$; hence

$$\delta(x - x') = \frac{2}{a}\left[\sum_0^\infty \cos\frac{(2n+1)\pi x}{a}\cos\frac{(2n+1)\pi x'}{a} + \sum_0^\infty \sin\frac{2n\pi x}{a}\sin\frac{2n\pi x'}{a}\right]. \tag{8.67b}$$

In the first series we substitute $\alpha_n = (2n + 1)\pi/a$ and in the second $\alpha_n = 2n\pi/a$. In each case $\Delta\alpha_n = \alpha_{n+1} - \alpha_n = 2\pi/a$ and letting $a \to \infty$, we obtain

$$\delta(x - x') = \frac{1}{\pi}\int_0^\infty \cos\alpha x \cos\alpha x'\, d\alpha + \frac{1}{\pi}\int_0^\infty \sin\alpha x \sin\alpha x'\, d\alpha$$

$$= (2\pi)^{-1}\int_{-\infty}^\infty \exp[i\alpha(x - x')]\, d\alpha, \tag{8.67c}$$

which is an appropriate delta function representation for the x coordinate of the parallel plate configuration shown in Fig. 8.6.

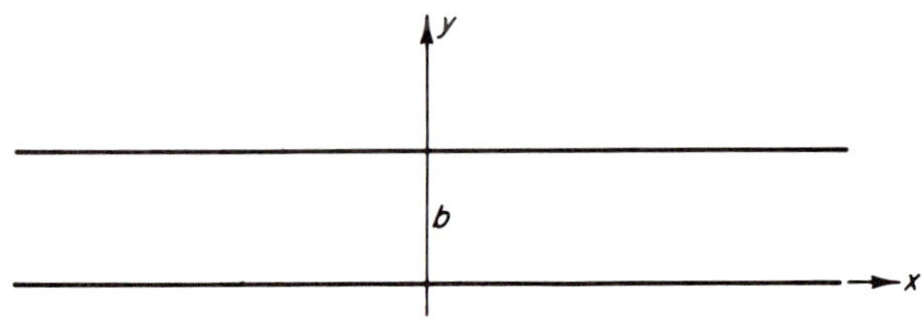

Fig. 8.6. Parallel plate configuration.

Since the delta function representation for the y coordinate of (8.51b) is still appropriate to the present problem, we have in this case

$$\delta(\boldsymbol{\rho} - \boldsymbol{\rho}') = \frac{1}{\pi b}\sum_1^\infty \sin\frac{n\pi y}{b}\sin\frac{n\pi y'}{b}\int_{-\infty}^\infty \exp[i\alpha(x - x')]\, d\alpha \tag{8.68a}$$

and, consequently, Green's function is found to be

$$G^{(e)}(\boldsymbol{\rho}, \boldsymbol{\rho}') = \frac{1}{\pi b}\sum_1^\infty \sin\frac{n\pi y}{b}\sin\frac{n\pi y'}{b}\int_{-\infty}^\infty \frac{\exp[i\alpha(x - x')]}{\alpha^2 + (n\pi/b)^2 - k^2}\, d\alpha. \tag{8.68b}$$

It can be shown that in the case corresponding to the axial magnetic field, we find for this configuration

$$\delta(\boldsymbol{\rho} - \boldsymbol{\rho}') = \frac{1}{2\pi b}\sum_0^\infty \varepsilon_n \cos\frac{n\pi y}{b}\cos\frac{n\pi y'}{b}\int_{-\infty}^\infty \exp[i\alpha(x - x')]\, d\alpha, \tag{8.69a}$$

$$G^{(m)}(\boldsymbol{\rho}, \boldsymbol{\rho}') = \frac{1}{2\pi b}\sum_0^\infty \varepsilon_n \cos\frac{n\pi y}{b}\cos\frac{n\pi y'}{b}\int_{-\infty}^\infty \frac{\exp[i\alpha(x - x')]}{\alpha^2 + (n\pi/b)^2 - k^2}\, d\alpha. \tag{8.69b}$$

8.2.5 Semi-Infinite Region

Appropriate delta function and Green's function expansions for the semi-infinite region shown in Fig. 8.7 can be obtained from the parallel plate configuration of Fig. 8.6 by letting $b \to \infty$. It is clear that in the electric case

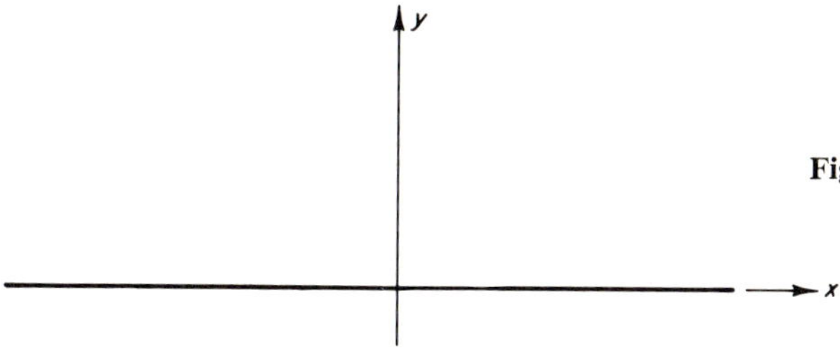

Fig. 8.7. Semi-infinite region.

the expansion of $\delta(x - x')$ in this case will be the same as in (8.67), while the expansion of $\delta(y - y')$ will be analogous to (8.61). Thus, we obtain

$$\delta(\boldsymbol{\rho} - \boldsymbol{\rho}') = \frac{1}{\pi^2} \int_{-\infty}^{\infty} \exp[i\alpha_1(x - x')]\, d\alpha_1 \int_0^{\infty} \sin \alpha_2 y \sin \alpha_2 y'\, d\alpha_2 . \tag{8.70a}$$

Hence, the Green function becomes

$$G^{(e)}(\boldsymbol{\rho}, \boldsymbol{\rho}') = \frac{1}{\pi^2} \int_{-\infty}^{\infty} d\alpha_1 \exp[i\alpha_1(x - x')] \int_0^{\infty} \frac{\sin \alpha_2 y \sin a_2 y'}{{\alpha_1}^2 + {\alpha_2}^2 - k^2}\, d\alpha_2 . \tag{8.70b}$$

In the corresponding magnetic case we find

$$\delta(\boldsymbol{\rho} - \boldsymbol{\rho}') = \frac{1}{\pi^2} \int_{-\infty}^{\infty} \exp[i\alpha_1(x - x')]\, d\alpha_1 \int_0^{\infty} \cos \alpha_2 y \cos \alpha_2 y'\, d\alpha_2 , \tag{8.71a}$$

$$G^{(m)}(\boldsymbol{\rho}, \boldsymbol{\rho}') = \frac{1}{\pi^2} \int_{-\infty}^{\infty} d\alpha_1 \exp[i\alpha_1(x - x')] \int_0^{\infty} \frac{\cos \alpha_2 y \cos \alpha_2 y'}{{\alpha_1}^2 + {\alpha_2}^2 - k^2}\, d\alpha_2 . \tag{8.71b}$$

8.2.6 Infinite Region

Appropriate delta function and Green's function expansions for an infinite region with no z dependence shown in Fig. 8.8 can be obtained from the parallel plate region by first shifting the origin to the mid-position and subsequently letting $b \to \infty$. We obtain for either the electric or magnetic case

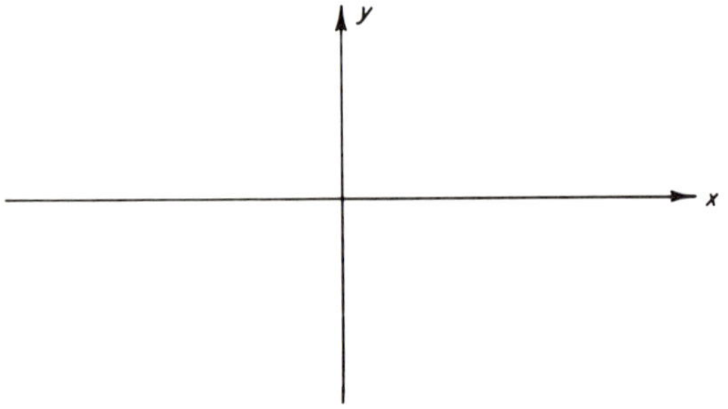

Fig. 8.8. Infinite region.

$$\delta(\boldsymbol{\rho} - \boldsymbol{\rho}') = \frac{1}{4\pi^2} \iint_{-\infty}^{\infty} \exp\{i[\alpha_1(x - x') + \alpha_2(y - y')]\}\, d\alpha_1\, d\alpha_2 \tag{8.72a}$$

and

$$G(\boldsymbol{\rho}, \boldsymbol{\rho}') = \frac{1}{4\pi^2} \iint_{-\infty}^{\infty} \frac{\exp\{i[\alpha_1(x - x') + \alpha_2(y - y')]\}}{\alpha_1{}^2 + \alpha_2{}^2 - k^2}\, d\alpha_1\, d\alpha_2\,. \tag{8.72b}$$

8.2.7 Finite Angular Sector

Consider a finite angular sector shown in Fig. 8.9 whose walls are assumed to be perfectly conducting. The axial component of either the electric or the magnetic field must satisfy the differential equation

$$\left(\frac{1}{\rho}\frac{\partial}{\partial\rho}\rho\frac{\partial}{\partial\rho} + \frac{1}{\rho^2}\frac{\partial^2}{\partial\varphi^2} + k^2\right)\psi(\boldsymbol{\rho}) = 0. \tag{8.73}$$

Assuming a product solution of the form $\psi(\boldsymbol{\rho}) = R(\boldsymbol{\rho})\Phi(\varphi)$, we obtain two ordinary differential equations

$$\Phi'' + \nu^2\Phi = 0, \tag{8.74a}$$

$$(\rho R')' + (k^2\rho - \nu^2/\rho)R = 0, \tag{8.74b}$$

where ν is the separation constant.

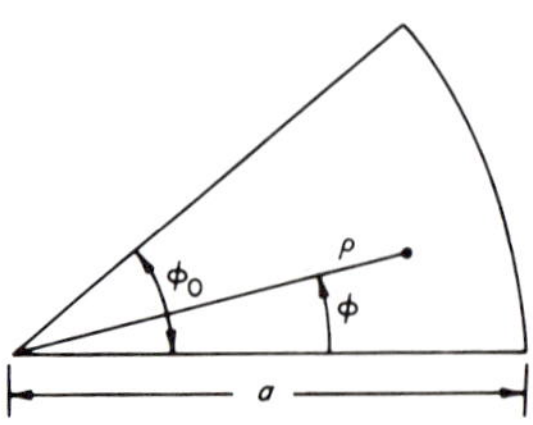

Fig. 8.9. Finite angular sector.

If the wave function $\psi(\boldsymbol{\rho})$ represents the axial component of the electric field, then $\psi^{(e)}(\rho, 0) = \psi^{(e)}(\rho, \varphi_0) = 0$. Consequently, the angular eigenfunctions $\Phi_m^{(e)}$ must be of the form

$$\Phi_m^{(e)} = A_m \sin \nu\varphi, \tag{8.75a}$$

$$\nu = \frac{m\pi}{\varphi_0}. \tag{8.75b}$$

Since

$$\int_0^{\varphi_0} \sin^2 \frac{m\pi\varphi}{\varphi_0}\, d\varphi = \frac{\varphi_0}{2}, \tag{8.76}$$

the normalized eigenfunctions are

$$\Phi_m^{(e)} = \left(\frac{2}{\varphi_0}\right)^{1/2} \sin \nu\varphi. \tag{8.77}$$

Hence, using (8.16), the angular delta function for the electric case becomes

$$\delta(\varphi - \varphi') = \frac{2}{\varphi_0} \sum_{m=1}^{\infty} \sin \nu\varphi \sin \nu\varphi'. \tag{8.78}$$

Next, let us consider the radial eigenfunctions, which must satisfy the differential equation (8.74b) and must vanish at $\rho = a$. A comparison of (8.74b) with the standard form of the Liouville equation in (8.2) reveals that $P = \rho$, $Q = -\nu^2/\rho$, $R = \rho$, and $\lambda = k_{\nu n}^2$. Since (8.74b) is the Bessel differential equation, the appropriate solution will be

$$R_n = B_n J_\nu(k_{\nu n}\rho), \tag{8.79}$$

which is finite at $\rho = 0$. Since the radial eigenfunctions must vanish at $\rho = a$, we must demand that

$$J_\nu(k_{\nu n} a) = 0, \tag{8.80a}$$

$$k_{\nu n} = p_{\nu n}/a, \tag{8.80b}$$

where $p_{\nu n}$ is the nth zero of the Bessel function of the first kind and the order ν. The radial eigenfunctions must now be normalized over the interval $0 \leq \rho \leq a$ by setting

$$B_n = \left[\int_0^{\infty} \rho J_\nu{}^2(k_{\nu n}\rho)\, d\rho\right]^{-1/2}$$

$$= \frac{\sqrt{2}}{aJ_{\nu+1}(k_{\nu n} a)}. \tag{8.81}$$

Hence, using the foregoing results together with (8.16), we obtain for the radial delta function representation

$$\frac{\delta(\rho-\rho')}{\rho'} = \frac{2}{a^2}\sum_{n=1}^{\infty}\frac{J_\nu(k_{\nu n}\rho)J_\nu(k_{\nu n}\rho')}{J_{\nu+1}^2(k_{\nu n}a)}. \tag{8.82}$$

Finally, combining the results of (8.78) and (8.82), we obtain the complete delta function expansion appropriate to the axial electric field

$$\begin{aligned}\delta(\boldsymbol{\rho}-\boldsymbol{\rho}') &= \frac{\delta(\rho-\rho')\delta(\varphi-\varphi')}{\rho'}\\ &= \frac{4}{a^2\varphi_0}\sum_{m=1}^{\infty}\sum_{n=1}^{\infty}\frac{J_\nu(k_{\nu n}\rho)J_\nu(k_{\nu n}\rho')\sin\nu\varphi\sin\nu\varphi'}{J_{\nu+1}^2(k_{\nu n}a)}.\end{aligned} \tag{8.83}$$

The Green function is found by using the prescription of (8.21) and the results above. Hence,

$$G^{(e)}(\boldsymbol{\rho},\boldsymbol{\rho}') = \frac{4}{a^2\varphi_0}\sum_{m=1}^{\infty}\sum_{n=1}^{\infty}\frac{J_\nu(k_{\nu n}\rho)J_\nu(k_{\nu n}\rho')\sin\nu\varphi\sin\nu\varphi'}{(k_{\nu n}^2-k^2)J_{\nu+1}^2(k_{\nu n}a)}. \tag{8.84}$$

In the corresponding case pertinent to the axial magnetic field we find

$$\delta(\boldsymbol{\rho}-\boldsymbol{\rho}') = \frac{2}{\varphi_0}\sum_{m=0}^{\infty}\sum_{n=1}^{\infty}\frac{\varepsilon_m\cos\nu\varphi\cos\nu\varphi'(k'_{\nu n})^2J_\nu(k'_{\nu n}\rho)J_\nu(k'_{\nu n}\rho')}{[(k'_{\nu n}a)^2-\nu^2]J_\nu{}^2(k'_{\nu n}a)}, \tag{8.85a}$$

$$G^{(m)}(\boldsymbol{\rho},\boldsymbol{\rho}') = \frac{2}{\varphi_0}\sum_{m=0}^{\infty}\sum_{n=1}^{\infty}\frac{\varepsilon_m\cos\nu\varphi\cos\nu\varphi'(k'_{\nu n})^2J_\nu(k'_{\nu n}\rho)J_\nu(k'_{\nu n}\rho')}{[(k'_{\nu n})^2-k^2][(k'_{\nu n}a)^2-\nu^2]J_\nu{}^2(k'_{\nu n}a)}, \tag{8.85b}$$

where $k'_{\nu n} = p'_{\nu n}/a$ and $J'_\nu(p'_{\nu n}) = 0$.

8.2.8 Infinite Angular Sector

If the radius of the finite angular sector in Fig. 8.9 is allowed to approach infinity, $a \to \infty$, we obtain an infinite angular sector as shown in Fig. 8.10. The transition from a finite to an infinite sector can be accomplished by first noting the asymptotic form

$$J_\nu(k_{\nu n}a) \xrightarrow[k_{\nu n}a\to\infty]{} \left(\frac{2}{\pi k_{\nu n}a}\right)^{1/2}\cos\left(k_{\nu n}a - \frac{\nu\pi}{2} - \frac{\pi}{4}\right) \tag{8.86}$$

so that the zeros of $J_\nu(k_{\nu n}a) = 0$ in the limit as $k_{\nu n}a \to \infty$ will be given by $k_{\nu n}a = (n + 3/4 + \nu/2)\pi$. Hence,

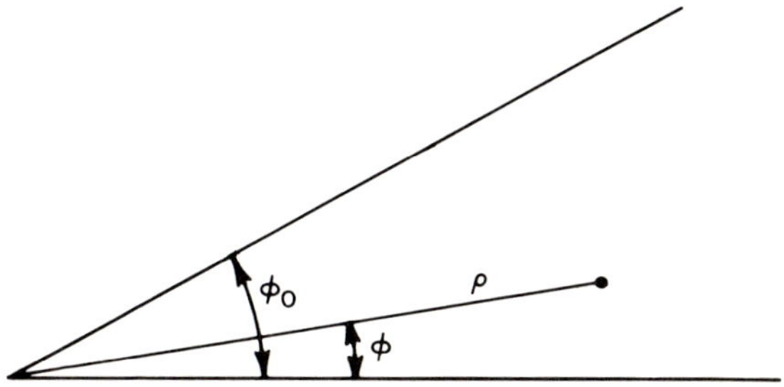

Fig. 8.10. Infinite angular sector.

$$\Delta k_{\nu n} = k_{\nu n+1} - k_{\nu n} = \frac{\pi}{a} \tag{8.87}$$

and

$$J_{\nu+1}(k_{\nu n} a) \xrightarrow[k_{\nu n}a \to \infty]{} (-1)^n \left(\frac{2}{\pi k_{\nu n} a}\right)^{1/2}. \tag{8.88}$$

Substituting the foregoing results into (8.82), we obtain

$$\begin{aligned}\frac{\delta(\rho - \rho')}{\rho'} &= \lim_{\Delta k_{\nu n} \to 0} \sum_{1}^{\infty} k_{\nu n} J_\nu(k_{\nu n}\rho) J_\nu(k_{\nu n}\rho') \, \Delta k_{\nu n} \\ &= \int_0^\infty \alpha J_\nu(\alpha\rho) J_\nu(\alpha\rho') \, d\alpha,\end{aligned} \tag{8.89}$$

which together with (8.78) gives for the electric case

$$\delta(\boldsymbol{\rho} - \boldsymbol{\rho}') = \frac{2}{\varphi_0} \sum_{m=1}^{\infty} \sin \nu\varphi \sin \nu\varphi' \int_0^\infty \alpha J_\nu(\alpha\rho) J_\nu(\alpha\rho') \, d\alpha. \tag{8.90}$$

The Green function pertinent to the electric case can be found by the same limiting process, starting with (8.84). We obtain

$$G^{(e)}(\boldsymbol{\rho}, \boldsymbol{\rho}') = \frac{2}{\varphi_0} \sum_{m=1}^{\infty} \sin \nu\varphi \sin \nu\varphi' \int_0^\infty \frac{\alpha J_\nu(\alpha\rho) J_\nu(\alpha\rho') \, d\alpha}{\alpha^2 - k^2}. \tag{8.91}$$

In the corresponding case pertinent to the axial magnetic field we find

$$\delta(\boldsymbol{\rho} - \boldsymbol{\rho}') = \varphi_0^{-1} \sum_{m=0}^{\infty} \varepsilon_m \cos \nu\varphi \cos \nu\varphi' \int_0^\infty \alpha J_\nu(\alpha\rho) J_\nu(\alpha\rho') \, d\alpha, \tag{8.92a}$$

$$G^{(m)}(\boldsymbol{\rho}, \boldsymbol{\rho}') = \varphi_0^{-1} \sum_{m=0}^{\infty} \varepsilon_m \cos \nu\varphi \cos \nu\varphi' \int_0^\infty \frac{\alpha J_\nu(\alpha\rho) J_\nu(\alpha\rho') \, d\alpha}{\alpha^2 - k^2}. \tag{8.92b}$$

8.2.9 Circular Region

A circular region containing the origin is shown in Fig. 8.11. This case is very similar to the case of the finite angular sector, with the exception that the angular eigenvalues must be equal to integers of π. Thus, in this case

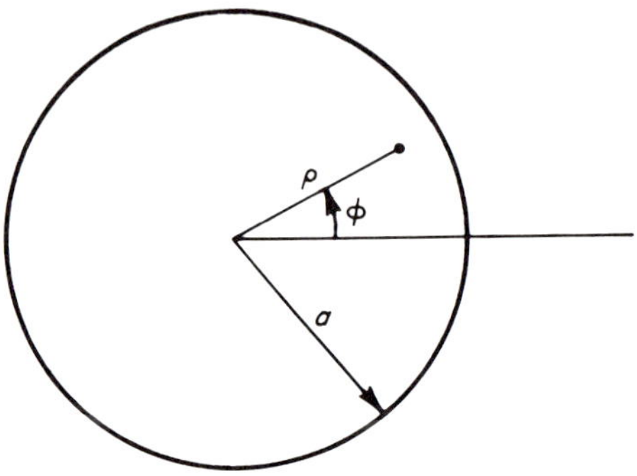

Fig. 8.11. Circular region.

$$\Phi_m = (2\pi)^{-1/2} \exp(im\varphi) \tag{8.93}$$

for either the electric or the magnetic case. Consequently,

$$\delta(\varphi - \varphi') = (2\pi)^{-1} \sum_{-\infty}^{\infty} \exp[im(\varphi - \varphi')], \tag{8.94}$$

while the radial delta function must be the same as in (8.82) for the electric case. Hence,

$$\delta(\boldsymbol{\rho} - \boldsymbol{\rho}') = \frac{1}{\pi a^2} \sum_{m=-\infty}^{\infty} \sum_{n=1}^{\infty} \frac{J_m(k_{mn}\rho)J_m(k_{mn}\rho')}{J_{m+1}^2(k_{mn}a)} \exp[im(\varphi - \varphi')], \tag{8.95a}$$

$$G^{(e)}(\boldsymbol{\rho}, \boldsymbol{\rho}') = \frac{1}{\pi a^2} \sum_{m=-\infty}^{\infty} \sum_{n=1}^{\infty} \frac{J_m(k_{mn}\rho)J_m(k_{mn}\rho')}{(k_{mn}^2 - k^2)J_{m+1}^2(k_{mn}a)}. \tag{8.95b}$$

In the corresponding case pertinent to the axial magnetic field we find

$$\delta(\boldsymbol{\rho}, \boldsymbol{\rho}') = \frac{1}{\pi} \sum_{m=-\infty}^{\infty} \sum_{n=1}^{\infty} \frac{(k'_{mn})^2 J_m(k'_{mn}\rho)J_m(k'_{mn}\rho')}{[(k'_{mn}a)^2 - m^2]J_m{}^2(k'_{mn}a)} \exp[im(\varphi - \varphi')], \tag{8.96a}$$

$$G(\boldsymbol{\rho}, \boldsymbol{\rho}') = \frac{1}{\pi} \sum_{m=-\infty}^{\infty} \sum_{n=1}^{\infty} \frac{(k'_{mn})^2 J_m(k'_{mn}\rho)J_m(k'_{mn}\rho') \exp[im(\varphi - \varphi')]}{[(k'_{mn}a)^2 - m^2][(k'_{mn})^2 - k^2]J_m{}^2(k'_{mn}a)}. \tag{8.96b}$$

8.3 Formal Green's Function Procedure

In the previous sections we came upon the concept of Green's function as a useful tool in the formulation of solutions in source-excited regions. Furthermore, in the last section, we were able to find various Green's function representations from its expansion in a series of eigenfunctions of the associated homogeneous equation. Even though this method is a very useful one in closed regions where eigenvalues are discrete, in open regions its usefulness is limited. In this section we shall study the properties of Green's function and a more direct method of finding it.

8.3.1 Introductory Remarks

Let us rewrite the nonhomogeneous Liouville equation (8.22) with $Q(x) + \lambda R(x) = S(x)$. Thus,

$$\frac{d}{dx}\left[P(x)\frac{dF}{dx}\right] + S(x)F = -f(x) \tag{8.97}$$

and as before we shall concern ourselves with the following boundary value problem. Find the solution $F(x)$, $a \leq x \leq b$, that satisfies the differential equation (8.97) and prescribed homogeneous boundary conditions at $x = a$ and $x = b$.

The physical meaning of (8.97) can be interpreted as a condition for equilibrium of a physical system under the influence of a time-independent disturbance distributed throughout the interval $a \leq x \leq b$. We can visualize a limiting process from a continuously distributed disturbance to a point disturbance, i.e., a disturbance acting at a single point $x = x'$ with a given intensity.

Let $G(x, x')$ denote the response of the system at a point x as a result of the action of a point disturbance of unity intensity applied at the point $x = x'$. Then the effect at any point in the interval $a \leq x \leq b$ of a continuously distributed disturbance $f(x)$ can be thought of as a superposition of the effects of the continuously distributed point disturbances whose linear density at each point $x = x'$ is equal to $f(x')$. Thus, we can expect the desired solution to be of the form

$$F(x) = \int_a^b G(x, x')f(x')\,dx'. \tag{8.98}$$

The function $G(x, x')$ is called the influence function or Green's function for the differential equation (8.97) and it satisfies the boundary conditions at $x = a$ and $x = b$.

8.3.2 Properties of Green's Function

Let us assume for the time being that the solution to the nonhomogeneous differential equation

$$\frac{d}{dx}\left[P(x)\frac{dF}{dx}\right] + S(x)F = -f(x), \qquad [8.97]$$

where $P(x)$ and $S(x)$ are continuous in the interval $a \leq x \leq b$ and $P(x) > 0$ therein, can indeed be written in the form

$$F(x) = \int_a^b G(x, x')f(x')\,dx'. \qquad [8.98]$$

We postulate the following properties of Green's function.

(i) For a fixed x', $G(x, x')$ is a continuous function of x and it satisfies the prescribed boundary conditions.

(ii) Except at the point $x = x'$, the first- and the second-order derivatives of $G(x, x')$ are continuous with respect to x in the interval $a \leq x \leq b$. At the point $x = x'$, the first derivative has a jump discontinuity given by

$$\frac{\partial}{\partial x} G(x, x')\bigg|_{x=x'-0}^{x=x'+0} = -\frac{1}{P(x')}.$$

(iii) The Green function $G(x, x')$ considered as a function of x satisfies the differential equation

$$\frac{d}{dx}\left(P(x)\frac{dG(x, x')}{dx}\right) + S(x)G(x, x') = 0$$

throughout the interval $a \leq x \leq b$ except at the point $x = x'$.

We shall now show that if $G(x, x')$ has the above-stated properties, then (8.98) is indeed the solution to the differential equation (8.97). To carry out the proof we must substitute (8.98) back into (8.97). To this end we shall need the first and second derivatives of $G(x, x')$ with respect to x. Since, as postulated, the first derivative of $G(x, x')$ is not continuous at $x = x'$ we must distinguish between the regions $x < x'$ and $x > x'$ when differentiating. Now

$$F'(x) = \frac{d}{dx}\int_a^{x-\varepsilon} G(x, x')f(x')\,dx' + \frac{d}{dx}\int_{x+\varepsilon}^b G(x, x')f(x')\,dx'. \qquad (8.99a)$$

We recall the *Leibnitz rule*. If

$$\Phi(x) = \int_{a(x)}^{b(x)} \varphi(t, x)\, dt,$$

where $a(x)$ and $b(x)$ are differentiable functions of x and $\varphi(t, x)$ and $\partial\varphi/\partial x$ are continuous in both x and t, then

$$\frac{d\Phi}{dx} = \int_{a(x)}^{b(x)} \frac{\partial\varphi}{\partial x}\, dt + \varphi(b(x), x)\frac{db}{dx} - \varphi(a(x), x)\frac{da}{dx}.$$

Applying the last result to (8.99a) and noting the property (i) of $G(x, x')$, we obtain

$$F'(x) = \int_a^{x-\varepsilon} f(x')\frac{\partial G}{\partial x}\, dx' + \int_{x+\varepsilon}^{b} f(x')\frac{\partial G}{\partial x}\, dx'. \tag{8.99b}$$

The second derivative of $F(x)$ can be obtained in a similar manner to give

$$\begin{aligned} F''(x) = {} & \int_a^{x-\varepsilon} f(x')\frac{\partial^2 G}{\partial x^2}\, dx' + \int_{x+\varepsilon}^{b} f(x')\frac{\partial^2 G}{\partial x^2}\, dx' \\ & - f(x)\left[\frac{\partial}{\partial x} G(x, x+\varepsilon) - \frac{\partial}{\partial x} G(x, x-\varepsilon)\right]. \end{aligned} \tag{8.100}$$

Substituting the above results into (8.97), we obtain

$$\begin{aligned} & -f(x) + P(x)f(x)\left[\frac{\partial}{\partial x} G(x, x+\varepsilon) - \frac{\partial}{\partial x} G(x, x-\varepsilon)\right] \\ & = \int_a^b dx'\, f(x')\left[P(x)\frac{\partial^2 G(x, x')}{\partial x^2} + P'(x)\frac{\partial G(x, x')}{\partial x} + S(x)G(x, x')\right], \end{aligned} \tag{8.101}$$

where the last integral has been written with the understanding that the point $x = x'$ is excluded from the integration. Since the quantity in the square brackets in the integrand of (8.101) is zero according to property (iii), of the Green function it follows that

$$\frac{\partial}{\partial x} G(x, x')\Big|_{x'=x-\varepsilon}^{x'=x+\varepsilon} = \frac{1}{P(x)}. \tag{8.102}$$

The limits on x' in (8.102) can be translated into the limits on x by noting that in the upper limit $x = x' - \varepsilon$ and in the lower limit $x = x' + \varepsilon$ and, consequently, the entire expression in (8.102) becomes a function of x' only. Hence,

$$\frac{\partial}{\partial x} G(x, x')\Big|_{x=x'-\varepsilon}^{x=x'+\varepsilon} = \frac{-1}{P(x')}, \tag{8.103}$$

which is identical with the (ii) property postulated earlier.

It is also worthwhile to note that properties (ii) and (iii) of the Green function $G(x, x')$ are implied by the differential equation

$$\frac{d}{dx}\left[P(x)\frac{dG(x, x')}{dx}\right] + S(x)G(x, x') = -\delta(x - x') \tag{8.104}$$

valid for all points $a \leq x \leq b$ including $x = x'$. For integrating formally both sides of (8.104), we obtain immediately

$$-1 = \int_{x'-\varepsilon}^{x'+\varepsilon}\left[\frac{d}{dx}\left(P\frac{dG}{dx}\right) + SG\right] dx \tag{8.105}$$

since the integral over the delta function is zero in regions not containing the point $x = x'$ and the only nonzero contribution results from a small interval $x' - \varepsilon \leq x \leq x' + \varepsilon$. Moreover, since the functions $S(x)$ and $G(x, x')$ are continuous at $x = x'$, the integration of $S(x)G(x, x')$ in (8.105) is zero and, consequently, we obtain (8.103) as before. This completes the proof.

8.4 Examples of Green's Function Construction

The knowledge of Green's function properties outlined in the preceding section are useful in the construction of Green's function. In fact, using these properties, it is possible to construct Green's function for many configurations in which the eigenfunction expansion method fails. Moreover, it will follow that the present method will yield Green's function in a form that is more explicit than before; i.e., there will be one summation or integration less than before appearing in the final result.

8.4.1 Finite Rectangular Region

Let us return to the problem of finding the Green function for a rectangular region shown in Fig. 8.1 and pertinent to the axial electric field. This Green function $G^{(e)}(\rho, \rho')$ must satisfy the differential equation (8.55) and must vanish at $x = 0$, a and $y = 0$, b.

As the first step toward finding $G^{(e)}(\boldsymbol{\rho}, \boldsymbol{\rho}')$ we must reduce the partial differential equation (8.55) to an ordinary one to which the rules developed earlier apply.

For the present case it will be appropriate to seek a solution in the form

$$G^{(e)}(\boldsymbol{\rho}, \boldsymbol{\rho}') = \sum_{1}^{\infty} A_m(x', y, y') \sin\frac{m\pi x}{a}, \tag{8.106}$$

which ensures its vanishing at $x = 0$, a. Substitution of (8.106) into (8.55) gives

$$\sum_{1}^{\infty} \sin \frac{m\pi x}{a} \left(\frac{d^2}{dy^2} + k_m{}^2 \right) A_m = -\delta(x - x')\, \delta(y - y'), \tag{8.107}$$

where $k_m{}^2 = k^2 - (m\pi/a)^2$. Multiplying both sides of (8.107) by $\sin(n\pi x/a)$ and integrating from 0 to a, where we use the orthogonality property of the eigenfunctions and the integral property of the delta function, we obtain

$$\left(\frac{d^2}{dy^2} + k_m{}^2 \right) g = -\delta(y - y'), \tag{8.108}$$

where

$$A_m = \frac{2}{a} \sin \frac{m\pi x'}{a}\, g(y, y'). \tag{8.109}$$

Equation (8.108) defines the Green function $g(y, y')$ in the interval $0 \le y \le b$, $0 \le y' \le b$, which must satisfy the boundary conditions $g(0, y') = g(b, y') = 0$.

It will be necessary to define the intervals $0 \le y < y'$ and $y' < y \le b$ where we correspondingly denote the Green function $g_<$ and $g_>$. Since the point $y = y'$ is now excluded, the functions $g_<$ and $g_>$ must satisfy the corresponding homogeneous equation of (8.108). Moreover, noting that $g_<$ must vanish at $y = 0$ and $g_>$ at $y = b$, we write

$$g_< = C_1 \sin k_m y, \tag{8.110a}$$

$$g_> = C_2 \sin k_m(b - y). \tag{8.110b}$$

To determine the coefficients C_1 and C_2, we use the properties of the Green function at $y = y'$

$$g_< = g_>, \tag{8.111a}$$

$$\frac{\partial g_>}{\partial y} - \frac{\partial g_<}{\partial y} = -1, \tag{8.111b}$$

which gives a pair of simultaneous algebraic equations as follows.

$$\begin{pmatrix} \cos k_m y' & \cos k_m(b - y') \\ \sin k_m y' & -\sin k_m(b - y') \end{pmatrix} \begin{pmatrix} C_1 \\ C_2 \end{pmatrix} = \begin{pmatrix} 1/k_m \\ 0 \end{pmatrix}. \tag{8.112}$$

Hence,

$$C_1 = \frac{\sin k_m(b - y')}{k_m \sin k_m b}, \tag{8.113a}$$

$$C_2 = \frac{\sin k_m y'}{k_m \sin k_m b}. \tag{8.113b}$$

Finally, substituting these results back into (8.109) and (8.106), we obtain

$$G_<^{(e)}(\boldsymbol{\rho}, \boldsymbol{\rho}') = \frac{2}{a}\sum_1^\infty \frac{\sin(m\pi x/a)\sin(m\pi x'/a)\sin k_m y \sin k_m(b-y')}{k_m \sin k_m b}, \tag{8.114a}$$

$$G_>^{(e)}(\boldsymbol{\rho}, \boldsymbol{\rho}') = \frac{2}{a}\sum_1^\infty \frac{\sin(m\pi x/a)\sin(m\pi x'/a)\sin k_m y' \sin k_m(b-y)}{k_m \sin k_m b}, \tag{8.114b}$$

where $G_<$ is appropriate to the interval $0 \le y \le y'$ and $G_>$ to the interval $y' \le y \le b$. Note that $G_>$ can be obtained from $G_<$ by interchanging the roles of x and x' and y and y', which is the basic symmetry property of the Green function associated with the Sturm–Liouville problem.

Looking back at (8.114) and the corresponding results obtained by the eigenfunction expansion in (8.58), we conclude that

$$\frac{2}{b}\sum_1^\infty \frac{\sin(n\pi y/b)\sin(n\pi y'/b)}{(n\pi/b)^2 - k_m{}^2} = \begin{cases} \dfrac{\sin k_m y' \sin k_m(b-y)}{k_m \sin k_m b}, & y' \le y, \\[2ex] \dfrac{\sin k_m y \sin k_m(b-y')}{k_m \sin k_m b}, & y \le y', \end{cases} \tag{8.115}$$

a fact that can be proven independently.

Using the same method, it can be readily shown that the Green function related to the axial magnetic field can be written as follows.

$$G_<^{(m)}(\boldsymbol{\rho}, \boldsymbol{\rho}') = -\frac{1}{a}\sum_0^\infty \frac{\varepsilon_m \cos(m\pi x/a)\cos(m\pi x'/a)\cos k_m y \cos k_m(b-y')}{k_m \sin k_m b} \tag{8.116}$$

and $G_>^{(m)}$ can be obtained from $G_<^{(m)}$ by interchanging the roles of y and y'.

8.4.2 Semi-Infinite Rectangular Region

To find the Green function of the electric type for a semi-infinite rectangular region as shown in Fig. 8.3, it will be convenient to start with the expansion

$$G^{(e)}(\boldsymbol{\rho}, \boldsymbol{\rho}') = \sum_1^\infty A_n(x, x', y')\sin(n\pi y/b), \tag{8.117}$$

which vanishes at $y = 0, b$, as it should, and will satisfy the differential equation (8.55) if

$$\left(\frac{d^2}{dx^2} + k_n{}^2\right) g = -\delta(x - x'), \tag{8.118}$$

where $k_n{}^2 = k^2 - (n\pi/b)^2$, and

$$A_n = \frac{2}{b} \sin\left(\frac{n\pi y'}{b}\right) g(x, x'). \tag{8.119}$$

To find $g(x, x')$, we divide the interval $0 \leq x \leq \infty$ into $0 \leq x < x'$ and $x' < x \leq \infty$ and denote the corresponding solutions with $g_<$ and $g_>$, respectively. In their respective regions $g_<$ and $g_>$ must satisfy the homogeneous differential equation corresponding to (8.118); hence we can write

$$g_< = C_1 \sin k_n x, \tag{8.120a}$$

$$g_> = C_2 \exp(ik_n x), \qquad \mathrm{Im}(k_n) > 0. \tag{8.120b}$$

The unknown coefficients C_1 and C_2 are readily found by imposing the required conditions on $g_<$ and $g_>$ and their derivatives at $x = x'$. Hence, we obtain

$$G_<^{(e)}(\boldsymbol{\rho}, \boldsymbol{\rho}') = \frac{2}{b} \sum_1^\infty \frac{\sin k_n x \exp(ik_n x') \sin(n\pi y/b) \sin(n\pi y'/b)}{k_n}, \tag{8.121}$$

while $G_<$ can be obtained from the foregoing by interchanging x and x'.

Comparison of (8.121) with the corresponding result derived from eigenfunction expansion in (8.63) shows that

$$\frac{2}{\pi} \int_0^\infty \frac{\sin \alpha x \sin \alpha x'}{\alpha^2 - k_n{}^2} \, d\alpha = \begin{cases} \dfrac{\sin k_n x'}{k_n} \exp(ik_n x), & x > x', \\[2ex] \dfrac{\sin k_n x}{k_n} \exp(ik_n x'), & x < x', \end{cases} \tag{8.122}$$

a fact that can be proven independently.

A still different representation of the Green function $G^{(e)}(\boldsymbol{\rho}, \boldsymbol{\rho}')$ can be obtained in the following way. Using the Fourier sine transform pair

$$\begin{aligned} \tilde{G}(\alpha) &= \left(\frac{2}{\pi}\right)^{1/2} \int_0^\infty G(x) \sin \alpha x \, dx, \\ G(x) &= \left(\frac{2}{\pi}\right)^{1/2} \int_0^\infty \tilde{G}(\alpha) \sin \alpha x \, d\alpha, \end{aligned} \tag{8.123}$$

the partial differential equation (8.55) can be reduced to an ordinary one as follows.

$$\left(\frac{d^2}{dy^2} + \tau^2\right)g = -\delta(y - y'), \tag{8.124a}$$

$$\tilde{G}^{(e)} = \left(\frac{2}{\pi}\right)^{1/2} \sin \alpha x' g(y, y'), \tag{8.124b}$$

where $\tau = (k^2 - \alpha)^{1/2}$, providing that

$$\sin \alpha x \frac{\partial G^{(e)}}{\partial x} - \alpha \cos \alpha x G^{(e)} \Big|_{x=0}^{x=\infty} = 0. \tag{8.125}$$

Since $g(y, y')$ in the present case is completely analogous to (8.108), we can write at once, using (8.123),

$$G_<^{(e)}(\boldsymbol{\rho}, \boldsymbol{\rho}') = \frac{2}{\pi} \int_0^\infty \frac{\sin \alpha x \sin \alpha x' \sin \tau y \sin \tau(b - y')}{\tau \sin \tau b} \, d\alpha, \tag{8.126}$$

and $G_>$ can be obtained from $G_<$ by interchanging the roles of y and y'.

It is not difficult to show that the condition (8.125) is satisfied by $G_{\gtrless}^{(e)}(\boldsymbol{\rho}, \boldsymbol{\rho}')$ in (8.126).

The Green function $G^{(m)}$ related to the axial magnetic field can be found in a similar manner. We obtain

$$G_<^{(m)}(\boldsymbol{\rho}, \boldsymbol{\rho}') = \frac{i}{b} \sum_0^\infty \frac{\varepsilon_n \cos k_n x \exp(ik_n x') \cos(n\pi y/b) \cos(n\pi y'/b)}{k_n}, \tag{8.127}$$

or

$$G_<^{(m)}(\boldsymbol{\rho}, \boldsymbol{\rho}') = -\frac{2}{\pi} \int_0^\infty \frac{\cos \alpha x \cos \alpha x' \cos \tau y \cos \tau(b - y')}{\tau \sin \tau b} \, d\alpha. \tag{8.128}$$

8.4.3 Quarter-Space Region

The partial differential equation can be reduced to an ordinary one by using the sine Fourier transform for the electric case and the cosine Fourier transform for the magnetic case. However, the same result can be obtained more simply by using the known delta function representation. Thus, for instance, for the electric case we have from (8.65a)

$$\delta(y - y') = \frac{2}{\pi} \int_0^\infty \sin \alpha y \sin \alpha y' \, d\alpha,$$

which upon substitution into (8.55) immediately suggests the expansion

$$G^{(e)}(\boldsymbol{\rho}, \boldsymbol{\rho}') = \frac{2}{\pi} \int_0^\infty g(x, x', \alpha) \sin \alpha y \sin \alpha y' \, d\alpha. \tag{8.129}$$

It is easy to show that $g(x, x', \alpha)$ must satisfy the differential equation

$$g'' + \tau^2 g = -\delta(x - x') \text{ with } \tau = (k^2 - \alpha^2)^{1/2}.$$

Hence, in this case

$$G_<^{(e)}(\boldsymbol{\rho}, \boldsymbol{\rho}') = \frac{2}{\pi}\int_0^\infty \frac{\sin \tau x \exp(i\tau x') \sin \alpha y \sin \alpha y'}{\tau}\, d\alpha. \tag{8.130}$$

Similarly, it can be shown that in the magnetic case

$$G_<^{(m)}(\boldsymbol{\rho}, \boldsymbol{\rho}') = \frac{2i}{\pi}\int_0^\infty \frac{\cos \tau x \exp(i\tau x') \cos \alpha y \cos \alpha y'}{\tau}\, d\alpha. \tag{8.131}$$

Alternately, carrying out the indicated integration in (8.130) and (8.131), we obtain the well-known "image" representations

$$\begin{aligned} G^{\binom{e}{m}}(\boldsymbol{\rho}, \boldsymbol{\rho}') = \frac{i}{4}\big(&H_0^{(1)}\{k[(x - x')^2 + (y - y')^2]^{1/2}\} \\ &\mp H_0^{(1)}\{k[(x - x')^2 + (y + y')^2]^{1/2}\} \\ &\mp H_0^{(1)}\{k[(x + x')^2 + (y - y')^2]^{1/2}\} \\ &+ H_0^{(1)}\{k[(x + x')^2 + (y + y')^2]^{1/2}\}\big), \end{aligned} \tag{8.132}$$

where the upper signs are pertinent to the electric case and the lower signs are pertinent to the magnetic case.

8.4.4 Half-Space Region

Following either one of the methods previously outlined, we find

$$G_{y<y'}^{(e)}(\boldsymbol{\rho}, \boldsymbol{\rho}') = \frac{1}{2\pi}\int_{-\infty}^\infty \frac{\exp[i\alpha(x - x')] \sin \tau y \exp(i\tau y')}{\tau}\, d\alpha, \tag{8.133}$$

$$G_{y<y'}^{(m)}(\boldsymbol{\rho}, \boldsymbol{\rho}') = \frac{i}{2\pi}\int_{-\infty}^\infty \frac{\exp[i\alpha(x - x')] \cos \tau y \exp(i\tau y')}{\tau}\, d\alpha, \tag{8.134}$$

or

$$G_{x<x'}^{(e)}(\boldsymbol{\rho}, \boldsymbol{\rho}') = \frac{i}{\pi}\int_0^\infty \frac{\exp[-i\tau(x - x')] \sin \alpha y \sin \alpha y'}{\tau}\, d\alpha, \tag{8.135}$$

and

$$G_{x<x'}^{(m)}(\boldsymbol{\rho}, \boldsymbol{\rho}') = \frac{i}{\pi}\int_0^\infty \frac{\exp[-i\tau(x - x')] \cos \alpha y \cos \alpha y'}{\tau}\, d\alpha. \tag{8.136}$$

The image representation can be obtained by carrying out the integration. Thus

$$G^{\binom{e}{m}}(\boldsymbol{\rho}, \boldsymbol{\rho}') = \frac{i}{4}\,(H_0^{(1)}\{k[(x - x')^2 + (y - y')^2]^{1/2}\}$$
$$\mp H_0^{(1)}\{k[(x - x')^2 + (y + y')^2]^{1/2}\}). \tag{8.137}$$

8.4.5 Free Space Region

In this case the polarization plays no role and we obtain

$$G(\boldsymbol{\rho}, \boldsymbol{\rho}') = \frac{i}{4\pi} \int_{-\infty}^{\infty} \frac{\exp(i\tau\,|x - x'|\exp[i\alpha(y - y')]}{\tau}\,d\alpha \tag{8.138}$$

or

$$G(\boldsymbol{\rho}, \boldsymbol{\rho}') = \frac{i}{4\pi} \int_{-\infty}^{\infty} \frac{\exp[i\alpha(x - x')]\exp(i\tau\,|y - y'|)}{\tau}\,d\alpha, \tag{8.139}$$

and in either case $G(\boldsymbol{\rho}, \boldsymbol{\rho}') = (i/4)H_0^{(1)}(k\,|\boldsymbol{\rho} - \boldsymbol{\rho}'|)$.

8.4.6 Finite Angular Sector

The geometrical configuration of the problem is shown in Fig. 8.9. The Green function of either kind must in this case satisfy the partial differential equation

$$\left(\frac{1}{\rho}\frac{\partial}{\partial \rho}\rho\frac{\partial}{\partial \rho} + \frac{1}{\rho^2}\frac{\partial^2}{\partial \varphi^2} + k^2\right)G(\boldsymbol{\rho}, \boldsymbol{\rho}') = -\frac{\delta(\rho - \rho')\,\delta(\varphi - \varphi')}{\rho}. \tag{8.140}$$

In the electric case $G^{(e)}(\boldsymbol{\rho}, \boldsymbol{\rho}')$ must vanish at $\rho = a$ and $\varphi = 0, \varphi_0$. Inasmuch as the appropriate expansion of the angular delta function has already been found in (8.78), we can seek the expansion of the Green function $G^{(e)}$ in a similar form, i.e.,

$$G^{(e)}(\boldsymbol{\rho}, \boldsymbol{\rho}') = \frac{2}{\varphi_0}\sum_{m=1}^{\infty} g(\rho, \rho')\sin\nu\varphi\,\sin\nu\varphi', \tag{8.141}$$

which is nothing else but a Fourier expansion of the Green function $G^{(e)}$ in terms of the angular eigenfunctions. Substitution of (8.141) together with (8.78) into (8.140) gives

$$\left(\frac{1}{\rho}\frac{\partial}{\partial \rho}\rho\frac{\partial}{\partial \rho} + k^2 - \frac{\nu^2}{\rho^2}\right)g(\rho, \rho') = -\frac{\delta(\rho - \rho')}{\rho}. \tag{8.142}$$

The region of interest must now be divided into $0 \leq \rho < \rho'$ and $\rho' < \rho \leq a$.

If $g_<$ and $g_>$ denote the respective solutions that must satisfy the homogeneous differential equation corresponding to (8.142), then we write

$$g_< = AJ_\nu(k\rho); \tag{8.143a}$$

$$g_> = BJ_\nu(k\rho) + CH_\nu^{(1)}(k\rho), \tag{8.143b}$$

which ensures finiteness of g at $\rho = 0$. At $\rho = \rho'$, the functions $g_>$ and $g_<$ must satisfy

$$g_> = g_<, \tag{8.144a}$$

$$\frac{\partial g_>}{\partial \rho} - \frac{\partial g_<}{\partial \rho} = -\frac{1}{\rho'}, \tag{8.144b}$$

since $P(\rho) = \rho$ in this case. Hence, we obtain two simultaneous algebraic equations as follows.

$$\begin{pmatrix} J_\nu(k\rho') & H_\nu^{(1)}(k\rho') \\ J_\nu'(k\rho') & H_\nu^{(1)\prime}(k\rho') \end{pmatrix} \begin{pmatrix} A - B \\ -C \end{pmatrix} = \begin{pmatrix} 0 \\ 1/k\rho' \end{pmatrix}. \tag{8.145}$$

Noting that the Wronskian determinant

$$W\{J_\nu(x), H_\nu^{(1)}(x)\} = \begin{pmatrix} J_\nu(x) & H_\nu^{(1)}(x) \\ J_\nu'(x) & H_\nu^{(1)\prime}(x) \end{pmatrix} = \frac{2i}{\pi x}, \tag{8.146}$$

we can find the coefficients A, B, and C by using Cramer's rule. Hence, we obtain, upon substituting back into (8.141),

$$G_<^{(e)}(\boldsymbol{\rho}, \boldsymbol{\rho}') = \frac{i\pi}{\varphi_0} \sum_{m=1}^{\infty} J_\nu(k\rho) \left[H_\nu^{(1)}(k\rho') - \frac{H_\nu^{(1)}(ka)}{J_\nu(ka)} J_\nu(k\rho') \right] \sin \nu\varphi \sin \nu\varphi' \tag{8.147}$$

and $G_>^{(e)}$ can be obtained from the foregoing by interchanging ρ and ρ'.

In the case corresponding to the axial magnetic field we obtain

$$G_<^{(m)}(\boldsymbol{\rho}, \boldsymbol{\rho}') = \frac{i\pi}{2\varphi_0} \sum_{m=0}^{\infty} \varepsilon_m \left[H_\nu^{(1)}(k\rho') - \frac{H_\nu^{(1)\prime}(ka)}{J_\nu'(ka)} J_\nu(k\rho') \right] J_\nu(k\rho) \cos \nu\varphi \cos \nu\varphi'. \tag{8.148}$$

8.4.7 Infinite Angular Sector

The geometry of the configuration is shown in Fig. 8.10 and it is clear that the Green function of either kind must satisfy the partial differential equation of (8.140).

In the electric case, $G^{(e)}(\boldsymbol{\rho}, \boldsymbol{\rho}')$ must vanish at $\varphi = 0$, φ_0 for all ρ and, consequently, its representation can be sought in the form of (8.141), where

$g(\rho, \rho')$ must still satisfy (8.142). In this case, however, we must choose

$$g_< = AJ_\nu(k\rho), \qquad g_> = BH_\nu^{(1)}(k\rho), \tag{8.149}$$

which ensures the finiteness of g at $\rho = 0$ and $\rho = \infty$. The functions $g_<$ and $g_>$ must satisfy (8.144) at $\rho = \rho'$; hence, we obtain

$$\begin{pmatrix} J_\nu(k\rho') & H_\nu^{(1)}(k\rho') \\ J_\nu'(k\rho') & H_\nu^{(1)\prime}(k\rho') \end{pmatrix} \begin{pmatrix} A \\ -B \end{pmatrix} = \begin{pmatrix} 0 \\ 1/k\rho' \end{pmatrix}. \tag{8.150}$$

Noting the Wronskian determinant in (8.146) we can easily determine the unknown coefficients A and B by using Cramer's rule. Substituting back into (8.141), we find

$$G_<^{(e)}(\boldsymbol{\rho}, \boldsymbol{\rho}') = \frac{i\pi}{\varphi_0} \sum_{m=1}^{\infty} H_\nu^{(1)}(k\rho')J_\nu(k\rho) \sin \nu\varphi' \sin \nu\varphi, \tag{8.151}$$

where $G_>^{(e)}(\boldsymbol{\rho}, \boldsymbol{\rho}')$ can be obtained from the foregoing by interchanging the roles of $\boldsymbol{\rho}$ and $\boldsymbol{\rho}'$.

In the magnetic case, $\partial G^{(m)}/\partial\varphi$ must vanish at $\varphi = 0$, φ_0 for all ρ and, using a procedure similar to the one above, we obtain

$$G_<^{(m)}(\boldsymbol{\rho}, \boldsymbol{\rho}') = \frac{i\pi}{2\varphi_0} \sum_{m=0}^{\infty} \varepsilon_m H_\nu^{(1)}(k\rho')J_\nu(k\rho) \cos \nu\varphi' \cos \nu\varphi. \tag{8.152}$$

An alternative approach to finding the Green function for an infinite angular sector when either the function itself or its normal derivative vanishes on the walls makes use of the integral transform of Lebedev–Kontorovich defined by (see Erdelyi *et al.*, 1954).

$$\tilde{f}(\alpha) = \int_0^\infty f(\rho)K_\alpha(-i\lambda\rho)\,\frac{d\rho}{\rho}, \tag{8.153a}$$

$$f(\rho) = (i\pi)^{-1} \int_{c-i\infty}^{c+i\infty} \alpha\tilde{f}(\alpha)I_\alpha(-i\lambda\rho)\,d\alpha, \tag{8.153b}$$

where $I_\alpha(\ )$ and $K_\alpha(\ )$ are the modified Bessel functions of the first and second kind, respectively, satisfying the differential equation

$$\left[\frac{1}{z}\frac{d}{dz} z \frac{d}{dz} - \left(1 + \frac{\alpha^2}{z^2}\right)\right] \begin{matrix} I_\alpha(z) \\ K_\alpha(z) \end{matrix} = 0. \tag{8.154}$$

The modified Bessel functions are related to the ordinary Bessel functions as follows.

$$I_\alpha(z) = \exp\left(-\frac{i\alpha\pi}{2}\right)J_\alpha(iz), \tag{8.155a}$$

$$K_\alpha(z) = \frac{i\pi}{2} \exp\left(\frac{i\alpha\pi}{2}\right)H_\alpha^{(1)}(iz), \tag{8.155b}$$

where z is such that $-\pi < \arg z \le \pi/2$.

Because of the singular character of the Bessel function of the second kind at the origin $\rho = 0$, since $K_\alpha(-i\lambda\rho) \sim \rho^{-\alpha}$ as $\rho \to 0$, the integral representation of $\tilde{f}(\alpha)$ in (8.153a) has meaning only if $f(\rho)$ approaches zero sufficiently fast and the integral exists at $\rho = 0$. Suppose that $f(\rho) \to \rho^\beta$ as $\rho \to 0$, and since $K_\alpha(-i\lambda\rho) \to \rho^{-\alpha}$ in this case, the integral in (8.153a) approaches $\rho^{(\beta-\alpha)}/(\beta - \alpha)$ as $\rho \to 0$. Hence, the integral will converge at $\rho = 0$ if $\beta > \alpha$ and since on the path of integration in (8.153b) $\alpha \geq 0$, β must be positive or $\beta > \alpha \geq 0$.

In the case of the Green function of the electric type, $G^{(e)} = 0$ at the edge $\rho = 0$, and it is clear that the convergence of the integral in (8.153a) is ensured at $\rho = 0$. In the case of the Green function of the magnetic type $G^{(m)}$, which does not have to vanish at the edge $\rho = 0$, the transform (8.153a) cannot be applied directly but it can still be used with a slight modification.

To illustrate the procedure of finding the Green function using the integral transform of Lebedev–Kontorovich, let us consider the Green function of the magnetic type $G^{(m)}$, which must satisfy the differential equation

$$\left(\frac{1}{\rho}\frac{\partial}{\partial\rho}\rho\frac{\partial}{\partial\rho} + \frac{1}{\rho^2}\frac{\partial^2}{\partial\varphi^2} + k^2\right)G^{(m)} = -\frac{\delta(\rho - \rho')\,\delta(\varphi - \varphi')}{\rho} \tag{8.156}$$

and the boundary condition $\partial G^{(m)}/\partial\varphi = 0$ at $\varphi = 0, \varphi_0$. Since $G^{(m)}$ does not necessarily vanish at the edge $\rho = 0$, we consider instead

$$f(\rho, \varphi) = G^{(m)}(\rho, \varphi) - G^{(m)}(0)\exp(ik\rho) \tag{8.157}$$

because $f(0)$ vanishes. Substituting from (8.157) into (8.156), we obtain

$$\left(\frac{1}{\rho}\frac{\partial}{\partial\rho}\rho\frac{\partial}{\partial\rho} + \frac{1}{\rho^2}\frac{\partial^2}{\partial\varphi^2} + k^2\right)f(\rho, \varphi) = -\frac{\delta(\rho - \rho')\,\delta(\varphi - \varphi')}{\rho} - \frac{ik}{\rho}G^{(m)}(0)e^{ik\rho}. \tag{8.158}$$

The Lebedev–Kontorovich transform can now be applied directly to (8.158). Multiplying the entire equation by $K_\alpha(-ik\rho)\rho\,d\rho$ and integrating from 0 to ∞, we get, from the first term on the right in (8.158), $-\delta(\varphi - \varphi')K(-ik\rho')$. To evaluate the integral corresponding to the second term on the right in (8.158), we note that (see, e.g., Gradshteyn and Ryzhik, 1965, p. 707)

$$\int_0^\infty e^{-\alpha x}K_\nu(\beta x)\,dx = \frac{\pi}{\beta\sin\nu\pi}\cdot\frac{\sin\nu\theta}{\sin\theta},$$
$$\cos\theta = \frac{\alpha}{\beta}; \qquad \operatorname{Re}(\alpha + \beta) > 0. \tag{8.159}$$

Hence, we obtain immediately

$$\int_0^\infty e^{ik\rho}K_\alpha(-ik\rho)\,d\rho = \frac{i\pi\alpha}{k\sin\alpha\pi}. \tag{8.160}$$

Going over to the left side of (8.158), we note that the first term gives

$$\int_0^\infty \frac{\partial}{\partial\rho}\left(\rho\frac{\partial f}{\partial\rho}\right)K_\alpha(-ik\rho)\,d\rho = -ik\int_0^\infty f\frac{\partial}{\partial\rho}[\rho K_\alpha'(-ik\rho)]\,d\rho, \quad (8.161a)$$

providing that the integrated part vanishes, i.e.,

$$\rho\left[\frac{\partial f}{\partial\rho}K_\alpha(-ik\rho) + ikfK_\alpha'(-ik\rho)\right]_{\rho=0}^{\rho=\infty} = 0. \quad (8.161b)$$

Moreover, from (8.154) we note that

$$\frac{\partial}{\partial\rho}[\rho K_\alpha'(-ik\rho)] = \frac{\alpha^2 - k^2\rho^2}{-ik\rho}K_\alpha(-ik\rho). \quad (8.161c)$$

Using the results above, we obtain the transformed differential equation as

$$\left(\frac{\partial^2}{\partial\varphi^2} + \alpha^2\right)\tilde{f}(\alpha, \varphi) = -\delta(\varphi - \varphi')K_\alpha(-ik\rho') + \frac{\alpha\pi}{\sin\alpha\pi}G^{(m)}(0). \quad (8.162)$$

Now we can define a new function

$$F(\alpha, \varphi) = \frac{1}{K_\alpha(-ik\rho')}\left[\tilde{f}(\alpha, \varphi) - \frac{\pi G^{(m)}(0)}{\alpha\sin\alpha\pi}\right], \quad (8.163)$$

which evidently satisfies the inhomogeneous differential equation

$$\frac{d^2F}{d\varphi^2} + \alpha^2F = -\delta(\varphi - \varphi') \quad (8.164)$$

that can be readily solved. Before proceeding with the solution, however, we must determine the boundary conditions to be imposed on $F(\alpha, \varphi)$ as a function of φ. Looking back at (8.163) and (8.157), we conclude that the boundary condition $\partial G^{(m)}/\partial\varphi = 0$ on $\varphi = 0, \varphi_0$ will be satisfied if $\partial F/\partial\varphi = 0$ on $\varphi = 0, \varphi_0$. Thus, we construct the appropriate solution to (8.164) by writing

$$F_< = A\cos\alpha\varphi, \quad (8.165a)$$

$$F_> = B\cos\alpha(\varphi_0 - \varphi), \quad (8.165b)$$

which satisfies the prescribed boundary conditions. The coefficients A and B are now determined from the requirements $F_> = F_<$ and $\partial F_>/\partial\varphi - \partial F_</\partial\varphi = -1$ at $\varphi = \varphi'$. Hence,

$$F_< = -\frac{\cos\alpha(\varphi_0 - \varphi')\cos\alpha\varphi}{\alpha\sin\alpha\varphi_0}, \quad (8.166a)$$

$$F_> = -\frac{\cos\alpha\varphi'\cos\alpha(\varphi_0 - \varphi)}{\alpha\sin\alpha\varphi_0}. \quad (8.166b)$$

The function $\tilde{f}(\alpha, \varphi)$ can now be determined by using (8.163) and it can be inverted according to (8.153b). Thus, we obtain

$$\underset{\varphi>\varphi'}{f}(\rho, \varphi) = \frac{i}{\pi}\int_{-i\infty}^{i\infty}\left[\frac{\cos\alpha\varphi'\cos\alpha(\varphi_0-\varphi)}{\text{s n }\alpha\varphi_0}K_\alpha(-ik\rho') - \frac{\pi G^{(m)}(0)}{\sin\alpha\pi}\right] \times I_\alpha(-ik\rho)\,d\alpha, \tag{8.167a}$$

$$\underset{\varphi<\varphi'}{f}(\rho, \varphi) = \frac{i}{\pi}\int_{-i\infty}^{i\infty}\left[\frac{\cos\alpha(\varphi_0-\varphi')\cos\alpha\varphi}{\sin\alpha\varphi_0}K_\alpha(-ik\rho') - \frac{\pi G^{(m)}(0)}{\sin\alpha\pi}\right] \times I_\alpha(-ik\rho)\,d\alpha. \tag{8.167b}$$

Next we shall dispose of the second integral on the right in (8.167a) and (8167b). Using contour integration, it will follow that (problem)

$$\int_{-i\infty}^{i\infty}\frac{I_\alpha(-ik\rho)}{\sin\alpha\pi}\,d\alpha = -ie^{ik\rho}. \tag{8.168}$$

Hence, the second term on the right in (8.167a) and (8.167b) is $-G^{(m)}(0)$ $\exp(ik\rho)$ and, consequently, the first term must be equal to $G^{(m)}(\rho, \varphi)$, according to (8.157). Hence, we secure

$$\underset{\varphi<\varphi'}{G^{(m)}}(\boldsymbol{\rho}, \boldsymbol{\rho}') = -\frac{1}{2}\int_{-i\infty}^{i\infty}\frac{\cos\alpha(\varphi_0-\varphi')\cos\alpha\varphi}{\sin\alpha\varphi_0}H_\alpha^{(1)}(k\rho')J_\alpha(k\rho)\,d\alpha, \tag{8.169a}$$

$$\underset{\varphi>\varphi'}{G^{(m)}}(\boldsymbol{\rho}, \boldsymbol{\rho}') = -\frac{1}{2}\int_{-i\infty}^{i\infty}\frac{\cos\alpha\varphi'\cos\alpha(\varphi_0-\varphi)}{\sin\alpha\varphi_0}H_\alpha^{(1)}(k\rho')J_\alpha(k\rho)\,d\alpha, \tag{8.169b}$$

where we also used (8.155) to express the modified Bessel functions in terms of $H_\alpha^{(1)}(k\rho')$ and $J_\alpha(k\rho)$. Noting the identity

$$H_\alpha^{(1)}(x')J_\alpha(x) = H_\alpha^{(1)}(x)J_\alpha(x') - \frac{i}{\sin\alpha\pi}[J_\alpha(x')J_{-\alpha}(x) - J_{-\alpha}(x')J_\alpha(x)], \tag{8.170}$$

and since the expression in the square brackets is an odd function of α, it will follow that its contribution to the value of the integral in (8.169) vanishes and, consequently, we conclude that $G^{(m)}(\boldsymbol{\rho}, \boldsymbol{\rho}') = G^{(m)}(\boldsymbol{\rho}', \boldsymbol{\rho})$, as it should.

The formulas in (8.169) constitute a formal answer to the problem we have set forth to solve. In order to show that the present representation is equivalent to the infinite series representation of (8.152), we must perform the indicated integration.

The path of integration in (8.169) runs along the axis of the imaginaries in the complex α plane and we shall now close it by adding a large semicircle in the right-hand plane, as shown in Fig. 8.12. Note that the contour is indented to the right in the vicinity of the pole at $\alpha = 0$, in accordance with the inversion formula in (8.153b).

Next we shall show that in the limit as the radius of the large semicircle approaches infinity, its contribution to the integral vanishes. For concreteness

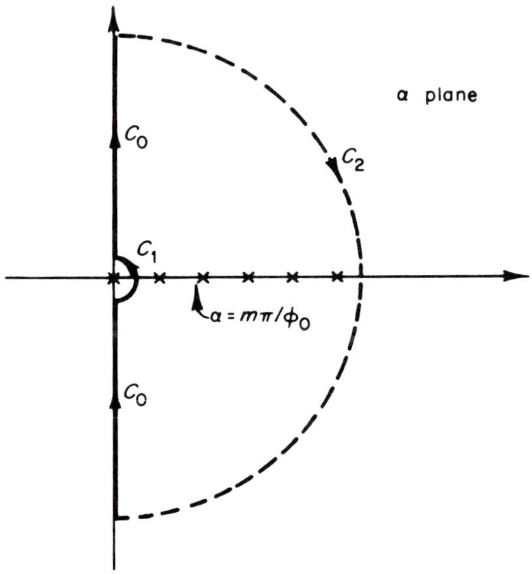

Fig. 8.12. The complex α plane for evaluation of the integrals in (8.169).

let us focus our attention on the integral in (8.169a). Setting $\alpha = R \exp(i\theta)$, $-\pi/2 < \theta < \pi/2$, we note immediately that

$$|\cos \alpha(\varphi_0 - \varphi') \cos \alpha\varphi/\sin \alpha\varphi_0| \xrightarrow[R\to\infty]{} \exp[-R \sin|\theta|(\varphi' - \varphi)]. \quad (8.171)$$

But we must also obtain an estimate on the product $H_\alpha^{(1)}(k\rho')J_\alpha(k\rho)$ as $|\alpha| \to \infty$ in the right plane. This can be obtained as follows.

Using the uniform asymptotic expansion for Bessel functions (Abramowitz and Stegun (1964), pp. 368, 448–449) we can write for $|\alpha| \to \infty$

$$J_\alpha(x) \sim (4\zeta)^{1/4}\alpha^{-1/3}Ai(w), \quad (8.172a)$$

$$Y_\alpha(x) \sim -(4\zeta)^{1/4}\alpha^{-1/3}Bi(w), \quad (8.172b)$$

$$H_\alpha^{(1)}(x) \sim (4\zeta)^{1/4}\alpha^{-1/3}[Ai(w) - iBi(w)], \quad (8.172c)$$

$$H_\alpha^{(2)}(x) \sim (4\zeta)^{1/4}\alpha^{-1/3}[Ai(w) + iBi(w)], \quad (8.172d)$$

where $Ai(w)$ and $Bi(w)$ are two linearly independent solutions of the Airy differential equation $W''(w) - wW(w) = 0$ and

$$w = \zeta\alpha^{2/3}, \tag{8.172e}$$

$$\tfrac{2}{3}\zeta^{3/2} \sim \ln(2\alpha/ex). \tag{8.172f}$$

Since $|w| \to \infty$ as $|\alpha| \to \infty$, we can use the asymptotic forms of the Airy functions $Ai(w)$ and $Bi(w)$ as follows.

$$Ai(w) \sim (2\sqrt{\pi})^{-1} w^{-1/4} e^{-\xi}, \qquad |\arg w| < \pi, \tag{8.173a}$$

$$Bi(w) \sim (\pi)^{-1/2} w^{-1/4} e^{\xi}, \qquad |\arg w| < \pi/3, \tag{8.173b}$$

where

$$\xi = (\tfrac{2}{3}) w^{3/2} \sim \ln(2\alpha/ex)^{\alpha}. \tag{8.173c}$$

Substituting (8.173) into (8.172) and noting that the contribution arising from $Ai(w)$ can be neglected in the expression for $H_\alpha^{(1)}(x)$, we obtain

$$J_\alpha(x) \sim (2\pi\alpha)^{-1/2} \left(\frac{ex}{2\alpha}\right)^{\alpha} \tag{8.174a}$$

$$H_\alpha^{(1)}(x) \sim -i\left(\frac{2}{\pi\alpha}\right)^{1/2} \left(\frac{2\alpha}{ex}\right)^{\alpha}, \tag{8.174b}$$

both valid in the range $|\arg \alpha| < \pi/2 - \delta$, $\delta \geq 0$. Hence, with $\alpha = R\exp(i\theta)$ it will follow that

$$|J_\alpha(k\rho)H_\alpha^{(1)}(k\rho')| \xrightarrow[R\to\infty]{} (\pi R)^{-1} \exp[-R\cos\theta \ln(\rho'/\rho)] \tag{8.175}$$

and the vanishing of the contribution from the large semicircle will, consequently, be ensured as long as $\rho' > \rho$.

Thus, the integral in (8.169a) is equal to $-2\pi i$ times the sum of the residues at the poles $\alpha = \nu = m\pi/\varphi_0$ plus the contribution from the small semicircle about the pole $\alpha = 0$. Hence, we obtain

$$\underset{\rho'>\rho}{G^{(m)}}(\boldsymbol{\rho}, \boldsymbol{\rho}') = \frac{i\pi}{2\varphi_0} \sum_{m=0}^{\infty} \varepsilon_m \cos\nu\varphi' \cos\nu\varphi J_\nu(k\rho) H_\nu^{(1)}(k\rho'), \tag{8.176}$$

which is identical with (8.152). Note that even though we have started with $G^{(m)}_{\varphi'>\varphi}$, the series representation in (8.176) is equally valid for $\varphi > \varphi'$ since Green's function is symmetrical. Moreover, since $G^{(m)}(\boldsymbol{\rho}', \boldsymbol{\rho}) = G^{(m)}(\boldsymbol{\rho}, \boldsymbol{\rho}')$, as remarked earlier, the series representation for $\rho > \rho'$ can be obtained from (8.176) by interchanging the roles of ρ and ρ'.

Following a similar procedure, it will follow (problem) that the Green function of the electric type can be written

$$G^{(e)}_{\varphi<\varphi'}(\boldsymbol{\rho}, \boldsymbol{\rho}') = \frac{1}{2}\int_{-i\infty}^{i\infty} \frac{\sin \alpha(\varphi_0 - \varphi') \sin \alpha\varphi}{\sin \alpha\varphi_0} H_\alpha^{(1)}(k\rho')J_\alpha(k\rho)\, d\alpha, \tag{8.177a}$$

$$G^{(e)}_{\varphi>\varphi'}(\boldsymbol{\rho}, \boldsymbol{\rho}') = \frac{1}{2}\int_{-i\infty}^{i\infty} \frac{\sin \alpha\varphi' \sin \alpha(\varphi_0 - \varphi)}{\sin \alpha\varphi_0} H_\alpha^{(1)}(k\rho')J_\alpha(k\rho)\, d\alpha. \tag{8.177b}$$

The integral representations of (8.169) and (8.177) can be expressed in a different form by noting the formula (Watson, 1962, p. 439)

$$H_\alpha^{(1)}(Z)J_\alpha(z) = \frac{1}{\pi i}\int_0^{c+i\infty} \exp\left[\frac{t}{2} - \frac{(Z^2 + z^2)}{2t^2}\right] I_\alpha\left(\frac{zZ}{t}\right) dt \tag{8.178a}$$

valid for $|z| < |Z|$. Moreover using (Watson, 1962, p. 181)

$$I_\alpha\left(\frac{zZ}{t}\right) = \frac{1}{2\pi i}\int_{\infty - i\pi}^{\infty + i\pi} \exp\left[\left(\frac{zZ}{t}\right) \cosh w - \alpha w\right] dw, \tag{8.178b}$$

we obtain by substituting (8.178b) into (8.178a) and changing the order of integration,

$$H_\alpha^{(1)}(Z)J_\alpha(z) = -\frac{1}{2\pi^2}\int_{\infty - i\pi}^{\infty + i\pi} dw \exp(-\alpha w)$$
$$\times \int_0^{c+i\infty} \exp\left\{\frac{1}{2}\left[t - \frac{(Z^2 + z^2 - 2zZ \cosh w)}{t}\right]\right\} \frac{dt}{t}. \tag{8.179}$$

With the substitution $x = (Z^2 + z^2 - 2zZ \cosh w)^{1/2}$ and $t = ux$ in the second integral, and noting the formula (Watson, 1962, p. 179)

$$H_0^{(1)}(x) = \frac{1}{\pi i}\int_0^{c+i\infty} \exp\left[\frac{x}{2}\left(u - \frac{1}{u}\right)\right] \frac{du}{u}, \qquad \operatorname{Im}(x) > 0, \tag{8.180}$$

we obtain

$$H_\alpha^{(1)}(Z)J_\alpha(z) = (2\pi i)^{-1}\int_{\infty - i\pi}^{\infty + i\pi} H_0^{(1)}(x)e^{-\alpha w}\, dw. \tag{8.181}$$

Substituting (8.181) into (8.169a) and (8.177a), we can write

$$G^{(e)}_{\varphi<\varphi'}(\boldsymbol{\rho}, \boldsymbol{\rho}') = -\frac{i}{4\pi}\int_{\infty - i\pi}^{\infty + i\pi} H_0^{(1)}[k(\rho^2 + \rho'^2 - 2\rho\rho' \cosh w)^{1/2}]Q^{(e)}(w)\, dw, \tag{8.182}$$

and

$$G^{(m)}_{\varphi<\varphi'}(\boldsymbol{\rho}, \boldsymbol{\rho}') = -\frac{1}{4\pi}\int_{\infty - i\pi}^{\infty + i\pi} H_0^{(1)}[k(\rho^2 + \rho'^2 - 2\rho\rho' \cosh w)^{1/2}]Q^{(m)}(w)\, dw, \tag{8.183}$$

where

$$Q^{(e)}(w) = \int_{-i\infty}^{i\infty} \frac{e^{-\alpha w} \sin \alpha(\varphi_0 - \varphi') \sin \alpha\varphi}{\sin \alpha\varphi_0}\, d\alpha, \tag{8.184a}$$

$$Q^{(m)}(w) = -\int_{-i\infty}^{i\infty} \frac{e^{-\alpha w} \cos \alpha(\varphi_0 - \varphi') \cos \alpha\varphi}{\sin \alpha\varphi_0}\, d\alpha. \tag{8.184b}$$

The integrals in (8.184) can be evaluated by closing the contour with a large semicircle of vanishing contribution in the right half plane and integrating over the poles $\alpha = m\pi/\varphi_0$. We obtain

$$Q^{(e)} = \frac{\pi i}{\varphi_0} \sum_{m=1}^{\infty} e^{-\nu w}[\cos \nu(\varphi - \varphi') - \cos \nu(\varphi + \varphi')], \tag{8.185a}$$

$$Q^{(m)} = \frac{\pi i}{2\varphi_0} \sum_{m=0}^{\infty} \varepsilon_m e^{-\nu w}[\cos \nu(\varphi - \varphi') + \cos \nu(\varphi + \varphi')]. \tag{8.185b}$$

The series on the right in (8.185) can be summed, using the well-known summation formula (Mangulis, 1965, p. 98)

$$\sum_{m=0}^{\infty} r^m \cos m\varphi = \frac{1 - r \cos \varphi}{1 - 2r \cos \varphi + r^2}, \qquad |r| < 1, \tag{8.186}$$

to obtain

$$Q^{(e,m)} = -\frac{i\pi}{4\varphi_0} \left[\frac{\sinh 2vw}{\sin v(iw + \varphi_1) \sin v(iw - \varphi_1)} \mp \frac{\sinh 2vw}{\sin v(iw + \varphi_2) \sin v(iw - \varphi_2)} \right], \tag{8.187}$$

where $v = \pi/2\varphi_0$, $\varphi_1 = \varphi' - \varphi$, and $\varphi_2 = \varphi' + \varphi$. Note that the requirement $|r| < 1$ in (8.186) now is translated into $\mathrm{Re}(w) > 0$. Substituting the results (8.187) into (8.182) and (8.183) and making a change of the variable $w = -iu$, where the requirement $\mathrm{Re}(w) > 0$ is now translated into $\mathrm{Im}(u) > 0$, we obtain

$$\underset{\varphi < \varphi'}{G^{(e)}} = -(16\varphi_0)^{-1}(W_1 - W_2) \tag{8.188}$$

and

$$\underset{\varphi < \varphi'}{G^{(m)}} = -(16\varphi_0)^{-1}(W_1 + W_2), \tag{8.189}$$

where

$$W_{1,2} = \int_{i\infty - \pi}^{i\infty + \pi} \frac{\sin 2vu H_0^{(1)}[k(\rho^2 + \rho'^2 - 2\rho\rho' \cos u)^{1/2}]}{\sin v(u + \varphi_{1,2}) \sin v(u - \varphi_{1,2})}\, du. \tag{8.190}$$

The integral in (8.190) can be reduced to a simpler form by integrating along the indicated contour with indentations above the poles $u_m = -\varphi_{1,2} \pm 2m\varphi_0$ and $u_n = \varphi_{1,2} \pm 2n\varphi_0$ in accordance with the requirement $\mathrm{Im}(u) > 0$ stated earlier and shown in Fig. 8.13a. Subsequently, the integral can be split as follows.

$$\int_{i\infty-\pi}^{i\infty+\pi} = \int_{i\infty-\pi}^{0} + \int_{0}^{i\infty+\pi}.$$

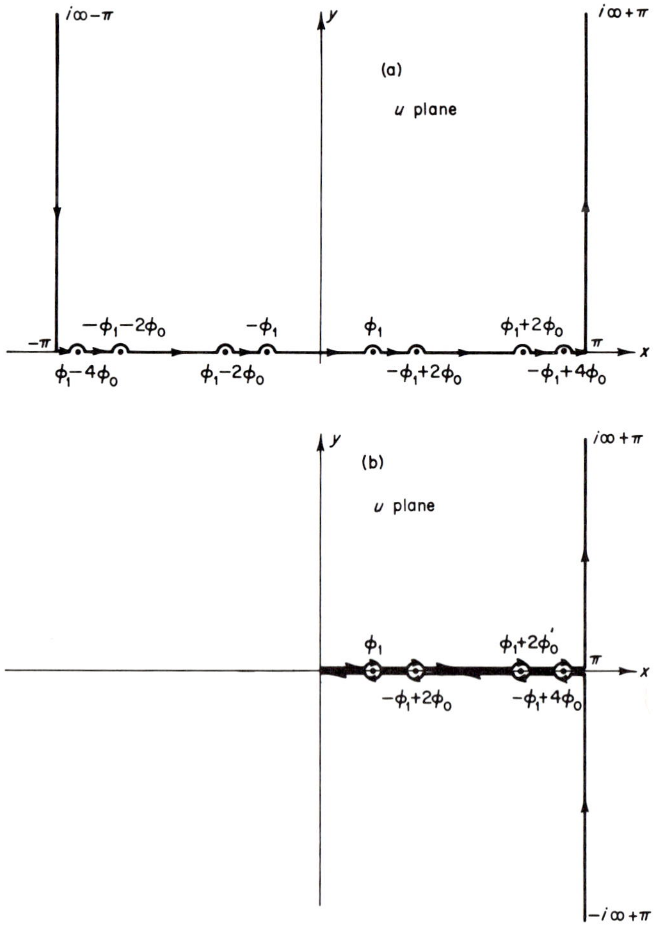

Fig. 8.13. The complex u plane pertinent to the evaluation of the integrals $W_{1,2}$ in (8.190).

In the first integral on the right we put $u = -u$, $\text{Im}(-u) < 0$, and hence obtain an equivalent contour, as shown in Fig. 8.13b.

It is clear by virtue of the contour in Fig. 8.13b that $W_{1,2} = -2\pi i$ times the sum of the residues at the poles $u_m = -\varphi_{1,2} + 2m\varphi_0$ and $u_n = \varphi_{1,2} + 2n\varphi_0$ plus the integral along the vertical path $-i\infty + \pi$, $i\infty + \pi$. The poles will be captured only if either

$$0 < 2M_{1,2}\varphi_0 - \varphi_{1,2} < \pi, \qquad M_{1,2} = 1, 2, 3, \ldots, \tag{8.191a}$$

or

$$0 < 2N_{1,2}\varphi_0 + \varphi_{1,2} < \pi, \qquad N_{1,2} = 0, 1, 2, 3, \ldots. \tag{8.191b}$$

Finally, the integral along the vertical path $-i\infty + \pi$, $i\infty + \pi$ can be simplified by putting $x = u - \pi$ and we obtain

$$\begin{aligned} W_{1,2} = -i4\varphi_0 \Bigg(& \sum_{m=1}^{M_{1,2}} H_0^{(1)}\{k[\rho^2 + \rho'^2 - 2\rho\rho' \cos(2m\varphi_0 - \varphi_{1,2})]^{1/2}\} \\ & + \sum_{n=0}^{N_{1,2}} H_0^{(1)}\{k[\rho^2 + \rho'^2 - 2\rho\rho' \cos(2n\varphi_0 + \varphi_{1,2})]^{1/2}\}\Bigg) \\ & + \int_{-i\infty}^{i\infty} \frac{\sin 2v(\pi + x) H_0^{(1)}\{k[\rho^2 + \rho'^2 + 2\rho\rho' \cos x]^{1/2}\}}{\sin v(x + \pi + \varphi_{1,2}) \sin v(x + \pi - \varphi_{1,2})}\, dx. \end{aligned} \tag{8.192}$$

It is important to note that the integral in (8.192) vanishes whenever $v = l/2$ where l is an integer other than zero so that $\varphi_0 = \pi/l$. In this case the solution can be represented solely in terms of images. Thus, for instance, in the case of an infinite plane $l = 1$, $\varphi_0 = \pi$, and since $\varphi_{1,2} > 0$, $N_{1,2} = 0$ while $M_{1,2}$ does not exist. Hence, only the summation over n will contribute and we obtain $W_{1,2} = -i4\pi H_0^{(1)}\{k[\rho^2 + \rho'^2 - 2\rho\rho' \cos(\varphi' \mp \varphi)]^{1/2}\}$.

The various integral representations of the Green function found in this section will prove useful in the discussion of diffraction by wedges in Chapter 10 and radiation from apertures on wedges in Chapter 11.

8.4.8 Cylindrically Tipped Angular Sector

The geometrical configuration of this problem is shown in Fig. 8.14. The Green function must satisfy (8.140), and in the electric case it must vanish at $\varphi = 0$, φ_0 and $\rho = a$. Consequently, we can seek its expansion in the form of (8.141) except that in the present case, since the origin is excluded, we must choose

$$g_< = AJ_\nu(k\rho) + BH_\nu^{(1)}(k\rho), \tag{8.193a}$$

$$g_> = CH_\nu^{(1)}(k\rho). \tag{8.193b}$$

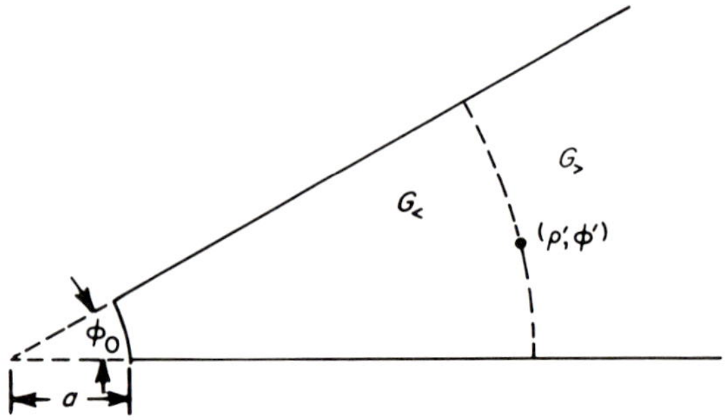

Fig. 8.14. Cylindrically tipped angular sector.

The coefficients A, B, and C are readily evaluated by applying the conditions (8.144) and the boundary conditions at $\rho = a$. Hence,

$$G_<^{(e)}(\boldsymbol{\rho}, \boldsymbol{\rho}') = \frac{i\pi}{\varphi_0} \sum_{m=1}^{\infty} \left[J_\nu(k\rho) - \frac{J_\nu(ka)}{H_\nu^{(1)}(ka)} H_\nu^{(1)}(k\rho) \right] H_\nu^{(1)}(k\rho') \sin \nu\varphi \sin \nu\varphi' \tag{8.194}$$

in the electric case.

For the magnetic case, $\partial G^{(m)}/\partial\varphi = 0$ at $\varphi = 0$, φ_0 and $\partial G^{(m)}/\partial\rho = 0$ at $\rho = a$ and we obtain

$$G_<^{(m)}(\boldsymbol{\rho}, \boldsymbol{\rho}') = \frac{i\pi}{2\varphi_0} \sum_{m=1}^{\infty} \varepsilon_m \left[J_\nu(k\rho) - \frac{J_\nu'(ka)}{H_\nu^{(1)\prime}(ka)} H_\nu^{(1)}(k\rho) \right] H_\nu^{(1)}(k\rho') \cos \nu\varphi \cos \nu\varphi'. \tag{8.195}$$

It is interesting to observe that (8.194) goes into (8.151), and (8.195) into (8.152), as $a \to 0$ as they should.

8.4.9 Internal Circular Cylindrical Region

In the case of a circular cylindrical region, as shown in Fig. 8.11, noting the delta function expansion of (8.94), the Green function expansion in the form

$$G(\boldsymbol{\rho}, \boldsymbol{\rho}') = (2\pi)^{-1} \sum_{-\infty}^{\infty} g(\rho, \rho') \exp[im(\varphi - \varphi')] \tag{8.196}$$

immediately suggests itself. Dividing the region $0 \le \rho \le a$ into $0 \le \rho < \rho'$ and $\rho' < \rho \le a$, we find

$$G_<^{(e)}(\boldsymbol{\rho}, \boldsymbol{\rho}') = \frac{i}{4} \sum_{-\infty}^{\infty} \exp[im(\varphi - \varphi')] \left[H_m^{(1)}(k\rho') - \frac{H_m^{(1)}(ka)}{J_m(ka)} J_m(k\rho') \right] J_m(k\rho) \tag{8.197}$$

in the electric case and

$$G_<^{(m)}(\boldsymbol{\rho}, \boldsymbol{\rho}') = \frac{i}{4} \sum_{-\infty}^{\infty} \exp[im(\varphi - \varphi')] \left[H_m^{(1)}(k\rho') - \frac{H_m^{(1)\prime}(ka)}{J_m'(ka)} J_m(k\rho') \right] J_m(k\rho) \tag{8.198}$$

in the magnetic case.

It is interesting to note that as $ka \to \infty$ (infinite region), the second term in the square brackets of (8.197) and (8.198) vanishes and we obtain in either case

$$G(\boldsymbol{\rho}, \boldsymbol{\rho}') = \frac{i}{4} \sum_{-\infty}^{\infty} \exp[im(\varphi - \varphi')] H_m^{(1)}(k\rho') J_m(k\rho). \tag{8.199}$$

The comparison of (8.199) with (5.41) demonstrates the special case of the addition theorem for the Bessel functions

$$\begin{aligned} H_0^{(1)}(k|\boldsymbol{\rho} - \boldsymbol{\rho}'|) &= \sum_{-\infty}^{\infty} \exp[im(\varphi - \varphi')] H_m^{(1)}(k\rho') J_m(k\rho), \quad \rho' < \rho, \\ &= \sum_{-\infty}^{\infty} \exp[im(\varphi - \varphi')] J_m(k\rho') H_m^{(1)}(k\rho), \quad \rho > \rho'. \end{aligned} \tag{8.200}$$

8.4.10 External Circular Cylindrical Region

The geometry of this problem is shown in Fig. 8.15. Again, we can seek an expansion of the Green function in the form of (8.196). In this case,

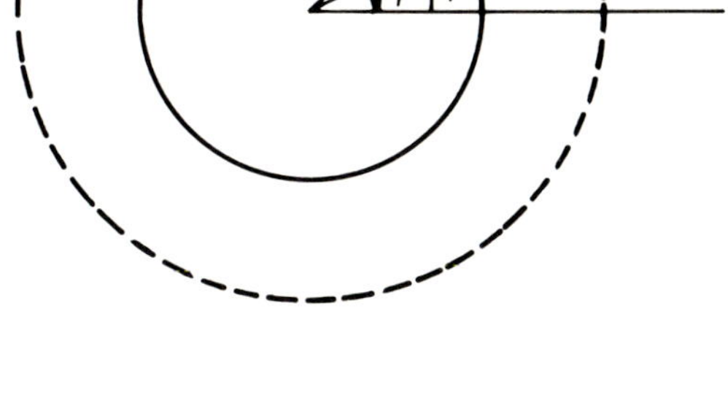

Fig. 8.15. External circular cylindrical region.

however, the point $\rho = 0$ is excluded, and we must choose

$$g_< = AJ_m(k\rho) + BH_m^{(1)}(k\rho), \tag{8.201a}$$

$$g_> = CH_m^{(1)}(k\rho), \tag{8.201b}$$

where the coefficients A, B, and C are evaluated in the usual manner. Thus, we obtain

$$G_<^{(e)}(\boldsymbol{\rho}, \boldsymbol{\rho}') = \frac{i}{4} \sum_{-\infty}^{\infty} \exp[im(\varphi - \varphi')] \left[J_m(k\rho) - \frac{J_m(ka)}{H_m^{(1)}(ka)} H_m^{(1)}(k\rho) \right] H_m^{(1)}(k\rho') \tag{8.202}$$

for the electric case and

$$G_<^{(m)}(\boldsymbol{\rho}, \boldsymbol{\rho}') = \frac{i}{4} \sum_{-\infty}^{\infty} \exp[im(\varphi - \varphi')] \left[J_m(k\rho) - \frac{J_m'(ka)}{H_m^{(1)\prime}(ka)} H_m^{(1)}(k\rho) \right] H_m^{(1)}(k\rho') \tag{8.203}$$

for the magnetic case.

Problems

8.1 Derive the orthogonality condition for complex eigenfunctions of the Sturm–Liouville problem. Also, show that the eigenvalues must be real.

8.2 If V is a closed region of space bounded by a regular surface S and ψ and φ are two scalar functions of position that, together with their first and second derivatives, are continuous throughout V and on the surface S, show that Green's theorem

$$\int_V (\psi \nabla^2 \varphi - \varphi \nabla^2 \psi)\, dv = \int_S \left(\psi \frac{\partial \varphi}{\partial n} - \varphi \frac{\partial \psi}{\partial n} \right) ds,$$

where $\partial/\partial n$ is the derivative in the direction of the outward normal, holds.

8.3 (a) Show that the solution to the boundary value problem

$$\nabla^2 \psi(\mathbf{r}) = -f(\mathbf{r}) \quad \text{in } V,$$
$$\psi(\mathbf{r}) = g(\mathbf{r}) \quad \text{on } S,$$

can be written

$$\psi(\mathbf{r}) = \int_{V'} f(\mathbf{r}') G(\mathbf{r}, \mathbf{r}')\, d\mathbf{r}' - \int_{S'} g(\mathbf{r}') \frac{\partial}{\partial n'} G(\mathbf{r}, \mathbf{r}')\, d\boldsymbol{\rho}'.$$

What requirements must be placed on the Green function $G(\mathbf{r}, \mathbf{r}')$?
(b) If the problem is modified so that

$$\nabla^2\psi(\mathbf{r}) = -f(\mathbf{r}) \quad \text{in } V,$$

$$\frac{\partial}{\partial n}\psi(\mathbf{r}) = \sigma(\mathbf{r}) \qquad \text{on } S,$$

show that

$$\psi(\mathbf{r}) = \int_{V'} f(\mathbf{r}')G(\mathbf{r}, \mathbf{r}')\, d\mathbf{r}' + \int_{S'} \sigma(\mathbf{r}')G(\mathbf{r}, \mathbf{r}')\, d\boldsymbol{\rho}'.$$

What are the requirements on the Green function in this case? *Hint*: Define a twice-differentiable function $\varphi(\mathbf{r})$ so that $\varphi(\mathbf{r}) = g(\mathbf{r})$ on S in the first case and $\partial\varphi(\mathbf{r})/\partial n = \sigma(\mathbf{r})$ on S in the second case, thus reducing the problem to a self-adjoint one.

8.4 Starting with (8.65b) and (8.66b), find the image representation of $G^{(e)}$ and $G^{(m)}$ for the quarter-space region shown in Fig. 8.4.

8.5 Show that the Green function representation for a free space cylindrical region

$$G(\boldsymbol{\rho}, \boldsymbol{\rho}') = \frac{1}{2\pi} \sum_{-\infty}^{\infty} \exp[im(\varphi - \varphi')] \int_0^{\infty} \frac{\alpha J_m(\alpha\rho)J_m(\alpha\rho')\, d\alpha}{\alpha^2 - k^2}$$

can be obtained directly from (8.72b).

8.6 Prove (8.115) by first showing that

$$\sum_{-\infty}^{\infty} \frac{\exp(im\theta)}{m^2 - a^2} = -\frac{\pi}{a} \frac{\cos a(\theta - \pi)}{\sin \pi a}, \qquad 0 \le \theta \le 2\pi.$$

Hint: Consider the integral

$$I = \int_C \frac{\exp(i\theta z)\, dz}{(z^2 - a^2)[\exp(i2\pi z) - 1]}$$

where C is a closed contour enclosing all of the poles of the integrand.

8.7 Show that the condition (8.125) is indeed satisfied by the Green function $G^{(e)}$ given in (8.126). What is the condition equivalent to (8.125) that must be satisfied by $G^{(m)}$ for the same configuration?

8.8 Prove (8.168).

8.9 Prove (8.177).

8.10 Starting with (8.190), prove (8.192).

8.11 Show that $G^{(m)}(\boldsymbol{\rho}, \boldsymbol{\rho}')$ in (8.169) satisfies the differential equation in (8.156).

References

Abramowitz, M., and Stegun, I. A., eds. (1964). "Handbook of Mathematical Functions." U.S. Gov't. Printing Office, Washington, D.C. (*Natl. Bur. Std. U.S. Appl. Math. Ser.* **55**.)

Arfken, G. (1966). "Mathematical Methods for Physicists." Academic Press, New York.

Courant, R., and Hilbert, D. (1953). "Methods of Mathematical Physics," Vol. 1. Wiley (Interscience), New York.

Dettman, J. W. (1962). "Mathematical Methods in Physics and Engineering." McGraw-Hill, New York.

Duff, G. F. D., and Naylor, D. (1966). "Differential Equations of Applied Mathematics." Wiley, New York.

Erdelyi, A., Magnus, W., Oberhettinger, F., and Tricomi, F. G. (1954). "Tables of Integral Transforms," Vol. II. McGraw-Hill, New York.

Felsen, L. B., and Marcuvitz, N. (1959). "Modal Analysis and Synthesis of Electromagnetic Fields." Polytech. Inst. Brooklyn Rept. No. R-726-59, PIB-654 (June).

Friedman, B. (1956). "Principles and Techniques of Applied Mathematics." Wiley, New York.

Gradshteyn, I. S., and Ryzhik, I. M. (1966). "Table of Integrals, Sums, Series, and Products," 4th ed. Academic Press, New York.

Mangulis, V. (1965). "Handbook of Series for Scientists and Engineers." Academic Press, New York.

Morse, M. M., and Feshbach, H. (1953). "Methods of Theoretical Physics." Vol. 1. McGraw-Hill, New York.

Sneddon, I. N. (1957). "Elements of Partial Differential Equations." McGraw-Hill, New York.

Sommerfeld, A. J. (1949). "Partial Differential Equations in Physics." Academic Press, New York.

Watson, G. N. (1962). "Theory of Bessel Functions." Cambridge Univ. Press, London and New York.

CHAPTER 9

AXIAL CURRENTS AND CYLINDRICAL BOUNDARIES

The Green function formalism developed in Chapter 8 will be used in this chapter to formulate and solve a class of boundary value problems dealing with axial current distributions in the presence of cylindrical conducting structures. The cylindrical structures to be considered will have one common property. They will be uniform and infinite in the axial direction.

9.1 General Considerations

Consider a cylindrical structure of an arbitrary shape that can be either open or closed in the presence of a bounded electric or magnetic axial current distribution as shown in Fig. 9.1. Since the current density vector has only an

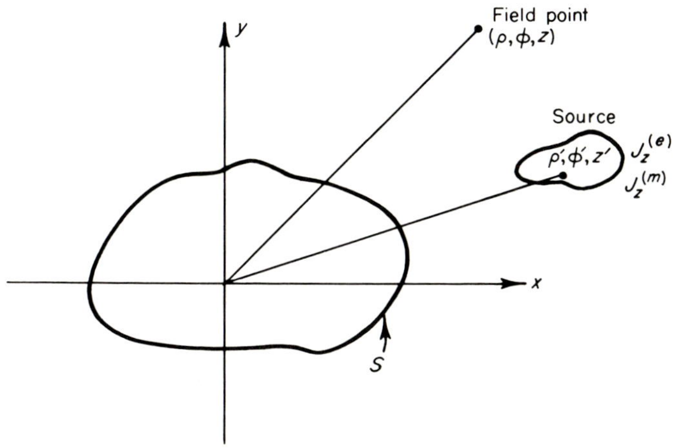

Fig. 9.1. A cylindrical structure in the presence of an axial electric or magnetic current distribution.

axial component, the electromagnetic fields can be derived from a single axial component of the Hertz vector. Thus, in the electric case we can write

$$\left(\frac{1}{\rho}\frac{\partial}{\partial\rho}\rho\frac{\partial}{\partial\rho}+\frac{1}{\rho^2}\frac{\partial^2}{\partial\varphi^2}+\frac{\partial^2}{\partial z^2}+k^2\right)\Pi_z^{(e)}=-\frac{i\mu}{\omega\varepsilon_0 n^2}J_z^{(e)}, \tag{9.1a}$$

where the magnetic and electric field components are given by

$$H_\rho=-\frac{i\omega\varepsilon_0 n^2}{\mu}\frac{1}{\rho}\frac{\partial\Pi_z^{(e)}}{\partial\varphi}, \tag{9.1b}$$

$$H_\varphi=\frac{i\omega\varepsilon_0 n^2}{\mu}\frac{\partial\Pi_z^{(e)}}{\partial\rho}, \tag{9.1c}$$

$$H_z=0, \tag{9.1d}$$

$$E_\rho=\frac{\partial^2\Pi_z^{(e)}}{\partial\rho\,\partial z}, \tag{9.1e}$$

$$E_\varphi=\frac{1}{\rho}\frac{\partial^2\Pi_z^{(e)}}{\partial\varphi\,\partial z}, \tag{9.1f}$$

$$E_z=\left(k^2+\frac{\partial^2}{\partial z^2}\right)\Pi_z^{(e)}. \tag{9.1g}$$

In the magnetic case we must have

$$\left(\frac{1}{\rho}\frac{\partial}{\partial\rho}\rho\frac{\partial}{\partial\rho}+\frac{1}{\rho^2}\frac{\partial^2}{\partial\varphi^2}+\frac{\partial^2}{\partial z^2}+k^2\right)\Pi_z^{(m)}=-\frac{i}{\omega\mu_0\mu}J_z^{(m)}, \tag{9.2a}$$

where the electric and magnetic field components are given by

$$E_\rho=i\omega\mu_0\mu\frac{1}{\rho}\frac{\partial\Pi_z^{(m)}}{\partial\varphi}, \tag{9.2b}$$

$$E_\varphi=-i\omega\mu_0\mu\frac{\partial\Pi_z^{(m)}}{\partial\rho}, \tag{9.2c}$$

$$E_z=0, \tag{9.2d}$$

$$H_\rho=\frac{\partial^2\Pi_z^{(m)}}{\partial\rho\,\partial z}, \tag{9.2e}$$

$$H_\varphi=\frac{1}{\rho}\frac{\partial^2\Pi_z^{(m)}}{\partial\varphi\,\partial z}, \tag{9.2f}$$

$$H_z=\left(k^2+\frac{\partial^2}{\partial z^2}\right)\Pi_z^{(m)}. \tag{9.2g}$$

If the structure is perfectly conducting, or if the impedance boundary conditions apply on the surface, then the problem is self-adjoint and the solution can be written in terms of the corresponding Green function as

$$\Pi_z^{(e)}(\mathbf{r}) = \frac{i\mu}{\omega\varepsilon_0 n^2} \iiint J_z^{(e)}(\mathbf{r}')G^{(e)}(\mathbf{r}, \mathbf{r}')\, d\mathbf{r}' \tag{9.3}$$

in the electric case, and

$$\Pi_z^{(m)}(\mathbf{r}) = \frac{i}{\omega\mu_0 \mu} \iiint J_z^{(m)}(\mathbf{r}')G^{(m)}(\mathbf{r}, \mathbf{r}')\, d\mathbf{r}' \tag{9.4}$$

in the magnetic case. The Green functions must satisfy the differential equation

$$\left(\frac{1}{\rho}\frac{\partial}{\partial\rho}\rho\frac{\partial}{\partial\rho} + \frac{1}{\rho^2}\frac{\partial^2}{\partial\varphi^2} + \frac{\partial^2}{\partial z^2} + k^2\right)G^{(e,\,m)} = -\frac{\delta(\rho-\rho')\,\delta(\varphi-\varphi')\,\delta(z-z')}{\rho} \tag{9.5}$$

and in the case of perfectly conducting boundaries $G^{(e)}(S, \mathbf{r}') = 0$ and $\partial G^{(m)}(S, \mathbf{r}')/\partial n = 0$.

Since the structure is uniform and of infinite extent in the axial direction, it will be appropriate to use the Fourier transform on the z coordinate as follows.

$$\tilde{G}(\rho, \varphi, \alpha; \mathbf{r}') = (2\pi)^{-1/2} \int_{-\infty}^{\infty} G(\mathbf{r}, \mathbf{r}')e^{-i\alpha z}\, dz, \tag{9.6a}$$

$$G(\mathbf{r}, \mathbf{r}') = (2\pi)^{-1/2} \int_{-\infty}^{\infty} \tilde{G}(\rho, \varphi, \alpha; \mathbf{r}')e^{i\alpha z}\, d\alpha \tag{9.6b}$$

in the differential equation of (9.5). We obtain in this case

$$\left(\frac{1}{\rho}\frac{\partial}{\partial\rho}\rho\frac{\partial}{\partial\rho} + \frac{1}{\rho^2}\frac{\partial^2}{\partial\varphi^2} + \tau^2\right)g = -\frac{\delta(\rho-\rho')\,\delta(\varphi-\varphi')}{\rho}, \tag{9.7}$$

where

$$\tilde{G} = \frac{e^{-i\alpha z'}}{(2\pi)^{1/2}}\, g, \tag{9.8a}$$

$$\tau = (k^2 - \alpha^2)^{1/2}, \qquad \mathrm{Im}(\tau) > 0, \tag{9.8b}$$

providing that the radiation condition

$$\left(\frac{\partial G}{\partial z} + i\alpha G\right)e^{-i\alpha z}\Bigg|_{z=-\infty}^{z=\infty} = 0 \tag{9.9}$$

is satisfied. It is easy to show that the boundary conditions to be satisfied by g are the same as those that must be satisfied by G, i.e., $g^{(e)}(S) = 0$ and $\partial g^{(m)}(S)/\partial n = 0$ in the case of perfectly conducting boundaries.

Finally, we can rewrite the solutions (9.3) and (9.4) in terms of $g^{(e)}$ and $g^{(m)}$ as follows.

$$\Pi_z^{(e)} = \frac{i\mu}{2\pi\omega\varepsilon_0 n^2} \int_{-\infty}^{\infty} d\alpha\, e^{i\alpha z} \iiint J_z^{(e)}(\mathbf{r}') g^{(e)}(\boldsymbol{\rho}, \boldsymbol{\rho}', \alpha) e^{-i\alpha z'}\, d\mathbf{r}' \tag{9.10}$$

and

$$\Pi_z^{(m)} = \frac{i}{2\pi\omega\mu_0 \mu} \int_{-\infty}^{\infty} d\alpha\, e^{i\alpha z} \iiint J_z^{(m)}(\mathbf{r}') g^{(m)}(\boldsymbol{\rho}, \boldsymbol{\rho}', \alpha) e^{-i\alpha z'}\, d\mathbf{r}'. \tag{9.11}$$

In the problems that follow, the function $g(\boldsymbol{\rho}, \boldsymbol{\rho}', \alpha)$ will be of the form $F(\phi, \alpha, \boldsymbol{\rho}')H_\nu^{(1)}(\tau\rho)$ and hence we will be faced with the approximate evaluation of the integral

$$I = \int_{-\infty}^{\infty} F(\alpha) H_\nu^{(1)}(\tau\rho) \exp[i\alpha(z - z')]\, d\alpha \tag{9.12}$$

when $k\rho \gg 1$. We shall now discuss the procedure for its evaluation in detail.

As the first step toward this goal we extract the asymptotic behavior of the Hankel function for a large value of the argument by writing

$$H_\nu^{(1)}(x) = \breve{H}_\nu^{(1)}(x) \exp[i(x - \nu\pi/2 - \pi/4)]$$

and noting that $\breve{H}_\nu^{(1)}(x) \to (2/\pi x)^{1/2}$ as $|x| \to \infty$. Hence, we obtain

$$I = -\exp[-i(\nu\pi/2 + \pi/4)] \int_{+\infty}^{-\infty} F(\alpha) \breve{H}_\nu^{(1)}(\tau\rho) \exp\{i[\tau\rho + \alpha(z - z')]\}\, d\alpha. \tag{9.13}$$

Let us now make the transformations

$$\begin{aligned} \alpha &= k \cos\beta, \qquad & \rho &= r \sin\theta, \\ \tau &= k \sin\beta, \qquad & z &= r\cos\theta, \end{aligned} \tag{9.14}$$

which recasts the integral in (9.13) into

$$\begin{aligned} I = k \exp[-i(\nu\pi/2 + \pi/4)] &\int_{\Gamma_2} F(\beta) \breve{H}_\nu^{(1)}(k\rho \sin\beta) \\ &\times \exp(-ik\cos\beta z') \sin\beta \exp[ikr\cos(\beta - \theta)]\, d\beta. \end{aligned} \tag{9.15}$$

We shall now discuss the consequences of the transformation $\alpha = k\cos\beta$ and the resulting new path of integration Γ_2.

Let us assume for the time being that k is real. Then it will follow that

$$\begin{aligned} \alpha' &= k \cos\beta' \cosh\beta'', \\ \alpha'' &= -k \sin\beta' \sinh\beta'', \\ \tau' &= k \sin\beta' \cosh\beta'', \\ \tau'' &= k \cos\beta' \sinh\beta'', \end{aligned} \tag{9.16}$$

and consequently, when $\alpha' > 0$, $-\pi/2 < \beta' < \pi/2$ and $3\pi/2 < \beta' < 5\pi/2$, and when $\alpha' < 0$, $\pi/2 < \beta' < 3\pi/2$ and $-3\pi/2 < \beta' < -\pi/2$. Moreover, when $\alpha'' > 0$ and $\beta'' > 0$, then $-\pi < \beta' < 0$ or $\pi < \beta' < 2\pi$; when $\alpha'' > 0$ and $\beta'' < 0$, then $0 < \beta' < \pi$ or $-2\pi < \beta' < -\pi$; when $\alpha'' < 0$ and $\beta'' < 0$, then $-\pi < \beta' < 0$ or $\pi < \beta' < 2\pi$; and finally, when $\alpha'' < 0$ and $\beta'' > 0$, then $0 < \beta' < \pi$ and $-2\pi < \beta' < -\pi$. Thus, it follows that the quadrants of the complex α plane are transformed into strips of $\pi/2$ width in the β plane, as shown in Fig. 9.2. Moreover, the upper sheet $\tau'' = k \cos\beta' \sinh\beta'' > 0$ will be transformed into $-\pi/2 < \beta' < \pi/2$ when $\beta'' > 0$ and $\pi/2 < \beta' < 3\pi/2$ when $\beta'' < 0$. The new path of integration Γ_2 will follow as shown in Fig. 9.2.

If k is complex, $k = k' + ik''$, then the transformation becomes

$$\begin{aligned} \alpha' &= k' \cos\beta' \cosh\beta'' + k'' \sin\beta' \sinh\beta'', \\ \alpha'' &= k'' \cos\beta' \cosh\beta'' - k' \sin\beta' \sinh\beta'', \\ \tau' &= k' \sin\beta' \cosh\beta'' - k'' \cos\beta' \sinh\beta'', \\ \tau'' &= k'' \sin\beta' \cosh\beta'' + k'' \cos\beta' \sinh\beta''; \end{aligned} \tag{9.17}$$

hence, it will follow that

$$\begin{aligned} &\cot\beta' \coth\beta'' \gtrless -\frac{k''}{k'}, && \alpha' \gtrless 0, \\ &\tan\beta' \tanh\beta'' \gtrless \frac{k''}{k'}, && \alpha'' \gtrless 0, \\ &\tan\beta' \coth\beta'' \gtrless \frac{k''}{k'}, && \tau' \gtrless 0, \\ &\cot\beta' \sinh\beta'' \gtrless -\frac{k''}{k'}, && \tau'' \gtrless 0. \end{aligned} \tag{9.18}$$

Moreover, it will be noted that the axis of the reals $\alpha'' = 0$ will be transformed into the β plane according to

$$\tan\beta' \tanh\beta'' = \frac{k''}{k'} = \tan\beta_0 \tag{9.19}$$

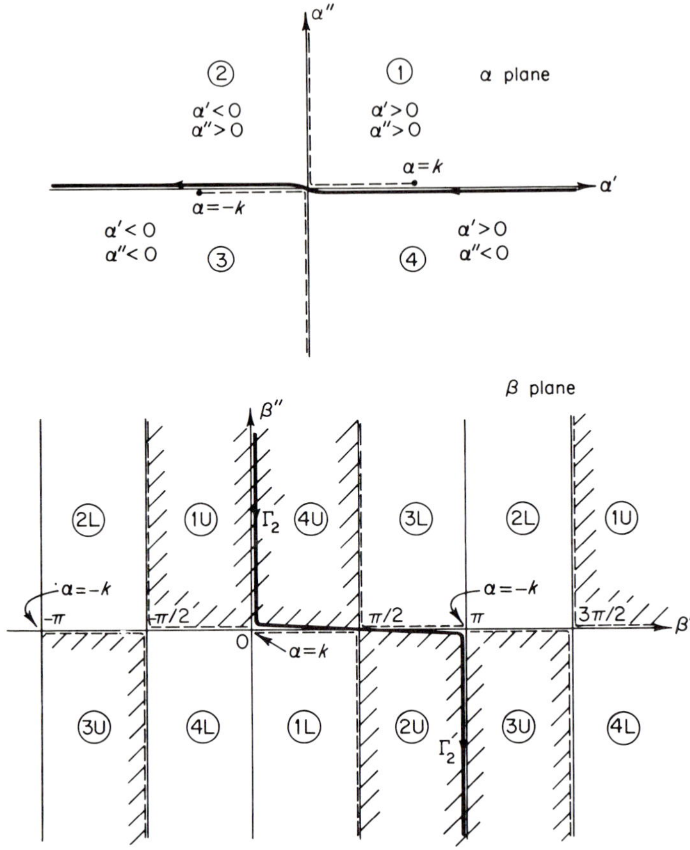

Fig. 9.2. The correspondence of the α and β planes due to a transformation $\alpha = k \cos \beta$ when k is real. The shaded area depicts the upper sheet $\tau'' > 0$.

so that on this line $\tan \beta' \to \tan \beta_0$ as $\beta'' \to \infty$ and hence, $\beta' \to \beta_0$, $\pi + \beta_0$, $-\pi + \beta_0$, and $\tan \beta' \to -\tan \beta_0$ as $\beta'' \to -\infty$ and hence, $\beta' \to -\beta_0$, $\pi - \beta_0$, $-\pi - \beta_0$. In the neighborhood of the real axis $\beta'' \to 0$, $\tan \beta' \to \tan \beta_0/\beta''$ as $\beta' = \pm\pi/2$, $\pm 3\pi/2$ and hence $\beta'' \sim -\tan \beta_0 \cdot (\beta' \pm \pi/2)$.

The axis of the imaginaries $\alpha' = 0$ will be transformed into the β plane according to

$$\cot \beta' \coth \beta'' = -\frac{k''}{k'} = -\tan \beta_0 \tag{9.20}$$

so that on this line $\cot \beta' \to -\tan \beta_0$ as $\beta'' \to \infty$ and hence, $\beta' = \pm\pi/2 + \beta_0$, $\pm 3\pi/2 + \beta_0$ and $\cot \beta' \to \tan \beta_0$ as $\beta'' \to -\infty$ and hence, $\alpha' = \pm\pi/2 - \beta_0$, $\pm 3\pi/2 - \beta_0$. In the neighborhood of the real axis $\beta'' \to 0$, $\cot \beta' \to -\beta'' \tan \beta_0$ as $\beta' = \pm\pi/2$, $\pm 3\pi/2$ and hence, $\beta'' \sim (\beta' \pm \pi/2)/\tan \beta_0$.

The branch cut $\tau'' = 0$ will be transformed as

$$\cot \beta' \tanh \beta'' = -\frac{k''}{k'} = -\tan \beta_0 \tag{9.21}$$

so that on this line $\cot \beta' \to -\tan \beta_0$ as $\beta'' \to \infty$ and hence $\beta' = \pm\pi/2 + \beta_0$, $\pm 3\pi/2 + \beta_0$ and $\cot \beta' \to \tan \beta_0$ as $\beta'' \to -\infty$ and hence $\beta' = \pm\pi/2 - \beta_0$, $\pm 3\pi/2 - \beta_0$. In the neighborhood of the real axis $\beta'' \to 0$, $\cot \beta' \to -\tan \beta_0/\beta''$ as $\beta' = 0, \pm\pi, \pm 2\pi$ and hence $\beta'' \sim -\beta' \cdot \tan \beta_0$.

Collecting the information above, we can construct the complex β plane as shown in Fig. 9.3.

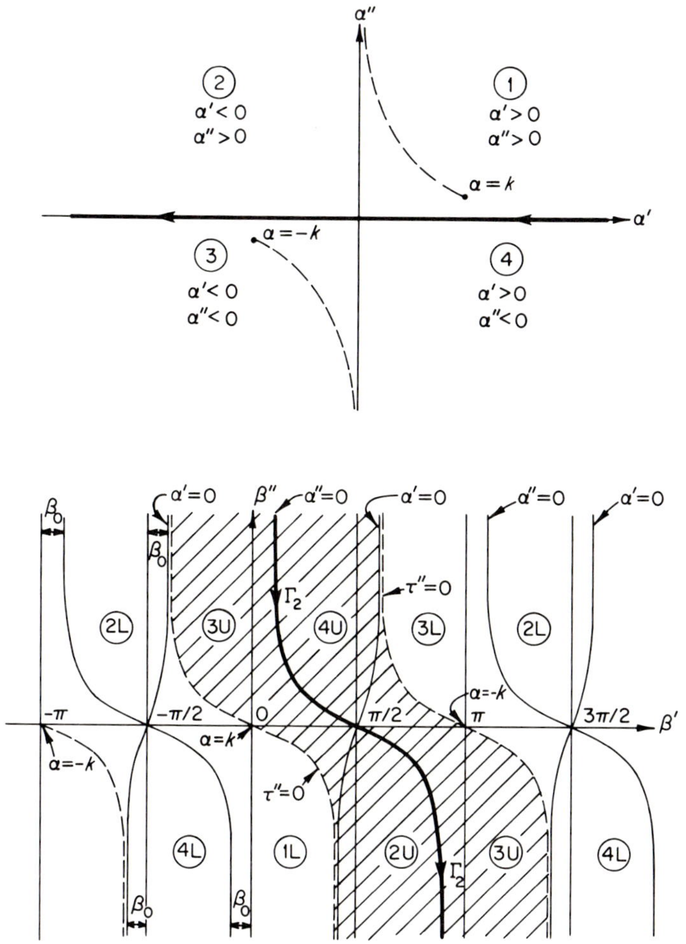

Fig. 9.3. The correspondence of the α and β planes under the transformation $\alpha = k \cos \beta$ when k is complex. The shaded area depicts the upper sheet $\tau'' > 0$ ($\beta_0 = \arctan(k''/k')$).

The saddle point of the integrand in (9.15) is at $\beta = \theta$ and the path of steepest descents runs as shown in Fig. 9.4. The deformation of the original path of integration Γ_2 into the path of steepest descents Γ can be accomplished taking proper account of the singularities of $F(\beta)$ that may be found between the two paths. The resulting saddle-point contribution is readily evaluated using (5.94). Hence

$$I_{SD} \sim -\frac{i2F(\theta)}{r} \exp\left[i\left(kr - \frac{\nu\pi}{2} - k\cos\theta z'\right)\right]. \tag{9.22}$$

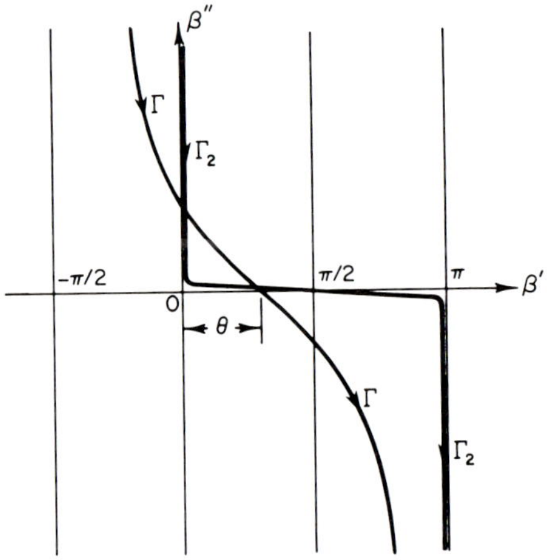

Fig. 9.4. The original path of integration Γ_2 and the path of steepest descents Γ.

9.2 Perfectly Conducting Wedge and an Axial Current Distribution

9.2.1 Electric Currents

The geometry of the problem is shown in Fig. 9.5. The function $g^{(e)}$ satisfying the differential equation (9.7a) and the boundary conditions $g^{(e)}(\rho, 0) = g^{(e)}(\rho, \phi_0) = 0$ was found in (8.151) and in (8.177). Using the representation of (8.151), we obtain, upon substitution into (9.10),

$$\underset{\rho>\rho'}{\Pi_z^{(e)}} = -\frac{\mu}{2\varphi_0\,\omega\varepsilon_0\,n^2} \sum_{m=1}^{\infty} \sin\nu\varphi \iiint d\mathbf{r}' J_z^{(e)}(\mathbf{r}') \sin\nu\varphi' I_>(\rho, \rho'; z, z'), \tag{9.23}$$

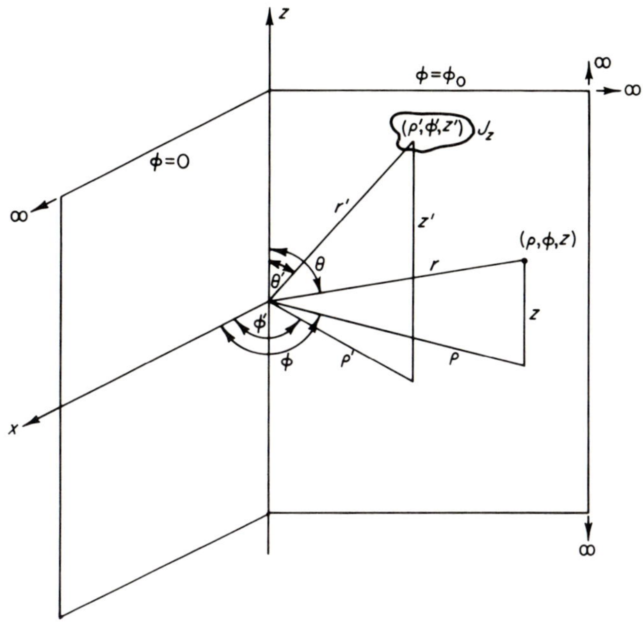

Fig. 9.5. Wedge and an axial current distribution.

where

$$I_{>} = \int_{-\infty}^{\infty} J_\nu(\tau\rho')H_\nu^{(1)}(\tau\rho) \exp[i\alpha(z - z')]\, d\alpha \tag{9.24}$$

and $\nu = m\pi/\varphi_0$ as before. When the distance from the apex of the wedge is large compared to the wavelength, i.e., when $k\rho \gg 1$, the integral $I_{>}$ can be readily evaluated using (9.22). The complete solution becomes, for $k\rho \gg 1$,

$$\underset{\rho>\rho'}{\Pi_z^{(e)}} = \frac{i\mu}{\varphi_0\,\omega\varepsilon_0\, n^2}\,\frac{\exp(ikr)}{r} \sum_{m=1}^{\infty} (-i)^\nu \sin\nu\varphi W^{(e)}(\theta, \nu), \tag{9.25a}$$

where

$$W^{(e)}(\theta, \nu) = \iiint J_z^{(e)}(\mathbf{r}') \sin\nu\varphi' J_\nu(k\rho'\sin\theta) \exp(-ik\cos\theta z')\, d\mathbf{r}'. \tag{9.25b}$$

When the wedge angle is an integral fraction of π, i.e., $\varphi_0 = \pi/l$ where l is an integer, a simple expression for the fields is possible in terms of images. Using (8.188) together with (8.192), where the integral vanishes for $v = l/2$, the Green function $g^{(e)}(\boldsymbol{\rho}, \boldsymbol{\rho}', \alpha)$ can be written

$$g^{(e)}(\boldsymbol{\rho}, \boldsymbol{\rho}', \alpha) = \frac{i}{4}\left(\sum_{m=1}^{M_1} H_0^{(1)}\{\tau[\rho^2 + \rho'^2 - 2\rho\rho' \cos(\varphi - \varphi' + 2m\pi/l)]^{1/2}\}\right.$$
$$- \sum_{m=1}^{M_2} H_0^{(1)}\{\tau[\rho^2 + \rho'^2 - 2\rho\rho' \cos(\varphi + \varphi' - 2m\pi/l)]^{1/2}\}$$
$$+ \sum_{n=0}^{N_1} H_0^{(1)}\{\tau[\rho^2 + \rho'^2 - 2\rho\rho' \cos(\varphi - \varphi' - 2n\pi/l)]^{1/2}\}$$
$$\left. - \sum_{n=0}^{N_2} H_0^{(1)}\{\tau[\rho^2 + \rho'^2 - 2\rho\rho' \cos(\varphi + \varphi' + 2n\pi/l)]^{1/2}\}\right), \tag{9.26a}$$

where $M_{1,2}$ and $N_{1,2}$ are integers such that

$$\begin{aligned} 0 &< 2M_1\pi/l + \varphi - \varphi' < \pi, \\ 0 &< 2M_2\pi/l - \varphi - \varphi' < \pi, \\ 0 &< 2N_1\pi/l - \varphi + \varphi' < \pi, \\ 0 &< 2N_2\pi/l + \varphi + \varphi' < \pi. \end{aligned} \tag{9.26b}$$

Substitution of (9.26a) into (9.10) gives

$$\Pi_z^{(e)} = \frac{i\mu}{4\pi\omega\varepsilon_0 n^2} \iiint d\mathbf{r}' J_z^{(e)}(\mathbf{r}')\left(\sum_{m=1}^{M_1} I_m^{(1)} - \sum_{m=1}^{M_2} I_m^{(2)} + \sum_{n=0}^{N_1} I_n^{(1)} - \sum_{n=0}^{N_2} I_n^{(2)}\right), \tag{9.27}$$

where, for instance

$$I_m^{(1)} = \frac{i}{2}\int_{-\infty}^{\infty} H_0^{(1)}\{\tau[\rho^2 + \rho'^2 - 2\rho\rho' \cos(\varphi - \varphi' + 2m\pi/l)]^{1/2}\} \times \exp[i\alpha(z - z')]\, d\alpha. \tag{9.28}$$

The integrals in (9.27) can be evaluated exactly if we note (see, e.g. Gradshteyn and Ryzhik (1966), p. 710)

$$\int_{-\infty}^{\infty} H_0^{(1)}[\rho(k^2 - \alpha^2)^{1/2}] \exp(i\alpha z)\, d\alpha = -2i\,\frac{\exp[ik(\rho^2 + z^2)^{1/2}]}{(\rho^2 + z^2)^{1/2}}. \tag{9.29}$$

Hence, we obtain for $I_m^{(1)}$

$$I_m^{(1)} = \frac{\exp\{ik[\rho^2 + \rho'^2 - 2\rho\rho' \cos(\varphi - \varphi' + 2m\pi/l) + (z - z')^2]^{1/2}\}}{[\rho^2 + \rho'^2 - 2\rho\rho' \cos(\varphi - \varphi' + 2m\pi/l) + (z - z')^2]^{1/2}}, \tag{9.30}$$

which upon substitution into (9.27) gives a more convenient representation when $\varphi_0 = \pi/l$.

If the distance from the apex to the field point is large and $k\rho'$ is moderate, so that $k\rho' \ll k\rho \to \infty$, then the radical in the exponent of (9.30) can be approximated by

$$r - \sin\theta\cos(\varphi - \varphi' + 2m\pi/l) - \cos\theta z',$$

whereas in the denominator we simply replace the radical with r. Hence, we obtain in this case

$$\begin{aligned}\Pi_z^{(e)} = \frac{i\mu\exp(ikr)}{4\pi\omega\varepsilon_0 n^2 r} & \iiint d\mathbf{r}' J_z^{(e)}(\mathbf{r}')\exp(-ikz'\cos\theta) \\ & \times \sum_{m=1}^{M_1} \exp[-ik\rho'\sin\theta\cos(\varphi - \varphi' + 2m\pi/l)] \\ & - \sum_{m=1}^{M_2} \exp[-ik\rho'\sin\theta\cos(\varphi + \varphi' - 2m\pi/l)] \\ & + \sum_{n=0}^{N_1} \exp[-ik\rho'\sin\theta\cos(\varphi - \varphi' - 2n\pi/l)] \\ & - \sum_{n=0}^{N_2} \exp[-ik\rho'\sin\theta\cos(\varphi + \varphi' + 2n\pi/l)]\Big\}. \end{aligned} \tag{9.31}$$

9.2.1.1 Axial Electric Dipole

If the electric current distribution is a single dipole oriented along the z axis at a point $\rho = \rho''$, $\varphi = \varphi''$, $z = z''$, then

$$J_z^{(e)} = p^{(e)} \frac{\delta(\rho' - \rho'')\,\delta(\varphi' - \varphi'')\,\delta(z' - z'')}{\rho'} \tag{9.32}$$

and the integral in (9.25b) can be readily evaluated to give

$$W^{(e)}(\theta, \nu) = p^{(e)} \sin\nu\varphi'' J_\nu(k\rho''\sin\theta)\exp(ik\cos\theta z'') \tag{9.33}$$

and consequently

$$\Pi_{z \atop \rho>\rho''}^{(e)} = \frac{i\mu p^{(e)}\exp[ik(r - z''\cos\theta)]}{\varphi_0\,\omega\varepsilon_0 n^2 r} \sum_{m=1}^{\infty} (-i)^\nu \sin\nu\varphi \sin\nu\varphi'' J_\nu(k\rho''\sin\theta). \tag{9.34}$$

In the case when $k\rho''\sin\theta \ll 1$, which occurs for small polar angles θ or for small separation of the dipole from the apex, the Bessel function can be replaced by its small argument approximation $J_\nu(x) \sim (x/2)^\nu/\Gamma(\nu + 1)$; consequently, the series will converge rapidly. Hence, retaining only the first term, we obtain

$$\Pi_{z \atop \rho>\rho''}^{(e)} = \frac{i\mu p^{(e)}\exp[ik(r - z''\cos\theta)]}{\pi\omega\varepsilon_0 n^2 r} \left(\frac{k\rho''\sin\theta}{2i}\right)^{\pi/\varphi_0} \sin\frac{\pi\varphi}{\varphi_0}\sin\frac{\pi\varphi''}{\varphi_0}. \tag{9.35}$$

If the angle of the wedge is an integral fraction of π, $\varphi_0 = \pi/l$, then we can use (9.31) and obtain in this case

$$\Pi_z^{(e)} = \frac{i\mu p^{(e)} \exp[ik(r - z'' \cos\theta)]}{4\pi r\omega\varepsilon_0 n^2}$$

$$\times \left\{ \sum_{m=1}^{M_1} \exp\left[-ik\rho'' \sin\theta \cos\left(\varphi - \varphi'' \quad \frac{2m\pi}{l}\right)\right] \right.$$

$$- \sum_{m=1}^{M_2} \exp\left[-ik\rho'' \sin\theta \cos\left(\varphi + \varphi'' - \frac{2m\pi}{l}\right)\right]$$

$$+ \sum_{n=0}^{N_1} \exp\left[-ik\rho'' \sin\theta \cos\left(\varphi - \varphi'' - \frac{2n\pi}{l}\right)\right]$$

$$\left. - \sum_{n=0}^{N_2} \exp\left[-ik\rho'' \sin\theta \cos\left(\varphi + \varphi'' + \frac{2n\pi}{l}\right)\right] \right\} \tag{9.36}$$

As an example of a case to which the formula in (9.36) can be applied, consider a wedge whose internal angle $\varphi_0 = \pi/4$ or $l = 4$. In this case $M_1 = 2$, $M_2 = 2$, $N_1 = 1$, and $N_2 = 1$. We obtain, after some manipulations (problem),

$$\Pi_z^{(e)} = \frac{i\mu p^{(e)} \exp[ik(r - z'' \cos\theta)]}{\pi r\omega\varepsilon_0 n^2}$$

$$\times [\cos(k\rho'' \cos\varphi'' \sin\theta \cos\varphi) \cos(k\rho'' \sin\varphi'' \sin\theta \sin\varphi)$$

$$+ \cos(k\rho'' \cos\varphi'' \sin\theta \sin\varphi) \cos(k\rho'' \sin\varphi'' \sin\theta \cos\varphi)], \tag{9.37}$$

which is relatively easy to use in numerical computations. To find the field components in spherical coordinates, we note that

$$\begin{aligned} \Pi_r &= \cos\theta \Pi_z, \\ \Pi_\theta &= -\sin\theta \Pi_z, \end{aligned} \tag{9.38}$$

so that the components of the magnetic field are

$$H_r = -\frac{i\omega\varepsilon_0 n^2}{\mu r} \frac{\partial \Pi_z^{(e)}}{\partial\varphi}, \tag{9.39a}$$

$$H_\theta = -\frac{i\omega\varepsilon_0 n^2 \cos\theta}{\mu r \sin\theta} \frac{\partial \Pi_z^{(e)}}{\partial\varphi}, \tag{9.39b}$$

$$H_\varphi = \frac{i\omega\varepsilon_0 n^2}{\mu r} \left[\sin\theta \frac{\partial}{\partial r} (r\Pi_z^{(e)}) + \frac{\partial}{\partial\theta} (\cos\theta \Pi_z^{(e)})\right], \tag{9.39c}$$

and

$$E_r = k^2 \cos \theta \Pi_z^{(e)} + \frac{\partial}{\partial r}(\nabla \cdot \mathbf{\Pi}^{(e)}), \tag{9.40a}$$

$$E_\theta = -k^2 \sin \theta \Pi_z^{(e)} + \frac{1}{r}\frac{\partial}{\partial \theta}(\nabla \cdot \mathbf{\Pi}^{(e)}), \tag{9.40b}$$

$$E_\varphi = \frac{1}{r \sin \theta}\frac{\partial}{\partial \varphi}(\nabla \cdot \mathbf{\Pi}^{(e)}), \tag{9.40c}$$

where

$$\nabla \cdot \mathbf{\Pi}^{(e)} = \frac{\cos \theta}{r^2}\frac{\partial}{\partial r}(r^2 \Pi_z^{(e)}) - \frac{1}{r \sin \theta}\frac{\partial}{\partial \theta}(\sin^2 \theta \Pi_z^{(e)}). \tag{9.40d}$$

Thus, to the first order in r

$$H_\varphi = -(k\omega\varepsilon_0 n^2/\mu) \sin \theta \Pi_z^{(e)}, \tag{9.41a}$$

$$E_\theta = -k^2 \sin \theta \Pi_z^{(e)}, \tag{9.41b}$$

and all other field components are zero. It will be noted that $E_\theta/H_\varphi = [\mu_0 \mu/(\varepsilon_0 \varepsilon + i\sigma/\omega)]^{1/2}$ is the intrinsic impedance of the surrounding medium.

9.2.1.2 Electric Current Line Source

If the electric current distribution is an infinite line source parallel to the z axis, i.e.,

$$J_z^{(e)} = I^{(e)} \frac{\delta(\rho' - \rho'')\,\delta(\varphi' - \varphi'')}{\rho'}, \tag{9.42}$$

then using this in (9.23) and recalling that

$$\int_{-\infty}^{\infty} \exp(-i\alpha z')\, dz' = 2\pi\, \delta(\alpha), \tag{9.43}$$

we obtain

$$\Pi_{z\ \rho<\rho''}^{(e)} = -\frac{\mu\pi I^{(e)}}{\omega\varepsilon_0 n^2 \varphi_0} \sum_{m=1}^{\infty} J_\nu(k\rho'')H_\nu^{(1)}(k\rho) \sin \nu\varphi \sin \nu\varphi''. \tag{9.44}$$

The far-field approximation can be obtained from the foregoing by using the asymptotic form of the Hankel function

$$H_\nu^{(1)}(x) \sim (2/\pi x)^{1/2}(-i)^\nu \exp[i(x - \pi/4)].$$

Hence, in the far field

$$\Pi_z^{(e)} = -\frac{\mu I^{(e)}}{\omega\varepsilon_0 n^2 \varphi_0}\left(\frac{2\pi}{k\rho}\right)^{1/2} \exp\left[i\left(k\rho - \frac{\pi}{4}\right)\right] \sum_{m=1}^{\infty} (-i)^\nu J_\nu(k\rho'') \sin \nu\varphi \sin \nu\varphi''. \tag{9.45}$$

The far-field components are found from $H_\varphi = -(k\omega\varepsilon_0 n^2/\mu)\Pi_z^{(e)}$ and $E_z = k^2\Pi_z^{(e)}$. It will be noted that in this case $-E_z/H_\varphi = [\mu_0\mu/(\varepsilon_0\varepsilon + i\sigma/\omega)]^{1/2}$, as before.

When the wedge angle is an integral fraction of π, i.e., $\varphi_0 = \pi/l$ where l is an integer, a simple expression for the fields in terms of images is possible. Using (8.188) together with (8.192) where the integral vanishes for $v = l/2$, and since in this case $\Pi_z^{(e)} = (i\mu I^{(e)}/\omega\varepsilon_0 n^2)g^{(e)}(\rho, \varphi; \rho'', \varphi'')$, we obtain

$$\begin{aligned}\Pi_z^{(e)} = -\frac{\mu I^{(e)}}{4\omega\varepsilon_0 n^2}\Bigg(&\sum_{m=1}^{M_1} H_0^{(1)}\left\{k\left[\rho^2 + \rho''^2 - 2\rho\rho''\cos\left(\varphi - \varphi'' + \frac{2m\pi}{l}\right)\right]^{1/2}\right\}\\ &-\sum_{m=1}^{M_2} H_0^{(1)}\left\{k\left[\rho^2 + \rho''^2 - 2\rho\rho''\cos\left(\varphi + \varphi'' - \frac{2m\pi}{l}\right)\right]^{1/2}\right\}\\ &+\sum_{n=0}^{N_1} H_0^{(1)}\left\{k\left[\rho^2 + \rho''^2 - 2\rho\rho''\cos\left(\varphi - \varphi'' - \frac{2n\pi}{l}\right)\right]^{1/2}\right\}\\ &-\sum_{n=0}^{N_2} H_0^{(1)}\left\{k\left[\rho^2 + \rho''^2 - 2\rho\rho''\cos\left(\varphi + \varphi'' + \frac{2n\pi}{l}\right)\right]^{1/2}\right\}\Bigg),\end{aligned} \tag{9.46}$$

which is the desired image representation of the solution.

9.2.2 Magnetic Currents

The procedure for obtaining the solution to the present problem is identical to the previous case, where now the Green function given in (8.152) may be used in (9.4). We obtain in this case

$$\Pi_{z\atop \rho>\rho'}^{(m)} = -(4\omega\mu_0\mu\varphi_0)^{-1}\sum_{m=0}^{\infty}\varepsilon_m\cos v\varphi\iiint d\mathbf{r}'\,J_z^{(m)}(\mathbf{r}')\cos v\varphi' I_> \tag{9.47}$$

with $I_>$ given by (9.24).

If the magnetic current distribution is a single magnetic dipole of moment $p^{(m)}$ oriented in the positive z direction and located at the point (ρ'', φ'', z''), the solution in the far zone simplifies to

$$\Pi_z^{(m)} = \frac{ip^{(m)}\exp[ik(r - \cos\theta z'')]}{2\omega\mu_0\mu\varphi_0 r}\sum_{m=0}^{\infty}\varepsilon_m(-i)^v\cos v\varphi\cos v\varphi'' J_v(k\rho''\sin\theta). \tag{9.48}$$

If the magnetic current distribution is a line source along the z axis at the point (ρ'', φ''), then the solution is

$$\Pi_{z\ \rho>\rho''}^{(m)} = -\frac{\pi I^{(m)}}{2\omega\mu_0 \mu\varphi_0} \sum_{m=0}^{\infty} \varepsilon_m J_\nu(k\rho'') H_\nu^{(1)}(k\rho) \cos \nu\varphi \cos \nu\varphi''. \tag{9.49}$$

When the wedge angle is an integral fraction of π, $\varphi_0 = \pi/l$, then simpler expressions for the fields are possible in terms of images. Using (8.189) together with (8.192), where the integral vanishes for $v = l/2$, the Green function $g^{(m)}(\boldsymbol{\rho}, \boldsymbol{\rho}', \alpha)$ can be written

$$\begin{aligned} g^{(m)}(\boldsymbol{\rho}, \boldsymbol{\rho}', \alpha) = \frac{i}{4}\Bigg(&\sum_{m=1}^{M_1} H_0^{(1)}\left\{\tau\left[\rho^2 + \rho'^2 - 2\rho\rho' \cos\left(\varphi - \varphi' + \frac{2m\pi}{l}\right)\right]^{1/2}\right\} \\ &+ \sum_{m=1}^{M_2} H_0^{(1)}\left\{\tau\left[\rho^2 + \rho'^2 - 2\rho\rho' \cos\left(\varphi + \varphi' - \frac{2m\pi}{l}\right)\right]^{1/2}\right\} \\ &+ \sum_{n=0}^{N_1} H_0^{(1)}\left\{\tau\left[\rho^2 + \rho'^2 - 2\rho\rho' \cos\left(\varphi - \varphi' - \frac{2n\pi}{l}\right)\right]^{1/2}\right\} \\ &+ \sum_{n=0}^{N_2} H_0^{(1)}\left\{\tau\left[\rho^2 + \rho'^2 - 2\rho\rho' \cos\left(\varphi + \varphi' + \frac{2n\pi}{l}\right)\right]^{1/2}\right\}\Bigg) \end{aligned} \tag{9.50}$$

with $M_{1,2}$ and $N_{1,2}$ given in (9.26b). Substitution of (9.50) into (9.11) gives the desired representation

$$\Pi_z^{(m)} = \frac{i}{4\pi\omega\mu_0 \mu} \iiint d\mathbf{r}' J_z^{(m)}(\mathbf{r}')\left(\sum_{m=1}^{M_1} I_m^{(1)} + \sum_{m=1}^{M_2} I_m^{(2)} + \sum_{n=0}^{N_1} I_n^{(1)} + \sum_{n=0}^{N_2} I_n^{(2)}\right), \tag{9.51}$$

where $I_{m,n}^{(1,2)}$ are integrals of the type shown in (9.28).

In the far zone, $k\rho' \ll k\rho \to \infty$, we can approximate as in the corresponding electric case and obtain

$$\begin{aligned} \Pi_z^{(m)} = \frac{i \exp(ikr)}{4\pi r\omega\mu_0 \mu} &\iiint d\mathbf{r}' J_z^{(m)}(\mathbf{r}') \exp(-ikz' \cos\theta) \\ \times \Bigg\{ &\sum_{m=1}^{M_1} \exp\left[-ik\rho' \sin\theta \cos\left(\varphi - \varphi' + \frac{2m\pi}{l}\right)\right] \\ &+ \sum_{m=1}^{M_2} \exp\left[-ik\rho' \sin\theta \cos\left(\varphi + \varphi' - \frac{2m\pi}{l}\right)\right] \\ &+ \sum_{n=0}^{N_1} \exp\left[-ik\rho' \sin\theta \cos\left(\varphi - \varphi' - \frac{2n\pi}{l}\right)\right] \\ &+ \sum_{n=0}^{N_2} \exp\left[-ik\rho' \sin\theta \cos\left(\varphi + \varphi' + \frac{2n\pi}{l}\right)\right]\Bigg\}. \end{aligned} \tag{9.52}$$

9.3 Cylindrically Tipped Wedge and an Axial Current Distribution

9.3.1 ELECTRIC CURRENTS

The geometry of the problem is shown in Fig. 9.6. The function $g^{(e)}(\rho, \varphi)$ satisfying the differential equation (9.7a) and the boundary conditions $g^{(e)}(a, \varphi) = g^{(e)}(\rho, 0) = g^{(e)}(\rho, \varphi_0) = 0$ was found in (8.194). Hence, the

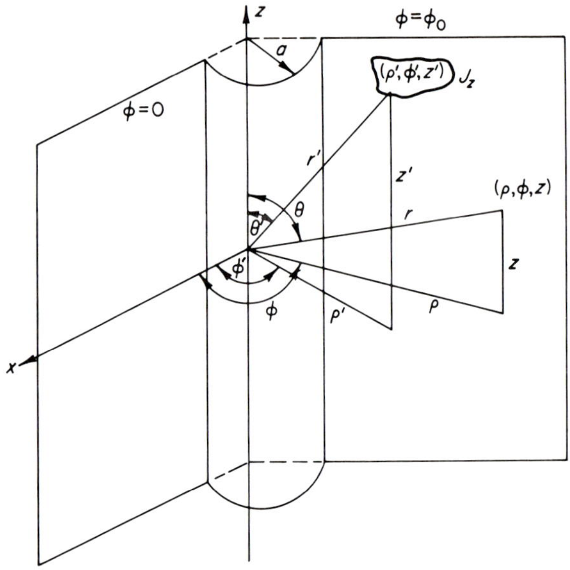

Fig. 9.6. Cylindrically tipped wedge and an axial current distribution.

solution to the present problem becomes

$$\Pi_{\substack{z \\ \rho > \rho'}}^{(e)} = -\frac{\mu}{2\varphi_0 \omega\varepsilon_0 n^2} \sum_{m=1}^{\infty} \sin \nu\varphi \iiint d\mathbf{r}' J_z^{(e)}(\mathbf{r}') \sin \nu\varphi' I_> , \tag{9.53a}$$

where in this case

$$I_> = \int_{-\infty}^{\infty} \left[J_\nu(\tau\rho') - \frac{J_\nu(\tau a)}{H_\nu^{(1)}(\tau a)} H_\nu^{(1)}(\tau\rho') \right] H_\nu^{(1)}(\tau\rho) \exp[i\alpha(z - z')]\, d\alpha \tag{9.53b}$$

and $\nu = m\pi/\varphi_0$ as before. For large distances compared to the wavelength, $k\rho \gg 1$, the integral in (9.53b) can be evaluated using (9.22). Hence, in this case

$$\Pi_{z\atop \rho>\rho'}^{(e)} = \frac{i\mu}{\varphi_0\,\omega\varepsilon_0\,n^2}\frac{\exp(ikr)}{r}\sum_{m=1}^{\infty}(-i)^{\nu}\sin\nu\varphi W^{(e)}(\theta,\nu) \tag{9.54a}$$

where

$$W^{(e)} = \iiint J_z^{(e)}(\mathbf{r}')\sin\nu\varphi'\left[J_\nu(k\rho'\sin\theta) - \frac{J_\nu(ka\sin\theta)}{H_\nu^{(1)}(ka\sin\theta)}H_\nu^{(1)}(k\rho'\sin\theta)\right]$$
$$\times\exp(-ik\cos\theta z')\,d\mathbf{r}'. \tag{9.54b}$$

9.3.1.1 Axial Electric Dipole

If the electric current distribution is a single dipole oriented along the z axis at a point $\rho = \rho''$, $\varphi = \varphi''$, and $z = z''$, then the integral in (9.54b) can be readily evaluated. Hence, the radiation field becomes

$$\Pi_{z\atop \rho>\rho''}^{(e)} = \frac{i\mu p^{(e)}\exp[ik(r - z''\cos\theta)]}{\varphi_0\,\omega\varepsilon_0\,n^2 r}\sum_{m=1}^{\infty}(-i)^{\nu}\sin\nu\varphi\sin\nu\varphi''$$
$$\times\left[J_\nu(k\rho''\sin\theta) - \frac{J_\nu(ka\sin\theta)}{H_\nu^{(1)}(ka\sin\theta)}H_\nu^{(1)}(k\rho''\sin\theta)\right]. \tag{9.55}$$

9.3.1.2 Electric Current Line Source

If the electric current distribution is an infinite line source parallel to the z axis and at a point (ρ'', φ'') in the ρ-φ plane, then noting (9.43), we obtain from (9.53)

$$\Pi_{z\atop \rho>\rho''}^{(e)} = -\frac{\mu\pi I^{(e)}}{\varphi_0\,\omega\varepsilon_0\,n^2}$$
$$\sum_{m=1}^{\infty}\left[J_\nu(k\rho'') - \frac{J_\nu(ka)}{H_\nu^{(1)}(ka)}H_\nu^{(1)}(k\rho'')\right]\times H_\nu^{(1)}(k\rho)\sin\nu\varphi\sin\nu\varphi''. \tag{9.56}$$

9.3.2 MAGNETIC CURRENTS

Using the results of (9.11) and (8.195), we obtain for the axial magnetic current distribution in the presence of a cylindrically tipped wedge of Fig. 9.6

$$\Pi_{z\atop \rho>\rho'}^{(m)} = -(4\varphi_0\,\omega\mu_0\,\mu)^{-1}\sum_{m=0}^{\infty}\varepsilon_m\cos\nu\varphi\iiint d\mathbf{r}' J_z^{(m)}(\mathbf{r}')\cos\nu\varphi'$$
$$\times\int_{-\infty}^{\infty}\left[J_\nu(\tau\rho') - \frac{J_\nu'(\tau a)}{H_\nu^{(1)\prime}(\tau a)}H_\nu^{(1)}(\tau\rho')\right]H_\nu^{(1)}(\tau\rho)\exp[i\alpha(z - z')]\,d\alpha. \tag{9.57}$$

If the magnetic current distribution is a single dipole of moment $p^{(m)}$ oriented in the positive z direction and situated at the point (ρ'', φ'', z''), then the radiation field becomes

$$\Pi_z^{(m)} = \frac{ip^{(m)} \exp[ik(r - z'' \cos\theta)]}{2\omega\mu_0 \mu\varphi_0 r} \sum_{m=0}^{\infty} \varepsilon_m(-i)^m \cos\nu\varphi \cos\nu\varphi''$$

$$\times \left[J_\nu(k\rho'' \sin\theta) - \frac{J_\nu'(ka \sin\theta)}{H_\nu^{(1)\prime}(ka \sin\theta)} H_\nu^{(1)}(k\rho'' \sin\theta) \right]. \tag{9.58}$$

If the magnetic current distribution is a line source parallel to the z axis at a point (ρ'', φ'') in the ρ-φ plane, then the solution is

$$\Pi_{z\ \rho>\rho''}^{(m)} = -\frac{\pi I^{(m)}}{2\omega\mu_0 \mu\varphi_0} \sum_{m=0}^{\infty} \varepsilon_m \cos\nu\varphi \cos\nu\varphi''$$

$$\times \left[J_\nu(k\rho'') - \frac{J_\nu'(ka)}{H_\nu^{(1)\prime}(ka)} H_\nu^{(1)}(k\rho'') \right] H_\nu^{(1)}(k\rho). \tag{9.59}$$

9.4 Perfectly Conducting Circular Cylinder and an Axial Current Distribution

9.4.1 Electric Currents

The geometry of the problem is shown in Fig. 9.7. The function $g^{(e)}$ satisfying the differential equation (9.7a) and the boundary conditions

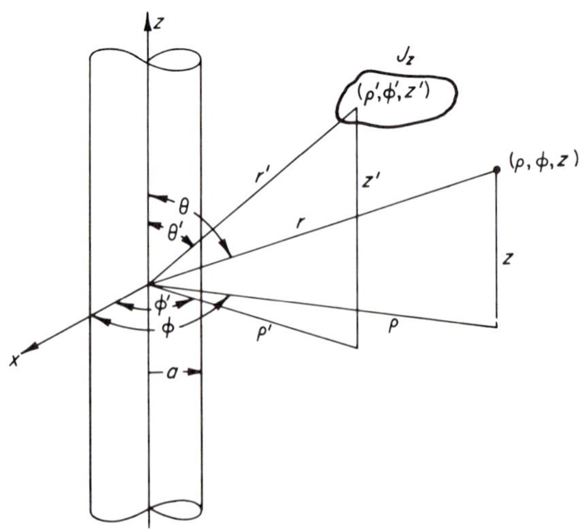

Fig. 9.7. Circular cylinder and an axial current distribution.

$g^{(e)}(a, \varphi) = 0$ was found in (8.202). Hence, the solution is

$$\underset{\rho>\rho'}{\Pi_z^{(e)}} = -\frac{\mu}{8\pi\omega\varepsilon_0 n^2} \sum_{-\infty}^{\infty} \exp(im\varphi) \iiint d\mathbf{r}' J_z^{(e)}(\mathbf{r}') \exp(-im\varphi')$$

$$\times \int_{-\infty}^{\infty} \left[J_m(\tau\rho') - \frac{J_m(\tau a)}{H_m^{(1)}(\tau a)} H_m^{(1)}(\tau\rho') \right] H_m^{(1)}(\tau\rho) \exp[i\alpha(z - z')]\, d\alpha. \tag{9.60}$$

For large distances compared to the wavelength, $k\rho \gg 1$, the last integral on the right in (9.60) can be evaluated using (9.22). Hence, the radiation field becomes

$$\Pi_z^{(e)} = \frac{i\mu \exp(ikr)}{4\pi\omega\varepsilon_0 n^2 r} \sum_{-\infty}^{\infty} (-i)^m \exp(im\varphi) \iiint d\mathbf{r}' J_z^{(e)}(\mathbf{r}') \exp[-i(m\varphi' + k\cos\theta z')]$$

$$\times \left[J_m(k\rho' \sin\theta) - \frac{J_m(ka\sin\theta)}{H_m^{(1)}(ka\sin\theta)} H_m^{(1)}(k\rho'\sin\theta) \right]. \tag{9.61}$$

9.4.1.1 Axial Electric Dipole

If the electric current distribution is a single dipole parallel to the z axis at a point (ρ'', φ'', z''), then the radiation field is

$$\Pi_z^{(e)} = \frac{i\mu p^{(e)} \exp[ik(r - z''\cos\theta)]}{4\pi\omega\varepsilon_0 n^2 r} \sum_{-\infty}^{\infty} (-i)^m \exp[im(\varphi - \varphi'')]$$

$$\times \left[J_m(k\rho''\sin\theta) - \frac{J_m(ka\sin\theta)}{H_m^{(1)}(ka\sin\theta)} H_m^{(1)}(k\rho''\sin\theta) \right]. \tag{9.62}$$

9.4.1.2 Electric Current Line Source

For an electric current line source parallel to the z axis and situated at a point (ρ'', φ'') in the ρ-φ plane, we obtain

$$\underset{\rho>\rho''}{\Pi_z^{(e)}} = -\frac{\mu I^{(e)}}{4\omega\varepsilon_0 n^2} \sum_{-\infty}^{\infty} e^{im(\varphi-\varphi'')} \left[J_m(k\rho'') - \frac{J_m(ka)}{H_m^{(1)}(ka)} H_m^{(1)}(k\rho'') \right] H_m^{(1)}(k\rho) \tag{9.63}$$

9.4.2 Magnetic Currents

Using the results of (9.11) and (8.203), we obtain for the axial magnetic current distribution in the presence of a circular cylinder of Fig. 9.7

$$\underset{\rho>\rho'}{\Pi_z^{(m)}} = -(8\pi\omega\mu_0\mu)^{-1} \sum_{-\infty}^{\infty} \exp(im\varphi) \iiint d\mathbf{r}' J_z^{(m)}(\mathbf{r}') \exp(-im\varphi')$$

$$\times \int_{-\infty}^{\infty} \left[J_m(\tau\rho') - \frac{J'_m(\tau a)}{H_m^{(1)\prime}(\tau a)} H_m^{(1)}(\tau\rho') \right] H_m^{(1)}(\tau\rho) \exp[i\alpha(z - z')]\, d\alpha. \tag{9.64}$$

If the magnetic current distribution is a single dipole of moment $p^{(m)}$ oriented in the positive z direction and situated at a point (ρ'', φ'', z''), then the radiation field is

$$\Pi_z^{(m)} = \frac{ip^{(m)} \exp[ik(r - z'' \cos\theta)]}{4\pi\omega\mu_0 \mu r} \sum_{-\infty}^{\infty} (-i)^m \exp[im(\varphi - \varphi'')]$$

$$\times \left[J_m(k\rho'' \sin\theta) - \frac{J_m'(ka \sin\theta)}{H_m^{(1)\prime}(ka \sin\theta)} H_m^{(1)}(k\rho'' \sin\theta)\right]. \qquad (9.65)$$

If the magnetic current distribution is a line source parallel to the z axis at a point $(\rho''\ \varphi'')$ in the ρ-φ plane, then the solution is

$$\Pi_{z\atop \rho>\rho''}^{(m)} = -\frac{I^{(m)}}{4\omega\mu_0 \mu} \sum_{-\infty}^{\infty} \exp[im(\varphi - \varphi'')]$$

$$\times \left[J_m(k\rho'') - \frac{J_m'(ka)}{H_m^{(1)\prime}(ka)} H_m^{(1)}(k\rho'')\right] H_m^{(1)}(k\rho). \qquad (9.66)$$

9.5 Dielectric Circular Cylinder and a Line Source

The problem of a circular cylinder composed of an arbitrary isotropic material in the presence of a line current source can also be solved. In this section we shall find solutions to electric and magnetic current lines and show that in the limit as the refractive index of the cylinder approaches infinity, $n \to \infty$, the solutions are identical to those that would be obtained by assuming the Leontovich boundary condition on the surface of the cylinder. As a consequence, we shall also show that the criterion for the application of the Leontovich boundary condition is that $nk_0 a \gg 1$ where a is the smallest radius of curvature on the surface.

9.5.1 Electric Line Current

Consider a circular cylinder in air whose electromagnetic properties are characterized by the refractive index $n = [\mu\varepsilon(1 + i\sigma/\omega\varepsilon_0 \varepsilon)]^{1/2}$ in the presence of an electric current line source carrying a uniform current $I^{(e)} \exp(-i\omega t)$, as shown in Fig. 9.8. Since the problem is two-dimensional $\partial/\partial z = 0$ and the axial component of the Hertzian vector in the region outside the cylinder must satisfy

$$\left(\frac{1}{\rho}\frac{\partial}{\partial\rho}\rho\frac{\partial}{\partial\rho} + \frac{1}{\rho^2}\frac{\partial^2}{\partial\varphi^2} + k_0{}^2\right)\Pi_{z0}^{(e)} = -\frac{iI^{(e)}}{\omega\varepsilon_0}\frac{\delta(\rho - \rho')\,\delta(\varphi - \varphi')}{\rho}, \qquad (9.67)$$

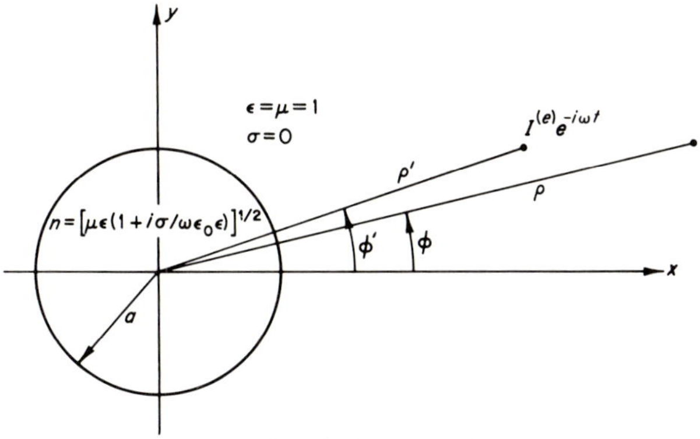

Fig. 9.8. Dielectric cylinder and a line source.

whereas inside the cylinder we must have

$$\left(\frac{1}{\rho}\frac{\partial}{\partial\rho}\rho\frac{\partial}{\partial\rho} + \frac{1}{\rho^2}\frac{\partial^2}{\partial\varphi^2} + k_0{}^2 n^2\right)\Pi_{z1}^{(e)} = 0. \tag{9.68}$$

The field components in the air are given by

$$H_{\rho 0} = -\frac{i\omega\varepsilon_0}{\rho}\frac{\partial\Pi_{z0}^{(e)}}{\partial\varphi}, \tag{9.69a}$$

$$H_{\varphi 0} = i\omega\varepsilon_0\frac{\partial\Pi_{z0}^{(e)}}{\partial\rho}, \tag{9.69b}$$

$$E_{z0} = k_0{}^2\Pi_{z0}^{(e)}, \tag{9.69c}$$

whereas inside the cylinder

$$H_{\rho 1} = -\frac{i\omega\varepsilon_0 n^2}{\mu\rho}\frac{\partial\Pi_{z1}^{(e)}}{\partial\varphi}, \tag{9.70a}$$

$$H_{\varphi 1} = \frac{i\omega\varepsilon_0 n^2}{\mu}\frac{\partial\Pi_{z1}^{(e)}}{\partial\rho}, \tag{9.70b}$$

$$E_{z1} = k_0{}^2 n^2\Pi_{z1}^{(e)}. \tag{9.70c}$$

We shall now seek solutions to $\Pi_{z0}^{(e)}$ and $\Pi_{z1}^{(e)}$ that are finite everywhere in their respective regions, except at the point $\boldsymbol{\rho} = \boldsymbol{\rho}'$ and ensure continuity of the tangential fields H_φ and E_z on the surface of the cylinder, i.e.

$$\Pi_{z0}^{(e)} = n^2\Pi_{z1}^{(e)}, \tag{9.71a}$$

$$\frac{\partial\Pi_{z0}^{(e)}}{\partial\rho} = \frac{n^2}{\mu}\frac{\partial\Pi_{z1}^{(e)}}{\partial\rho}, \tag{9.71b}$$

at $\rho = a$ for all φ.

It will be appropriate to seek solutions in the form

$$\Pi_{z0}^{(e)} = \frac{iI^{(e)}}{2\pi\omega\varepsilon_0} \sum_{-\infty}^{\infty} \exp[im(\varphi - \varphi')] g_0(\rho), \tag{9.72a}$$

$$\Pi_{z1}^{(e)} = \frac{iI^{(e)}}{2\pi\omega\varepsilon_0} \sum_{-\infty}^{\infty} \exp[im(\varphi - \varphi')] g_1(\rho), \tag{9.72b}$$

where $g_0(\rho)$ and $g_1(\rho)$ must satisfy the Bessel differential equations

$$\left[\frac{1}{\rho}\frac{\partial}{\partial\rho}\rho\frac{\partial}{\partial\rho} + \left(k_0{}^2 - \frac{m^2}{\rho^2}\right)\right] g_0(\rho) = -\frac{\delta(\rho - \rho')}{\rho}, \tag{9.73a}$$

$$\left[\frac{1}{\rho}\frac{\partial}{\partial\rho}\rho\frac{\partial}{\partial\rho} + \left(k^2 - \frac{m^2}{\rho^2}\right)\right] g_1(\rho) = 0, \tag{9.73b}$$

where $k = k_0 n$. Now we divide the region outside the cylinder into $a \leq \rho < \rho'$ and $\rho' < \rho \leq \infty$ and denote the respective solutions as $g_{0<}$ and $g_{0>}$. The appropriate solutions in each one of these regions are

$$g_{0<} = AJ_m(k_0 \rho) + BH_m^{(1)}(k_0 \rho), \tag{9.74a}$$

$$g_{0>} = CH_m^{(1)}(k_0 \rho), \tag{9.74b}$$

Moreover, in the region inside the cylinder we must put

$$g_1 = DJ_m(k\rho), \tag{9.75}$$

which is finite at $\rho = 0$.

The solutions $g_{0<}$ and $g_{0>}$ in (9.74) must be continuous at $\rho = \rho'$ and moreover $\partial g_{0>}/\partial\rho - \partial g_{0<}/\partial\rho = -1/\rho'$, which gives

$$A = \frac{i\pi}{2} H_m^{(1)}(k_0 \rho'), \tag{9.76a}$$

$$C - B = \frac{i\pi}{2} J_m(k_0 \rho'). \tag{9.76b}$$

Moreover, using (9.71), we obtain

$$\frac{B}{A} = -\left[\frac{\mu J_m(ka)J_m'(k_0 a) - nJ_m'(ka)J_m(k_0 a)}{\mu J_m(ka)H_m^{(1)\prime}(k_0 a) - nJ_m'(ka)H_m^{(1)}(k_0 a)}\right], \tag{9.77a}$$

$$D = \left[\frac{J_m(k_0 a) + (B/A)H_m^{(1)}(k_0 a)}{n^2 J_m(ka)}\right] A. \tag{9.77b}$$

Hence, all of the unknown coefficients are now uniquely determined.

Let us focus our attention on the fields in the region $\rho > \rho'$ outside the cylinder. After some manipulations, we obtain

$$\Pi_{\substack{z0 \\ \rho>\rho'}}^{(e)} = -\frac{I^{(e)}}{4\omega\varepsilon_0}\sum_{-\infty}^{\infty}\exp[im(\varphi-\varphi')]H_m^{(1)}(k_0\rho)$$
$$\times\left[J_m(k_0\rho') - \frac{J_m(k_0 a) - Z^{(e)}J_m'(k_0 a)}{H_m^{(1)}(k_0 a) - Z^{(e)}H_m^{(1)\prime}(k_0 a)}H_m^{(1)}(\ {}_0\rho')\right], \tag{9.78a}$$

where

$$Z^{(e)} = \frac{\mu J_m(ka)}{nJ_m'(ka)}. \tag{9.78b}$$

If $ka = k_0 na \gg 1$ and since $\mathrm{Im}(n) > 0$ we can replace the Bessel functions in (9.78b) by their asymptotic equivalents and show that

$$Z^{(e)} \xrightarrow[|ka|\to\infty]{} \frac{i\mu}{n}. \tag{9.79}$$

Hence in this case

$$\Pi_{\substack{z0 \\ \rho>\rho'}}^{(e)} = -\frac{I^{(e)}}{4\omega\varepsilon_0}\sum_{-\infty}^{\infty}\exp[im(\varphi-\varphi')]H_m^{(1)}(k_0\rho)$$
$$\times\left[J_m(k_0\rho') - \frac{nJ_m(k_0 a) - i\mu J_m'(k_0 a)}{nH_m^{(1)}(k_0 a) - i\mu H_m^{(1)\prime}(k_0 a)}H_m^{(1)}(k_0\rho')\right]. \tag{9.80}$$

We shall have an occasion to refer back to the solution in (9.80) shortly.

Now let us solve the problem of a circular cylindrical surface impedance and an electric line source as shown in Fig. 9.9. Note that in this problem we need not be concerned about the fields inside the cylinder. The axial component of the Hertzian vector pertinent to the region outside the cylinder must satisfy the partial differential equation of (9.67). On the surface of the

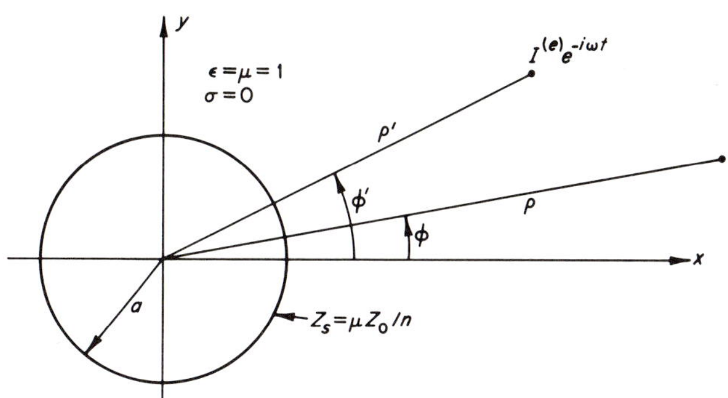

Fig. 9.9. Cylindrical surface impedance and a line source.

cylinder at $\rho = a$ the ratio of the tangential electric and magnetic field components must be equal to the surface impedance $Z_s = \mu Z_0/n$, which is nothing else but the plane wave impedance of the cylinder material (see (2.74)). In terms of the Hertzian vector, this boundary condition is readily translated into

$$\Pi_{z0}^{(e)} - \left(\frac{i\mu}{k}\right)\frac{\partial \Pi_{z0}^{(e)}}{\partial \rho} = 0, \qquad \rho = a. \tag{9.81}$$

Note that (9.81) specifies a homogeneous boundary condition of the type discussed in Section 8.1; hence we know that the present problem is still self-adjoint and, consequently, the eigenfunctions must be orthogonal.

To proceed with the solution, we use the expansion of (9.72a) together with (9.74). It is clear that the coefficient A is still given by (9.76a) and the relation between B and C in (9.76b) is still valid. Moreover, since $g_{0<}$ must satisfy the same boundary condition as $\Pi_{z0}^{(e)}$ in (9.81), we obtain

$$B = -\frac{i\pi}{2}\left[\frac{nJ_m(k_0 a) - i\mu J_m'(k_0 a)}{nH_m^{(1)}(k_0 a) - i\mu H_m^{(1)\prime}(k_0 a)}\right]H_m^{(1)}(k_0 \rho'), \tag{9.82a}$$

$$C = \frac{i\pi}{2}\left\{J_m(k_0 \rho') - \left[\frac{nJ_m(k_0 a) - i\mu J_m'(k_0 a)}{nH_m^{(1)}(k_0 a) - i\mu H_m^{(1)\prime}(k_0 a)}\right]H_m^{(1)}(k_0 \rho')\right\}. \tag{9.82b}$$

Substitution of (9.82b) into (9.74b) and hence using (9.72a) gives a solution identical to (9.80).

It is clear that the amount of work involved in the latter case was considerably less than in the former since in the latter case we did not have to be concerned about the fields inside the cylinder. This demonstrates the usefulness of the Leontovich or impedance boundary condition.

From what has been said we can deduce that the criterion for the applicability of the Leontovich boundary condition to an arbitrary surface is that $k_0 na \gg 1$ where n is the refractive index of the material, which is represented by its surface impedance $Z_s = \mu Z_0/n$, and a is to be understood as the smallest radius of curvature of that surface.

9.5.2 Magnetic Line Current

If a magnetic current line is substituted for the electric current line in Fig. 9.8, then in the region outside the cylinder the axial component of the Hertzian vector must satisfy

$$\left(\frac{1}{\rho}\frac{\partial}{\partial \rho}\rho\frac{\partial}{\partial \rho} + \frac{1}{\rho^2}\frac{\partial^2}{\partial \varphi^2} + k_0{}^2\right)\Pi_{z0}^{(m)} = -\frac{iI^{(m)}}{\omega\mu_0}\frac{\delta(\rho - \rho')\,\delta(\varphi - \varphi')}{\rho}, \tag{9.83}$$

whereas inside the cylinder

$$\left(\frac{1}{\rho}\frac{\partial}{\partial\rho}\rho\frac{\partial}{\partial\rho}+\frac{1}{\rho^2}\frac{\partial^2}{\partial\varphi^2}+k_0{}^2n^2\right)\Pi_{z1}^{(m)}=0. \tag{9.84}$$

The field components in the air are given by

$$E_{\rho 0}=\frac{i\omega\mu_0}{\rho}\frac{\partial\Pi_{z0}^{(m)}}{\partial\varphi}, \tag{9.85a}$$

$$E_{\varphi 0}=-i\omega\mu_0\frac{\partial\Pi_{z0}^{(m)}}{\partial\rho}, \tag{9.85b}$$

$$H_{z0}=k_0{}^2\Pi_{z0}^{(m)}, \tag{9.85c}$$

whereas inside the cylinder

$$E_{\rho 1}=\frac{i\omega\mu_0\,\mu}{\rho}\frac{\partial\Pi_{z1}^{(m)}}{\partial\varphi}, \tag{9.86a}$$

$$E_{\varphi 1}=-i\omega\mu_0\,\mu\frac{\partial\Pi_{z1}^{(m)}}{\partial\rho}, \tag{9.86b}$$

$$H_{z1}=k_0{}^2n^2\Pi_{z1}^{(m)}. \tag{9.86c}$$

On the surface of the cylinder, $\rho=a$, the tangential electric and magnetic fields must be continuous, which implies that

$$\Pi_{z0}^{(m)}=n^2\Pi_{z1}^{(m)}, \tag{9.87a}$$

$$\frac{\partial\Pi_{z0}^{(m)}}{\partial\rho}=\mu\frac{\partial\Pi_{z1}^{(m)}}{\partial\rho} \tag{9.87b}$$

at $\rho=a$.

As in the corresponding electric case, we now expand $\Pi_{z0}^{(m)}$ and $\Pi_{z1}^{(m)}$:

$$\Pi_{z0}^{(m)}=\frac{iI^{(m)}}{2\pi\omega\mu_0}\sum_{-\infty}^{\infty}\exp[im(\varphi-\varphi')]g_0(\rho), \tag{9.88a}$$

$$\Pi_{z1}^{(m)}=\frac{iI^{(m)}}{2\pi\omega\mu_0}\sum_{-\infty}^{\infty}\exp[im(\varphi-\varphi')]g_1(\rho), \tag{9.88b}$$

where $g_0(\rho)$ and $g_1(\rho)$ must satisfy (9.73a) and (9.73b), respectively, and their solutions can be written as in (9.74) and (9.75). It is clear that (9.76a) and (9.76b) still hold true in the present case; however, since $g_{0<}$ and g_1 must satisfy the same boundary conditions as $\Pi_{z0}^{(m)}$ and $\Pi_{z1}^{(m)}$, respectively, we obtain

$$\frac{B}{A} = -\left[\frac{J'_m(k_0 a) - Z^{(m)} J_m(k_0 a)}{H_m^{(1)\prime}(k_0 a) - Z^{(m)} H_m^{(1)}(k_0 a)}\right], \tag{9.89a}$$

where

$$Z^{(m)} = \frac{\mu J'_m(ka)}{n J_m(ka)}. \tag{9.89b}$$

Hence, the solution for the region $\rho > \rho'$ outside the cylinder becomes

$$\begin{aligned}\underset{\rho>\rho'}{\Pi_{z0}^{(m)}} = &-\frac{I^{(m)}}{4\omega\mu_0} \sum_{-\infty}^{\infty} \exp[im(\varphi - \varphi')] H_m^{(1)}(k_0 \rho) \\ &\times \left[J_m(k_0 \rho') - \frac{J'_m(k_0 a) - Z^{(m)} J_m(k_0 a)}{H_m^{(1)\prime}(k_0 a) - Z^{(m)} H_m^{(1)}(k_0 a)} H_m^{(1)}(k_0 \rho')\right].\end{aligned} \tag{9.90}$$

When $ka = k_0 na \gg 1$, we can replace the Bessel functions in (9.89b) by their asymptotic equivalents and show that

$$Z^{(m)} \xrightarrow[|ka| \to \infty]{} -\frac{i\mu}{n}, \tag{9.91}$$

which upon substitution into (9.90) gives

$$\begin{aligned}\underset{\rho>\rho'}{\Pi_{z0}^{(m)}} = &-\frac{I^{(m)}}{4\omega\mu_0} \sum_{-\infty}^{\infty} \exp[im(\varphi - \varphi')] H_m^{(1)}(k_0 \rho) \\ &\times \left[J_m(k_0 \rho') - \frac{n J'_m(k_0 a) + i\mu J_m(k_0 a)}{n H_m^{(1)\prime}(k_0 a) + i\mu H_m^{(1)}(k_0 a)} H_m^{(1)}(k_0 \rho')\right].\end{aligned} \tag{9.92}$$

It can be shown (problem) that the solution in (9.92) is identical to that which can be obtained by assuming that on the surface of the cylinder $-E_{\varphi 0}/H_{z0} = Z_s = \mu Z_0/n$ and by disregarding the fields inside the cylinder altogether.

Problems

9.1 Starting with (9.36), prove (9.37).

9.2 Obtain formulas equivalent to (9.37) and (9.46) for the magnetic case.

9.3 Show that (9.92) can be obtained by solving the problem depicted in Fig. 9.9 with a magnetic line source replacing the electric line source.

9.4 Consider the problem of a perfectly conducting circular cylinder coated with an isotropic homogeneous dielectric in the presence of a magnetic current line source in air as shown in Fig. 9.10.

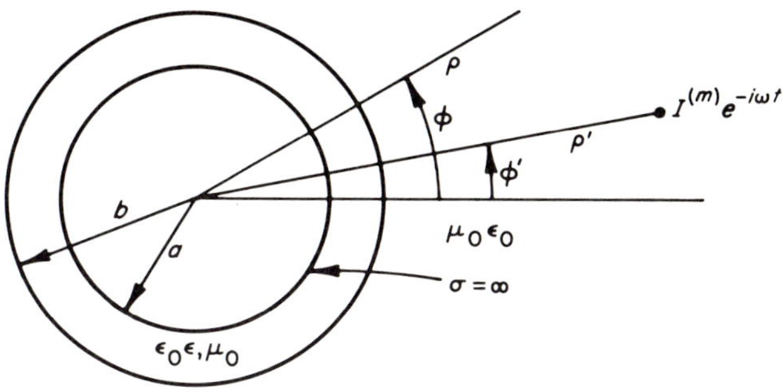

Fig. 9.10.

(a) Find the axial component of the Hertzian vector in the regions $a \le \rho \le b$ and $b \le \rho \le \infty$ when $\rho' \ge b$.
(b) Find the components of the electric and magnetic fields in the radiation zone ($k\rho \to \infty$).
(c) Find the approximate form of the radiation field when the thickness of the dielectric layer is amall, i.e. $k(b - a) \ll 1$.

9.5 Consider the problem of a perfectly conducting cylindrically tipped wedge with a dielectric coating in the presence of a magnetic current line source as shown in Fig. 9.11.
(a) Find the axial component of the Hertzian vector in the regions $a \le \rho \le b$ and $b \le \rho \le \infty$.

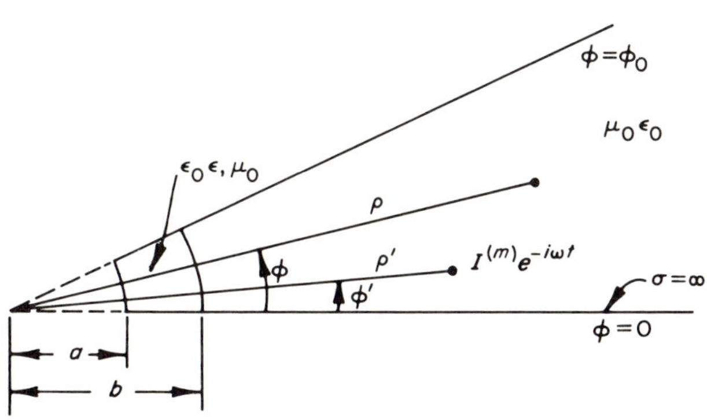

Fig. 9.11.

(b) Find the components of the electric and magnetic fields in the far field ($k\rho \to \infty$).

(c) Find the approximate form of the radiation field components when the thickness of the coating is small, i.e., $k(b - a) \ll 1$.

References

Gradshteyn, I. S., and Ryzhik, I. M. (1966). "Table of Integrals, Sums, Series, and Products," 4th ed. Academic Press, New York.

Harrington, R. F. (1961). "Time Harmonic Electromagnetic Fields." McGraw-Hill, New York.

Jones, D. S. (1964). "The Theory of Electromagnetism." Macmillan, New York.

Wait, J. R. (1959). "Electromagnetic Radiation from Cylindrical Structures." Pergamon, Oxford.

CHAPTER 10

DIFFRACTION BY CYLINDRICAL STRUCTURES

In this chapter we discuss the problem of diffraction by cylindrical structures. Specifically, we consider the circular cylinder and the wedge and its variants (i.e., a cylindrically tipped wedge and a half plane).

In the circular cylinder problem, after finding the customary "harmonic series" representation, which is suitable for numerical calculations in the range $ka < 1$, we make use of the Watson transformation to find the "residue series" representation, which is useful for numerical calculations in the range $ka \gg 1$. The concept of "creeping waves" is discussed in this connection.

The wedge problem is discussed from the standpoint of the separation of variables method and the Lebedev–Kontorovich transform. In the special case of a half plane, the Wiener–Hopf technique of solution is also illustrated.

10.1 Circular Cylinder

10.1.1 Harmonic Series Representation

In Chapter 8 we found the Green functions for regions outside a perfectly conducting cylinder as follows.

$$G_{<}^{(e)} = \frac{i}{4}\sum_{-\infty}^{\infty} \exp[im(\varphi - \varphi')]\left[J_m(k\rho) - \frac{J_m(ka)}{H_m^{(1)}(ka)} H_m^{(1)}(k\rho)\right] H_m^{(1)}(k\rho') \quad (10.1)$$

for the electric excitation, and

$$G_{<}^{(m)} = \frac{i}{4}\sum_{-\infty}^{\infty} \exp[im(\varphi - \varphi')]\left[J_m(k\rho) - \frac{J_m'(ka)}{H_m^{(1)\prime}(ka)} H_m^{(1)}(k\rho)\right] H_m^{(1)}(k\rho') \quad (10.2)$$

for the magnetic excitation. If the source point is allowed to be moved to infinity ($k\rho' \to \infty$), as shown in Fig. 10.1 then we can use the asymptotic form

$$H_m^{(1)}(k\rho') \xrightarrow[k\rho' \to \infty]{} \left(\frac{2}{\pi k\rho'}\right)^{1/2} (-i)^m \exp\left[i\left(k\rho' - \frac{\pi}{4}\right)\right] \quad (10.3)$$

to find the limiting form of the Green functions pertinent to waves incident on the cylinder from infinity.

Without loss of generality we can set $\varphi' = 0$. Moreover, since $E_z = iI^{(e)}\omega\mu_0\mu G^{(e)}$ and $H_z = i\omega\varepsilon_0 n^2 I^{(m)} G^{(m)}/\mu$, we obtain, using (10.3) in (10.1) and (10.2),

$$E_z = E_0 \sum_{-\infty}^{\infty} (-i)^m \exp(im\varphi)\left[J_m(k\rho) - \frac{J_m(ka)}{H_m^{(1)}(ka)} H_m^{(1)}(k\rho)\right], \quad (10.4a)$$

where

$$E_0 = -\frac{\omega\mu_0\mu I^{(e)}}{2(2\pi k\rho')^{1/2}} \exp\left[i\left(k\rho' - \frac{\pi}{4}\right)\right] \quad (10.4b)$$

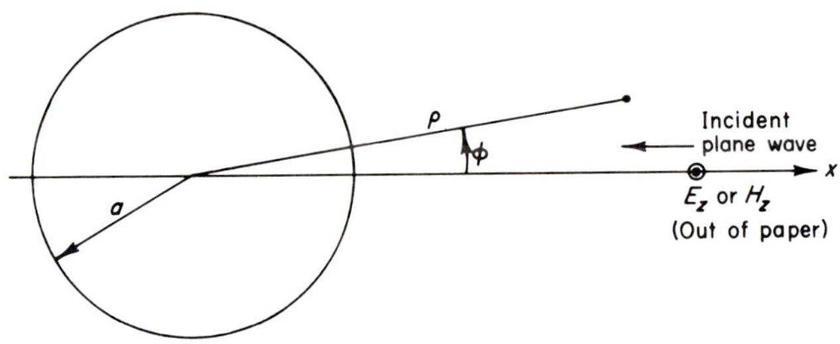

Fig. 10.1. Geometry of the problem of diffraction by a circular cylinder.

for the electric case, and

$$H_z = H_0 \sum_{-\infty}^{\infty} (-i)^m \exp(im\varphi)\left[J_m(k\rho) - \frac{J_m'(ka)}{H_m^{(1)\prime}(ka)} H_m^{(1)}(k\rho)\right], \quad (10.5a)$$

where

$$H_0 = -\frac{\omega\varepsilon_0 n^2 I^{(m)}}{2\mu(2\pi k\rho')^{1/2}} \exp\left[i\left(k\rho' - \frac{\pi}{4}\right)\right]. \quad (10.5b)$$

Using the well-known formula

$$\sum_{-\infty}^{\infty} (-i)^m J_m(k\rho) \exp(im\varphi) = \exp(-ik\rho\cos\varphi)$$
$$= \exp(-ikx), \quad (10.6)$$

we can write (10.4a) and (10.5a) as

$$E_z = E_0\left\{e^{-ikx} - \sum_{-\infty}^{\infty} (-i)^m e^{im\varphi} \frac{J_m(ka)}{H_m^{(1)}(ka)} H_m^{(1)}(k\rho)\right\} \quad (10.7)$$

and

$$H_z = H_0\left\{e^{-ikx} - \sum_{-\infty}^{\infty}(-i)^m e^{im\varphi}\,\frac{J_m'(ka)}{H_m^{(1)\prime}(ka)}\,H_m^{(1)}(k\rho)\right\}. \tag{10.8}$$

The first term in the curly brackets of (10.7) and (10.8) is easily identified as the phase factor of the incident wave traveling along the x axis in the negative x direction. The second term in the curly brackets of (10.7) and (10.8) must then represent the field diffracted by the cylinder. Far away from the cylinder $k\rho \to \infty$, and we can replace the Hankel function $H_m^{(1)}(k\rho)$ by its asymptotic equivalent to obtain

$$E_z = E_0\left\{\exp(-ikx) - \left(\frac{2}{\pi k\rho}\right)^{1/2}\exp\left[i\left(k\rho - \frac{\pi}{4}\right)\right]\right.$$
$$\left.\times \sum_{-\infty}^{\infty}(-1)^m \exp(im\varphi)\,\frac{J_m(ka)}{H_m^{(1)}(ka)}\right\} \tag{10.9}$$

and

$$H_z = H_0\left\{\exp(-ikx) - \left(\frac{2}{\pi k\rho}\right)^{1/2}\exp\left[i\left(k\rho - \frac{\pi}{4}\right)\right]\right.$$
$$\left.\times \sum_{-\infty}^{\infty}(-1)^m \exp(im\varphi)\,\frac{J_m'(ka)}{H_m^{(1)\prime}(ka)}\right\}. \tag{10.10}$$

Moreover, in the far zone $H_\varphi = -(n/\mu Z_0)E_z$ for the electric case and $E_\varphi = (Z_0 n/\varepsilon)H_z$ for the magnetic case. Hence, the Poynting vector is in the radial direction and

$$S_\rho^{(e)} = \frac{n^*}{2\mu Z_0}\,|E_z|^2 \tag{10.11}$$

for the electric case and

$$S_\rho^{(m)} = \frac{Z_0\, n}{2\varepsilon}\,|H_z|^2 \tag{10.12}$$

for the magnetic case. Hence, the ratio of the scattered to the incident power intensity will be proportional to $|E_z^s/E_z^i|^2$ in the electric case and $|H_z^s/H_z^i|^2$ in the magnetic case. Now we find, for the case when the surrounding medium is nonconducting ($\sigma = 0$),

$$\left|\frac{E_z^s}{E_z^i}\right|^2 = \frac{2}{\pi k\rho}\left|\sum_{-\infty}^{\infty}(-1)^m \varepsilon_m \cos m\varphi\,\frac{J_m(ka)}{H_m^{(1)}(ka)}\right|^2 \tag{10.13}$$

and

$$\left|\frac{H_z^{\,s}}{H_z^{\,i}}\right|^2 = \frac{2}{\pi k\rho}\left|\sum_{-\infty}^{\infty}(-1)^m \varepsilon_m \cos m\varphi \, \frac{J'_m(ka)}{H_m^{(1)\prime}(ka)}\right|^2. \tag{10.14}$$

If the frequency is sufficiently low or the radius of the cylinder is sufficiently small, so that $ka = 2\pi a/\lambda \ll 1$, the series will converge fast enough so we can retain only the first few terms. Moreover, in this case we can set

$$\frac{J_0(x)}{H_0^{(1)}(x)} \sim -\frac{i\pi}{2}\frac{1}{\ln x}, \qquad \frac{J'_0(x)}{H_0^{(1)\prime}(x)} \sim i\pi\left(\frac{x}{2}\right)^2,$$

$$\frac{J_m(x)}{H_m^{(1)}(x)} \sim \frac{i\pi m}{(m!)^2}\left(\frac{x}{2}\right)^{2m}, \qquad \frac{J'_m(x)}{H_m^{(1)\prime}(x)} \sim -\frac{i\pi m}{(m!)^2}\left(\frac{x}{2}\right)^{2m}, \tag{10.15}$$

and obtain

$$\left|\frac{E_z^{\,s}}{E_z^{\,i}}\right|^2 = \frac{\pi}{2k\rho}\left|\frac{1}{\ln(ka)} + (ka)^2\cos\varphi - \frac{(ka)^4}{8}\cos 2\varphi + \cdots\right|^2 \tag{10.16}$$

and

$$\left|\frac{H_z^{\,s}}{H_z^{\,i}}\right|^2 = \frac{\pi(ka)^4}{8k\rho}\left|1 + 2\cos\varphi - \left(\frac{ka}{2}\right)^2\cos 2\varphi + \cdots\right|^2. \tag{10.17}$$

It is significant that in the electric case the power is scattered almost uniformly around the cylinder when $ka \ll 1$. In the magnetic case the scattered power pattern is proportional to $(1 + 2\cos\varphi)^2$, which attains a maximum in the direction of the incident field and becomes zero at $\varphi = 180$ degrees.

If the radius of the cylinder is not small or if the frequency is high, so that $ka > 1$, then the series in (10.13) and (10.14) converge slowly and, consequently, many terms are needed for a fair degree of accuracy. In this case the convergence can be improved by using the Watson transformation, which we describe next.

10.1.2 The Watson Transformation

Consider the integral

$$I = \frac{i}{2}\int_C \frac{\exp[i\nu(\varphi - \pi)]}{\sin \nu\pi} B_\nu \, d\nu \tag{10.18}$$

taken around the closed contour C enclosing all of the poles due to the zeros of $\sin \nu\pi$, as shown in Fig. 10.2. Since the integral taken around the contour C is equal to $-2\pi i$ times the sum of the residues at the poles $\nu = m$, we obtain the identity

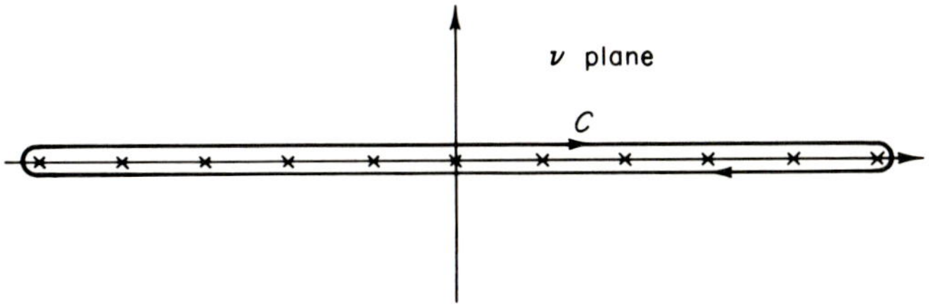

Fig. 10.2. The complex ν plane for the Watson transformation.

$$\frac{i}{2}\int_C \frac{\exp[i\nu(\varphi-\pi)]}{\sin \nu\pi} B_\nu \, d\nu = \sum_{-\infty}^{\infty} \exp(im\varphi) B_m \tag{10.19}$$

providing that B_m has no singularities on the real axis. It is clear that if we put

$$\begin{aligned} B_\nu^{(e)} &= E_0 \frac{\exp(-i\nu\pi/2)}{H_\nu^{(1)}(ka)} [H_\nu^{(1)}(ka)J_\nu(k\rho) - J_\nu(ka)H_\nu^{(1)}(k\rho)] \\ &= \frac{E_0}{2}\frac{\exp(-i\nu\pi/2)}{H_\nu^{(1)}(ka)} [H_\nu^{(1)}(ka)H_\nu^{(2)}(k\rho) - H_\nu^{(2)}(ka)H_\nu^{(1)}(k\rho)] \end{aligned} \tag{10.20}$$

for the electric case and

$$\begin{aligned} B_\nu^{(m)} &= H_0 \frac{\exp(-i\nu\pi/2)}{H_\nu^{(1)\prime}(ka)} [H_\nu^{(1)\prime}(ka)J_\nu(k\rho) - J_\nu'(ka)H_\nu^{(1)}(k\rho)] \\ &= \frac{H_0}{2}\frac{\exp(-i\nu\pi/2)}{H_\nu^{(1)\prime}(ka)} [H_\nu^{(1)\prime}(ka)H_\nu^{(2)}(k\rho) - H_\nu^{(2)\prime}(ka)H_\nu^{(1)}(k\rho)] \end{aligned} \tag{10.21}$$

for the magnetic case, then the correspondences

$$E_z = \frac{i}{2}\int_C \frac{\exp[i\nu(\varphi-\pi)]}{\sin \nu\pi} B_\nu^{(e)} \, d\nu, \tag{10.22}$$

$$H_z = \frac{i}{2}\int_C \frac{\exp[i\nu(\varphi-\pi)]}{\sin \nu\pi} B_\nu^{(m)} \, d\nu. \tag{10.23}$$

are immediately established. Moreover, since $H_{-\nu}^{(1)}(x) = \exp(i\nu\pi)H_\nu^{(1)}(x)$ and $H_{-\nu}^{(2)}(x) = \exp(-i\nu\pi)\, H_\nu^{(2)}(x)$ then $B_{-\nu}^{(e)} = B_\nu^{(e)}$ and $B_\nu^{(m)} = B_\nu^{(m)}$ and we obtain

$$E_z = i\int_{-\infty}^{\infty} \frac{\cos \nu(\varphi-\pi)}{\sin \nu\pi} B_\nu^{(e)} \, d\nu \tag{10.24}$$

and

$$H_z = i\int_{-\infty}^{\infty} \frac{\cos \nu(\varphi-\pi)}{\sin \nu\pi} B_\nu^{(m)} \, d\nu \tag{10.25}$$

where the path of integration is understood to run *slightly above* the real axis.

We now attempt to evaluate the integral in (10.24) and (10.25) by closing the contour with a large semicircle in the upper ν plane. To this end we must investigate the location of the zeros of

$$H^{(1)}_{\nu_n}(ka) = 0 \tag{10.26}$$

for the electric case and

$$H^{(1)\prime}_{\nu'_n}(ka) = 0 \tag{10.27}$$

for the magnetic case. Moreover, we must also investigate the behavior of the Hankel functions and their derivatives as $|\nu| \to \infty$.

Since ka is assumed to be large and in the radiation zone $k\rho$ is large as well, we are dealing with Hankel functions of large arguments.

When $|\nu| \ll |ka|$, we can use the well-known asymptotic forms

$$H^{(1)}_\nu(ka) \sim \left(\frac{2}{\pi ka}\right)^{1/2} \exp\left[i\left(ka - \frac{\nu\pi}{2} - \frac{\pi}{4}\right)\right], \tag{10.28a}$$

$$H^{(1)\prime}_\nu(ka) \sim i\left(\frac{2}{\pi ka}\right)^{1/2} \exp\left[i\left(ka - \frac{\nu\pi}{2} - \frac{\pi}{4}\right)\right], \tag{10.28b}$$

to ascertain that in this case neither $H^{(1)}_\nu(ka)$ nor $H^{(1)\prime}_\nu(ka)$ has any zeros in the complex ν plane.

When $|\nu|$ is comparable to or larger than $|ka|$, the asymptotic forms in (10.28) are not valid. Hence, we must investigate the behavior of Bessel functions with real and positive argument and an arbitrary complex index. (We henceforth assume that the medium surrounding the cylinder is lossless, so that k is real.)

Consider the Sommerfeld integral representation of the Bessel functions

$$\mathscr{C}_\nu(x) = (\pi)^{-1} \int_C \exp\{i[x \cos\beta + \nu(\beta - \pi/2)]\}\, d\beta \tag{10.29}$$

where C stands for C_1, C_2, or C_0, in which case $\mathscr{C}_\nu(x)$ represents $H^{(1)}_\nu(x)$, $H^{(2)}_\nu(x)$, or $2J_\nu(x)$, respectively. The various paths are shown in Fig. 10.3. Let us put

$$f(\beta) = x\cos\beta + \nu(\beta - \pi/2) \tag{10.30}$$

and assume that x and ν are real. Any saddle points are determined by setting $f'(\beta) = 0$; hence we find, for the two saddle points in the range $-\pi/2 < \beta < 3\pi/2$,

$$\beta_{s1} = \arcsin\frac{\nu}{x}, \qquad \nu < x, \tag{10.31a}$$

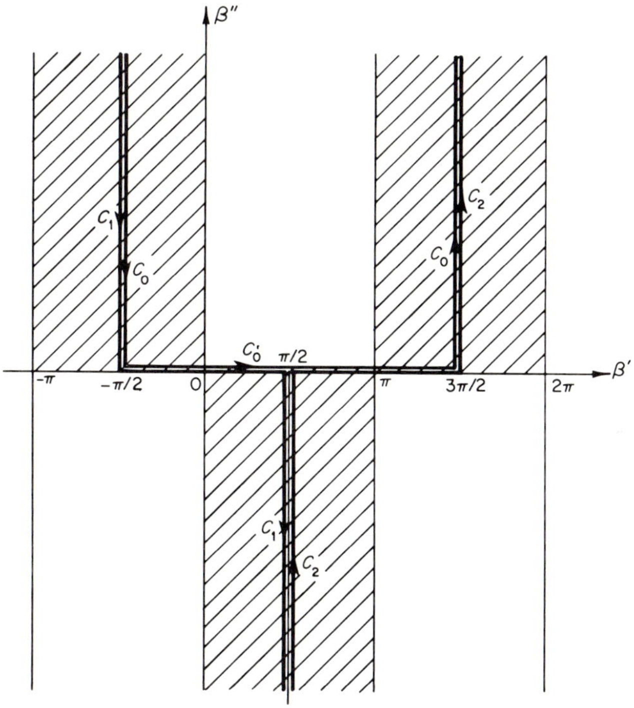

Fig. 10.3. Paths of integration for integral representation of the various Bessel functions.

$$\beta_{s1} = \frac{\pi}{2} + i \operatorname{arccosh} \frac{v}{x}, \qquad v > x, \tag{10.31b}$$

and

$$\beta_{s2} = \pi - \beta_{s1}. \tag{10.31c}$$

The location of the saddle points β_{s1} and β_{s2} is shown in Fig. 10.4.

The paths C_0, C_1, and C_2 can be deformed to pass through the saddle points somewhat as shown in Fig. 10.4. In the neighborhood of each of the saddle points we can expand the exponent in a Taylor series about $\beta = \beta_s$ to obtain

$$f(\beta) = f(\beta_s) - (x/2) \cos \beta_s(\beta - \beta_s)^2 + \cdots, \tag{10.32a}$$

which upon substitution into (10.29) gives

$$\mathscr{C}_v(x) = \frac{\exp(if(\beta_s))}{\pi} \int_C \exp\left[-i\left(\frac{x}{2}\right) \cos \beta_s(\beta - \beta_s)^2\right] d\beta. \tag{10.32b}$$

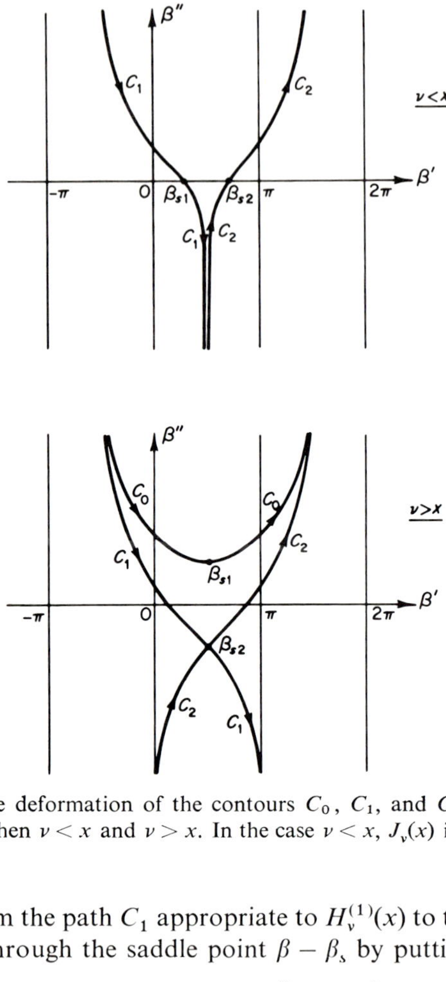

Fig. 10.4. Possible deformation of the contours C_0, C_1, and C_2 through the saddle points βs_1 and βs_2 when $\nu < x$ and $\nu > x$. In the case $\nu < x$, $J_\nu(x)$ is found from $J_\nu(x) = \frac{1}{2}[H_\nu^{(1)}(x) + H_\nu^{(2)}(x)]$.

We can now deform the path C_1 appropriate to $H_\nu^{(1)}(x)$ to the path of steepest descents passing through the saddle point $\beta - \beta_s$ by putting

$$(x/2)\cos\beta_s(\beta - \beta_s)^2 = -is^2 \tag{10.33}$$

and noting that $d\beta/ds = (2/x\cos\beta_s)^{1/2}\exp(-i\pi/4)$. The resulting integral can be evaluated by using (5.72) and we obtain

$$H_\nu^{(1)}(x) = \left(\frac{2}{\pi x\cos\beta_s}\right)^{1/2}\exp\left\{i\left[x\cos\beta_s + \nu\left(\beta_s - \frac{\pi}{2}\right) - \frac{\pi}{4}\right]\right\}. \tag{10.34}$$

The function $H_\nu^{(2)}(x)$ can be obtained in a similar manner or simply by inserting a minus sign in front of i in (10.34).

When $\nu < x$, the pertinent saddle point is β_{s1} (see Fig. 10.4), in which case $\sin\beta_{s1} = \nu/x$ and we obtain

$$H_\nu^{(1)}(x) \sim \left[\frac{2}{\pi(x^2-\nu^2)^{1/2}}\right]^{1/2} \exp\{i[(x^2-\nu^2)^{1/2} - \nu \arccos(\nu/x) - \pi/4]\}, \tag{10.35a}$$

$$H_\nu^{(1)\prime}(x) \sim i\left(\frac{x^2-\nu^2}{x}\right)\left[1 + \frac{ix^2}{2(x^2-\nu^2)^{3/2}}\right] H_\nu^{(1)}(x). \tag{10.35b}$$

When $\nu > x$, the pertinent saddle point is $\beta_{s2} = \pi/2 - i \operatorname{arccosh}(\nu/x)$, and we obtain

$$H_\nu^{(1)}(x) \sim -i\left[\frac{2}{\pi(\nu^2-x^2)^{1/2}}\right]^{1/2} \exp[\nu \operatorname{arccosh}(\nu/x) - (\nu^2-x^2)^{1/2}], \tag{10.36a}$$

$$H_\nu^{(1)\prime}(x) \sim -\frac{(\nu^2-x^2)}{x}\left[1 - \frac{x^2}{2(\nu^2-x^2)^{3/2}}\right] H_\nu^{(1} (x). \tag{10.36b}$$

Note that the asymptotic forms for $\nu > x$ can be obtained from those for $\nu < x$ by making the substitution $(x^2-\nu^2)^{1/2} = i(\nu^2-x^2)^{1/2}$. Moreover, the asymptotic forms for $H_\nu^{(2)}(x)$ and $H_\nu^{(2)\prime}(x)$ can be obtained from the foregoing by changing the sign in front of i.

The asymptotic forms of (10.35) or (10.36) give no clue regarding the zeros of $H_{\nu_n}^{(1)}(x) = 0$ or $H_{\nu'_n}^{(1)\prime}(x) = 0$. Note, however, that in the case when $\nu \sim x$, the foregoing formulas do not hold and we must obtain a better estimate. It is also clear that as $\nu \to x$, then $\beta_{s1} \to \beta_{s2}$ and the contribution of both saddle points must be considered. Thus we must write in this case

$$\begin{aligned} H_\nu^{(1)}(x) \sim {} & \left(\frac{2}{\pi x \cos \beta_{s1}}\right)^{1/2} \exp\left\{i\left[x \cos \beta_{s1} + \nu\left(\beta_{s1} - \frac{\pi}{2}\right) - \frac{\pi}{4}\right]\right\} \\ & + \left(\frac{2}{\pi x \cos \beta_{s2}}\right)^{1/2} \exp\left\{i\left[x \cos \beta_{s2} + \nu\left(\beta_{s2} - \frac{\pi}{2}\right) - \frac{\pi}{4}\right]\right\}. \end{aligned} \tag{10.37}$$

Now since $\beta_{s2} = \pi - \beta_{s1}$, so that $\cos \beta_{s2} = -\cos \beta_{s1}$, we obtain

$$H_\nu^{(1)}(x) \sim 2i\left(\frac{2}{\pi x \cos \beta_{s1}}\right)^{1/2} \sin\left[x \cos \beta_{s1} + \nu\left(\beta_{s1} - \frac{\pi}{2}\right) - \frac{\pi}{4}\right], \tag{10.38a}$$

$$H_\nu^{(1)\prime}(x) \sim 2i\left(\frac{2 \cos \beta_{s1}}{\pi x}\right)^{1/2} \cos\left[x \cos \beta_{s1} + \nu\left(\beta_{s1} - \frac{\pi}{2}\right) - \frac{\pi}{4}\right], \tag{10.38b}$$

which, presumably, could be used for finding the zeros of $H_{\nu_n}^{(1)}(x) = 0$ and $H_{\nu'_n}^{(1)\prime}(x) = 0$. Hence we set

$$x[\cos \beta_{s1} + (\beta_{s1} - \pi/2) \sin \beta_{s1}] - \pi/4 = \pm n\pi \tag{10.39a}$$

for the zeros of $H_{\nu_n}^{(1)}(x) = 0$ and

$$x[\cos \beta_{s1} + (\beta_{s1} - \pi/2) \sin \beta_{s1}] - \pi/4 = \pm(n + \tfrac{1}{2})\pi \qquad (10.39b)$$

for the zeros of $H_{\nu'_n}^{(1)\prime}(x) = 0$. It is clear that for $x \gg 1$ and n not too large the quantities in the square brackets in (10.39a) and (10.39b) must be small, which seems to indicate that the zeros in each case will be found in the vicinity of $\beta_{s1} = \arcsin(\nu/x) \sim \pi/2$. In the neighborhood of $\beta_{s1} = \pi/2$, however, the formula in (10.38a) diverges and the one in (10.38b) approaches zero independently, which suggests that the formulas in (10.38) are not accurate enough to determine the zeros of $H_{\nu_n}^{(1)}(x) = 0$ and $H_{\nu'_n}^{(1)\prime}(x) = 0$.

Inasmuch as the second term in the expansion of $f(\beta)$ in (10.32a) vanishes for $\beta_{s1} = \pi/2$, we next attempt to obtain a better approximation for $H_\nu^{(1)}(x)$ in the vicinity of $\beta_{s1} = \pi/2$ by retaining the next higher-order term. Hence we write

$$f(\beta) = f(\pi/2) + (\beta - \pi/2)f'(\pi/2) + \tfrac{1}{2}(\beta - \pi/2)^2 f''(\pi/2) + \tfrac{1}{6}(\beta - \pi/2)^3 f'''(\pi/2) + \cdots \qquad (10.40a)$$

and note that

$$f(\pi/2) = 0, \qquad f''(\pi/2) = 0, \qquad f'(\pi/2) = \nu - x, \qquad f'''(\pi/2) = x, \qquad (10.40b)$$

which upon substitution into (10.29) gives

$$\mathscr{C}_\nu(x) \sim \pi^{-1} \int_C \exp[i(\nu - x)(\beta - \pi/2) + i(x/6)(\beta - \pi/2)^3]\, d\beta. \qquad (10.40c)$$

Note that when $\beta \sim \pi/2$, the first term in the exponent plays the dominant role, and when β is far away from the point $\pi/2$, the cubic term plays the dominant role. Then if we set

$$\beta - \pi/2 = re^{i\theta}, \qquad (10.41)$$

it will follow that the lines $\mathrm{Im}(\beta - \pi/2)^3 = 0$ will be given by $\theta = m\pi/3$ where m is an integer including zero. Moreover, it will be found that in the shaded regions of the diagram of Fig. 10.5 $\mathrm{Im}(\beta - \pi/2)^3 > 0$ and in the unshaded regions $\mathrm{Im}(\beta - \pi/2)^3 < 0$.

Since the integral in (10.40c) will decrease rapidly for a large x in the shaded regions of Fig. 10.5, we can easily deform the original path of integration to run in the shaded regions and set for the limits

$H_\nu^{(1)}(x)$: $\quad -\infty \exp(-i\pi/3)$ to $\infty \exp(-i\pi/3)$;

$H_\nu^{(2)}(x)$: $\quad -\infty \exp(i\pi/3)$ to $\infty \exp(i\pi/3)$;

$J_\nu(x)$: $\quad -\infty$ to ∞.

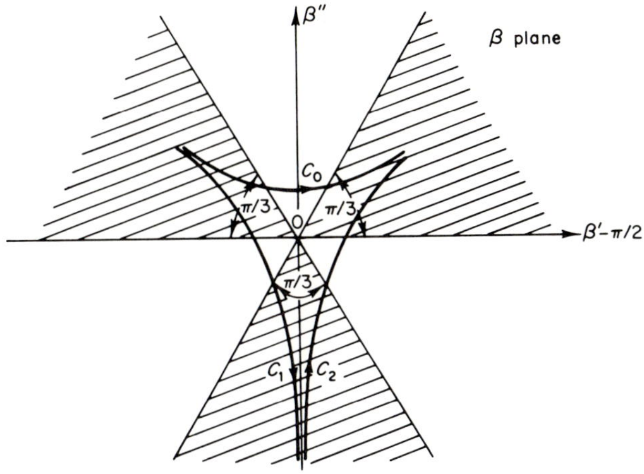

Fig. 10.5. Regions (shaded) in which Im $(\beta - \pi/2)^3 > 0$.

Now we make a change of variable

$$\tau = (x/6)^{1/3}(\beta - \pi/2)\begin{cases}\exp(i\pi/3)\\ \exp(-i\pi/3)\\ -1\end{cases} \tag{10.42}$$

for $H_\nu^{(1)}$, $H_\nu^{(2)}$, and J_ν, respectively. Hence, we obtain for $H_\nu^{(1)}$

$$H_\nu^{(1)}(x) \sim (\pi)^{-1}(6/x)^{1/3}\exp(-i\pi/3)\int_{-\infty}^{\infty}\exp[i(q_1\tau - \tau^3)]\,d\tau, \tag{10.43a}$$

where

$$q_1 = (\nu - x)(6/x)^{1/3}\exp(-i\pi/3). \tag{10.43b}$$

Using the definition of the Airy function (Abramowitz and Stegun (1964), p. 447)

$$\begin{aligned} Ai\left[-\frac{x}{(3a)^{1/3}}\right] &= \frac{(3a)^{1/3}}{\pi}\int_0^\infty \cos(at^3 - xt)\,dt \\ &= \frac{(3a)^{1/3}}{2\pi}\int_{-\infty}^{\infty}\exp[i(xt - at^3)]\,dt, \end{aligned} \tag{10.44}$$

we obtain after some manipulations

$$\int_{-\infty}^{\infty}\exp[i(q\tau - \tau^3)]\,d\tau = \frac{2\pi}{3^{1/3}}Ai\left(-\frac{q}{3^{1/3}}\right). \tag{10.45}$$

Using the result of (10.45) in (10.43a), we obtain the representation of the Hankel function of the first kind in terms of the Airy function as

$$H_\nu^{(1)}(x) \sim 2(2/x)^{1/3} \exp(-i\pi/3) Ai[-(\nu - x)(2/x)^{1/3} \exp(-i\pi/3)]. \tag{10.46}$$

In a similar manner we can easily show that the Hankel function of the second kind can be expressed as

$$H_\nu^{(2)}(x) \sim 2(2/x)^{(1/3)} \exp(i\pi/3) Ai[-(\nu - x)(2/x)^{1/3} \exp^{(i\pi/3)}]. \tag{10.47}$$

The Bessel function $J_\nu(x)$ can be obtained by using (10.46) and (10.47) to give

$$J_\nu(x) \sim (2/x)^{1/3} Ai[(\nu - x)(2/x)^{1/3}], \tag{10.48}$$

where the recurrence relation for the Airy functions

$$Ai(x) = \exp(-i\pi/3)\{Ai[-z \exp(-i\pi/3)] - \exp(-i\pi/3) Ai[-z \exp(-i\pi/3)]\} \tag{10.49}$$

has been used.

In what follows we shall also need the Airy function representation of the derivatives of the Hankel functions. These can be obtained from the formulas above by differentiation. We obtain

$$H_\nu^{(1)\prime}(x) \sim -2(2/x)^{2/3} \exp(i\pi/3) Ai'[-(\nu - x)(2/x)^{1/3} \exp(-i\pi/3)] \tag{10.50a}$$

$$H_\nu^{(2)\prime}(x) \sim -2(2/x)^{2/3} \exp(-i\pi/3) Ai'[-(\nu - x)(2/x)^{1/3} \exp(i\pi/3)], \tag{10.50b}$$

$$J_\nu'(x) \sim -(2/x)^{2/3} Ai'[(\nu - x)(2/x)^{1/3}]. \tag{10.50c}$$

The zeros of the Airy functions and its derivatives are tabulated (Abramowitz and Stegun (1964), p. 478). Some of them are listed for convenience in Table 10.1.

Table 10.1

ZEROS AND ASSOCIATED VALUES OF THE AIRY FUNCTIONS AND THEIR DERIVATIVES

n	$-a_n$	$Ai'(a_n)$	$-a_n'$	$Ai(a_n')$
1	2.338	+0.701	1.019	+0.536
2	4.088	−0.803	3.248	−0.419
3	5.521	+0.865	4.820	+0.380
4	6.787	−0.911	6.163	−0.358
5	7.944	+0.947	7.372	+0.342

The determination of the zeros of the Hankel function $H_\nu^{(1)}(x)$ and its derivative $H_\nu^{(1)\prime}(x)$ is now a straightforward procedure. We find

$$\nu_n = x - a_n(x/2)^{1/3} \exp(i\pi/3), \tag{10.51a}$$

$$\nu_n' = x - a_n'(x/2)^{1/3} \exp(i\pi/3) \tag{10.51b}$$

for the roots of $H_\nu^{(1)}(x)$ and $H_\nu^{(1)\prime}(x)$, respectively.

Using the values of $-a_n$ and $-a_n'$ from Table 10.1, we can write explicitly the following.

$\nu_n = x + x^{1/3} \exp(i\pi/3) \times 1.856$	$\nu_n' = x + x^{1/3} \exp(i\pi/3) \times 0.809$
3.245	2.578
4.381	3.826
5.387	4.892
6.305	5.851 (10.52)

In what follows we shall also need to know the behavior of $J_\nu(x)$ and $H_\nu^{(1)}(x)$ and their derivatives as $|\nu| \to \infty$ in the upper plane. These could be obtained, as before, by using the method of saddle points (see, e.g., Beckmann and Franz, 1957) with the additional complexities arising from the now complex values of ν. It is somewhat simpler, however, to start with the readily available "Uniform Asymptotic Expansions" formulas listed in the "Handbook of Mathematical Functions" (Abramowitz and Stegun (1964), p. 368).

When $|\nu| \to \infty$ and x is fixed, the uniform asymptotic expansions for $J_\nu(x)$ and $H_\nu^{(1)}(x)$ simplify to

$$J_\nu(x) \sim (4\zeta)^{1/4}\nu^{-1/3}Ai(w), \tag{10.53a}$$

$$H_\nu^{(1)}(x) \sim (4\zeta)^{1/4}\nu^{-1/3}[Ai(w) - iBi(w)], \tag{10.53b}$$

where $Ai(w)$ and $Bi(w)$ are two linearly independent solutions of the Airy differential equation $W''(w) - wW(w) = 0$ and

$$w = \zeta\nu^{2/3}, \tag{10.54a}$$

$$\tfrac{2}{3}\zeta^{3/2} \sim \ln(2\nu/ex). \tag{10.54b}$$

It can be seen that as $|\nu| \to \infty$, then $|w| \to \infty$ and, consequently, we can use the asymptotic forms of the Airy functions $Ai(w)$ and $Bi(w)$. However, we must first determine the range of arg(w) as $0 < \arg \nu < \pi$ in the upper ν plane.

Now it is clear that for a fixed real x, $\arg\{\ln(2\nu/ex)\}$ approaches zero as $|\nu| \to \infty$, hence $\arg(\zeta) \sim 0$ and

$$\arg(w) \sim \tfrac{2}{3}\arg(\nu). \tag{10.55}$$

Thus we must seek an asymptotic expansions for $Ai(w)$ and $Bi(w)$ when $|w| \to \infty$ in the range

$$0 < \arg(w) < 2\pi/3. \tag{10.56}$$

Now we have

$$Ai(w) \sim \frac{-1}{(2\sqrt{\pi})} w^{-1/4} e^{-\xi}, \qquad |\arg(w)| < \pi, \tag{10.57a}$$

$$Bi(w) \sim \pi^{-1/2} w^{-1/4} e^{\xi}, \qquad |\arg(w)| < \pi/3, \tag{10.57b}$$

where

$$\xi = \tfrac{2}{3} w^{3/2} \sim \ln(2v/ex)^{v}. \tag{10.57c}$$

We note that the approximation for $Ai(w)$ in (10.57a) is appropriate over the entire range of interest, whereas the approximation for $Bi(w)$ in (10.57b) is valid only for $\arg(v) < \pi/2$. To obtain an approximation for $Bi(w)$ for $\pi/2 < \arg(v)$, we set $w' = w\exp(-i\pi/3)$ and note

$$Bi\left[w' \exp\left(\frac{i\pi}{3}\right)\right] \sim \left(\frac{2}{\pi}\right)^{1/2} \exp\left(\frac{i\pi}{6}\right)(w')^{-1/4} \sin\left(\xi' + \frac{\pi}{4} - \frac{i}{2}\ln 2\right),$$

$$\arg(w') < \frac{2\pi}{3}, \tag{10.58a}$$

where

$$\xi' = \tfrac{2}{3}(w')^{3/2} = -i\xi. \tag{10.58b}$$

Since $\arg(w) = \arg(w') + \pi/3$, it is seen that (10.58a) is valid for $\arg(w) < \pi$ or $\arg(v) < 3\pi/2$, which includes the desired range. Moreover, in the region $\arg(v) > \pi/2$, $\operatorname{Re}\xi < 0$, and the sin( ) term in (10.58) can be approximated by $-(2i)^{-1}\exp(-\xi - \ln\sqrt{2} - i\pi/4)$. Hence we can write

$$Bi(w) \sim \frac{i}{2\sqrt{\pi}} w^{-1/2} e^{-\xi}, \qquad \frac{\pi}{2} < \arg(v) < \frac{3\pi}{2}. \tag{10.59}$$

Substituting (10.57) and (10.59) into (10.53), we secure

$$J_v(x) \sim (2\pi v)^{-1/2}\left(\frac{ex}{2v}\right)^{v}, \qquad 0 < \arg(v) < \pi, \tag{10.60a}$$

$$H_v^{(1)}(x) \sim -i\left(\frac{2}{\pi v}\right)^{1/2}\left(\frac{2v}{ex}\right)^{v}, \qquad |\arg(v)| < \frac{\pi}{2}, \tag{10.60b}$$

$$H_v^{(1)}(x) \sim \left(\frac{2}{\pi v}\right)^{1/2}\left(\frac{ex}{2v}\right)^{v}, \qquad \frac{\pi}{2} < \arg(v) < \frac{3\pi}{2}. \tag{10.60c}$$

Moreover, differentiating the results above with respect to x, we obtain for the derivatives

$$J_\nu'(x) \sim (\nu/x)J_\nu(x), \qquad 0 < \arg(\nu) < \pi, \tag{10.61a}$$

$$H_\nu^{(1)\prime}(x) \sim -(\nu/x)H_\nu^{(1)}(x), \qquad |\arg(\nu)| < \pi/2, \tag{10.61b}$$

$$(H_\nu^{(1)\prime}x) \sim (\nu/x)H_\nu^{(1)}(x), \qquad \pi/2 < \arg(\nu) < 3\pi/2. \tag{10.61c}$$

Before leaving this section it will be instructive to examine the location of any possible zeros of $H_{\nu'_n}^{(1)}(x) = 0$ and $H_{\nu'_n}^{(1)\prime}(x) = 0$ and $|\nu| \to \infty$ in the upper half plane. If we set $\nu = R\exp(i\theta)$, then it will follow that

$$|H_\nu^{(1)}(x)| \sim R^{-1/2}\exp[R\cos\theta\ln(2R/ex) - \theta R\sin\theta] \tag{10.62a}$$

for $-\pi/2 < \theta < \pi/2$ and

$$|H_\nu^{(1)}(x)| \sim R^{-1/2}\exp[-R\cos\theta\ln(2R/ex) + \theta R\sin\theta] \tag{10.62b}$$

for $\pi/2 < \theta < 3\pi/2$. Thus we note that there is no possibility of zeros in the region $\pi/2 < \theta < 3\pi/2$ since $\cos\theta < 0$ therein. In the region $0 < \theta < \pi/2$ zeros of $H_{\nu_n}^{(1)}(x) = 0$ are possible only in the vicinity of $\theta = \pi/2$ where $\cos\theta \sim 0$ and $|H_\nu^{(1)}(x)| \to R^{-1/2}\exp(-\theta R\sin\theta)$. Combining this information with that obtained earlier, we conclude that the distribution of zeros of $H_{\nu_n}^{(1)}(x) = 0$ and $H_{\nu'_n}^{(1)\prime}(x) = 0$ in the upper ν plane looks somewhat as shown in Fig. 10.6.

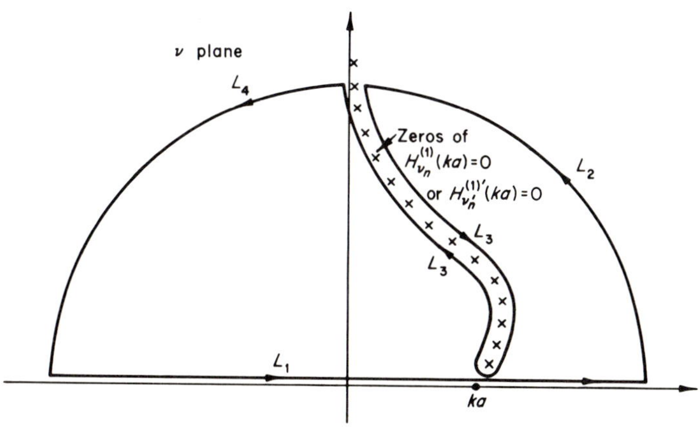

Fig. 10.6. Integration path for evaluation of the residue series.

10.1.3 Residue Series Representation of Solutions (Shadow Region)

We have now gathered all of the necessary information to evaluate the integrals in (10.24) and (10.25). Closing the contour with a large semicircle in the upper halfplane, as shown in Fig. 10.6, we can evaluate the integral by using the residue theorem. It is easy to show (problem), using (10.60) or (10.61), that the contribution from the semicircular arc vanishes as its radius approaches infinity. Hence, the integral along the real axis is equal to $2\pi i$ times the sum of residues at the poles due to the zeros of $H_{\nu_n}^{(1)}(ka) = 0$ or $H_{\nu'_n}^{(1)\prime}(ka) = 0$. Thus we obtain

$$E_z = \pi E_0 \sum_{n=1}^{\infty} \frac{H_{\nu_n}^{(2)}(ka)}{\partial/\partial\nu \, H_\nu^{(1)}(ka)|_{\nu=\nu_n}} \frac{\cos \nu_n(\varphi - \pi) \exp(-i\nu_n \pi/2)}{\sin \nu_n \pi} H_{\nu_n}^{(1)}(k\rho), \tag{10.63}$$

and

$$H_z = \pi H_0 \sum_{n=1}^{\infty} \frac{H_{\nu'_n}^{(2)\prime}(ka)}{\partial/\partial\nu \, H_\nu^{(1)\prime}(ka)|_{\nu=\nu'_n}} \frac{\cos \nu'_n(\varphi - \pi) \exp(-i\nu'_n \pi/2)}{\sin \nu'_n \pi} H_{\nu'_n}^{(1)}(k\rho). \tag{10.64}$$

Using the Wronskian

$$W\{H_\nu^{(1)}(x), H_\nu^{(2)}(x)\} = H_\nu^{(1)}(x)H_\nu^{(2)\prime}(x) - H_\nu^{(1)\prime}(x)H_\nu^{(2)}(x) = -\frac{4i}{\pi x}, \tag{10.65}$$

we find

$$H_{\nu_n}^{(2)}(ka) = \frac{4i}{\pi ka H_{\nu_n}^{(1)\prime}(ka)} \sim -\frac{\exp(i\pi/6)}{\pi Ai'(a_n)} \left(\frac{2}{ka}\right)^{1/3}, \tag{10.66a}$$

$$H_{\nu'_n}^{(2)\prime}(ka) = -\frac{4i}{\pi ka H_{\nu'_n}^{(1)}(ka)} \sim \frac{\exp(-i\pi/6)}{\pi Ai(a'_n)} \left(\frac{2}{ka}\right)^{2/3}, \tag{10.66b}$$

where we also used (10.46) and (10.50a). Moreover, we note that

$$\frac{\partial}{\partial\nu} H_\nu^{(1)}(ka)|_{\nu=\nu_n} \sim 2\left(\frac{2}{ka}\right)^{2/3} \exp(i\pi/3) Ai'(a_n) \tag{10.67a}$$

and

$$\frac{\partial}{\partial\nu} H_\nu^{(1)\prime}(ka)|_{\nu=\nu_n'} \sim \frac{4}{ka} Ai''(a'_n) = \frac{4a'_n}{ka} Ai(a'_n), \tag{10.67b}$$

where in (10.67b) we availed ourselves of the recurrence relation $Ai''(x) = xAi(x)$. Using the facts above in (10.63) and (10.64), we obtain

$$E_z = -\frac{E_0}{2} \exp\left(-\frac{i\pi}{6}\right)\left(\frac{ka}{2}\right)^{1/3} \sum_{n=1}^{\infty} \frac{\cos \nu_n(\varphi - \pi) \exp(-i\nu_n \pi/2)}{[Ai'(a_n)]^2 \sin \nu_n \pi} H_{\nu_n}^{(1)}(k\rho) \tag{10.68}$$

and

$$H_z = \frac{H_0}{2} \exp\left(-\frac{i\pi}{6}\right)\left(\frac{ka}{2}\right)^{1/3} \sum_{n=1}^{\infty} \frac{\cos \nu_n'(\varphi - \pi) \exp(-i\nu_n' \pi/2)}{a_n'[Ai(a_n')]^2 \sin \nu_n' \pi} H_{\nu_n'}^{(1)}(k\rho), \tag{10.69}$$

which is the residue series representation of the solutions. Moreover, noting the relation

$$\frac{\cos \nu(\varphi - \pi)}{\sin \nu\pi} \exp\left(-\frac{i\nu\pi}{2}\right) = -i \frac{\exp[i\nu(\varphi - \pi/2)] + \exp[i\nu(3\pi/2 - \varphi)]}{1 - \exp(i\nu 2\pi)}, \tag{10.70}$$

it can be seen that the residue series (10.68) and (10.69) converges rapidly in the shadow region of the cylinder, $\pi/2 < \varphi < 3\pi/2$.

The residue series representations of (10.68) and (10.69) also provide a physically interesting interpretation of how the waves are transmitted in the shadow region. Using the asymptotic form of $H_{\nu_n}^{(1)}(k\rho)$ and $H_{\nu_n'}^{(1)}(k\rho)$ as in (10.35a) and noting that $(k\rho)^2 - \nu_n{}^2 \sim (k\rho)^2 - \nu_n'^2 \sim k^2(\rho^2 - a^2)$ and $\nu_n/k\rho \sim \nu_n'/k\rho \sim a/\rho$, we obtain

$$\begin{aligned} E_z = {} & \frac{E_0 \exp(i\pi/12)}{[2\pi k(\rho^2 - a^2)^{1/2}]^{1/2}} \left(\frac{ka}{2}\right)^{1/3} \\ & \times \sum_{n=1}^{\infty} D_n^{(e)}(\exp\{i[k(\rho^2 - a^2)^{1/2} + \nu_n(\varphi - \pi/2 - \operatorname{arc}\cos(a/\rho)]\} \\ & + \exp\{i[k(\rho^2 - a^2)^{1/2} + \nu_n(3\pi/2 - \varphi - \operatorname{arc}\cos(a/\rho)]\}), \end{aligned} \tag{10.71}$$

and

$$\begin{aligned} H_z = {} & -\frac{H_0 \exp(i\pi/12)}{[2\pi k(\rho^2 - a^2)^{1/2}]^{1/2}} \left(\frac{ka}{2}\right)^{1/3} \\ & \times \sum_{n=1}^{\infty} D_n^{(m)}(\exp\{i[k(\rho^2 - a^2)^{1/2} + \nu_n'(\varphi - \pi/2 - \operatorname{arc}\cos(a/\rho)]\} \\ & + \exp\{i[k(\rho^2 - a^2)^{1/2} + \nu_n'(3\pi/2 - \varphi - \operatorname{arc}\cos(a/\rho)]\}), \end{aligned} \tag{10.72}$$

where

$$D_n^{(e)} = \{[1 - \exp(i2\pi\nu_n)][Ai'(a_n)]^2\}^{-1} \tag{10.73}$$

and

$$D_n^{(m)} = \{[1 - \exp(i2\pi\nu_n')]a_n'[Ai(a_n')]^2\}^{-1}. \tag{10.74}$$

The geometrical representation of (10.71) and (10.72) is shown in Fig. 10.7. It can be seen that the field in the shadow region is produced from the incident rays that strike the cylinder at the points $\varphi = \pi/2$ and $\varphi = 3\pi/2$. These rays travel along the surface of the cylinder through a distance $a[\varphi - \pi/2 - \arccos(a/\rho)]$ and $a[3\pi/2 - \varphi - \arccos(a/\rho)]$ with attenuation proportional to the imaginary part of v_n in the electric case and v_n' in the magnetic case. Subsequently, these rays leave the surface of the cylinder, evidently by radiation, and again travel in free space through a distance $(\rho^2 - a^2)^{1/2}$ to join at a point in the shadow region of the cylinder. Hence, the name "creeping waves."

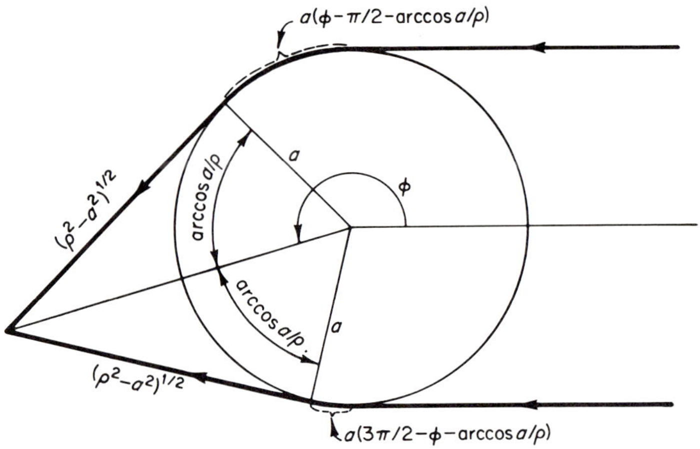

Fig. 10.7. Geometry of the "creeping" waves on a circular cylinder.

10.1.4 Geometrical Optics Approximation (Illuminated Region)

The formulas (10.71) and (10.72) are suitable for field calculations in the shadow region of the cylinder but they are useless in the illuminated region $-\pi/2 < \varphi < \pi/2$. A suitable representation of the solution for the illuminated region can be derived as follows.

We note the transformation

$$\cos v(\pi - \varphi) = e^{iv\pi} \cos v\varphi - ie^{iv\varphi} \sin v\pi, \tag{10.75a}$$

which upon substitution into (10.24) and (10.25) gives

$$E_z = I_1^{(e)} + iI_2^{(e)}, \tag{10.75b}$$

$$H_z = I_1^{(m)} + iI_2^{(m)}, \tag{10.75c}$$

where

$$I_1^{(e,m)} = \int_{-\infty}^{\infty} B_\nu^{(e,m)} e^{i\nu\varphi}\, d\nu \tag{10.75d}$$

and

$$I_2^{(e,m)} = \int_{-\infty}^{\infty} \frac{e^{i\nu\pi}\cos\nu\varphi}{\sin\nu\pi} B_\nu^{(e,m)}\, d\nu. \tag{10.75e}$$

The integral I_2 can be evaluated by closing the contour in the upper half plane and integrating about the poles due to $H_{\nu_n}^{(1)}(ka) = 0$ in the electric case and $H_{\nu_n'}^{(1)\prime}(ka) = 0$ in the magnetic case. We obtain

$$I_2^{(e)} = \frac{E_0}{2}\exp\left(\frac{i\pi}{3}\right)\left(\frac{ka}{2}\right)^{1/3} \sum_{n=1}^{\infty} \frac{\cos\nu_n\varphi\exp(i\nu_n\pi/2)}{[Ai'(a_n)]^2\sin\nu_n\pi} H_{\nu_n}^{(1)}(k\rho) \tag{10.76a}$$

and

$$I_2^{(m)} = -\frac{H_0}{2}\exp\left(\frac{i\pi}{3}\right)\left(\frac{ka}{2}\right)^{1/3} \sum_{n=1}^{\infty} \frac{\cos\nu_n'\varphi\exp(i\nu_n'\pi/2)}{a_n'[Ai(a_n')]^2\sin\nu_n'\pi} H_{\nu_n'}^{(1)}(k\rho). \tag{10.76b}$$

Moreover, using the transformation

$$\frac{\cos\nu\varphi}{\sin\nu\pi}\exp\left(\frac{i\nu\pi}{2}\right) = -i\,\frac{\exp[i\nu(3\pi/2+\varphi)] + \exp[i\nu(3\pi/2-\varphi)]}{1-\exp(i\nu 2\pi)} \tag{10.77}$$

and the asymptotic form for $H_\nu^{(1)}(k\rho)$ as $k\rho \to \infty$, we can show that

$$I_2^{(e)} \sim \frac{\exp(ik\rho)}{\sqrt{\rho}} \sum_{n=1}^{\infty} D_n^{(e)}\{\exp[i\nu_n(\pi+\varphi)] + \exp[i\nu_n(\pi-\varphi)]\} \tag{10.78a}$$

and

$$I_2^{(m)} \sim \frac{\exp(ik\rho)}{\sqrt{\rho}} \sum_{n=1}^{\infty} D_n^{(m)}\{\exp[i\nu_n'(\pi+\varphi)] + \exp[i\nu_n'(\pi-\varphi)]\}. \tag{10.78b}$$

Inasmuch as the imaginary part of ν_n and ν_n' is positive, it can be seen that the contribution of $I_2^{(e)}$ and $I_2^{(m)}$ will be small in the illuminated region $-\pi/2 < \varphi < \pi/2$.

Let us now concentrate on $I_1^{(e,m)}$ in (10.75d). Since $H_\nu^{(2)}(k\rho)$ has no zeros in the upper ν plane, we can write

$$I_1^{(e)} = -\frac{E_0}{2}\int_{L_3} \frac{H_\nu^{(2)}(ka)}{H_\nu^{(1)}(ka)} H_\nu^{(1)}(k\rho)\exp\left[i\nu\left(\varphi - \frac{\pi}{2}\right)\right] d\nu, \tag{10.79a}$$

$$I_1^{(m)} = -\frac{H_0}{2}\int_{L_3} \frac{H_\nu^{(2)\prime}(ka)}{H_\nu^{(1)\prime}(ka)} H_\nu^{(1)}(k\rho)\exp\left[i\nu\left(\varphi - \frac{\pi}{2}\right)\right] d\nu, \tag{10.79b}$$

where the path L_3 is shown in Fig. 10.6. In what follows we shall evaluate the foregoing integrals by the method of stationary phase. Hence, we must determine the location of the stationary points of the integrands in (10.79a) and (10.79b) in the ν plane. It follows from the formulas in (10.35) and (10.36) and from the accompanied remarks that

$$\frac{H_\nu^{(2)}(x)}{H_\nu^{(1)}(x)} \sim +i \exp\left\{-i2\left[(x^2-\nu^2)^{1/2} - \nu \arccos\left(\frac{\nu}{x}\right)\right]\right\}, \qquad \nu < x, \tag{10.80a}$$

$$\frac{H_\nu^{(2)}(x)}{H_\nu^{(1)}(x)} \sim -1, \qquad \nu > x, \tag{10.80b}$$

and

$$\frac{H_\nu^{(2)\prime}(x)}{H_\nu^{(1)\prime}(x)} \sim -i \exp\left\{-i2\left[(x^2-\nu^2)^{1/2} - \nu \arccos\left(\frac{\nu}{x}\right)\right]\right\}, \qquad \nu < x, \tag{10.81a}$$

$$\frac{H_\nu^{(2)\prime}(x)}{H_\nu^{(1)\prime}(x)} \sim -1, \qquad \nu > x. \tag{10.81b}$$

Moreover, using the asymptotic approximation (10.35a) for $H_\nu^{(1)}(k\rho)$ as $k\rho \to \infty$, we obtain

$$I_1^{(e)} \sim -\frac{E_0 \exp(i\pi/4)}{(2\pi)^{1/2}}(W_< + W_>), \tag{10.82a}$$

$$I_1^{(m)} \sim -\frac{H_0 \exp(i\pi/4)}{(2\pi)^{1/2}}(-W_< + iW_>), \tag{10.82b}$$

where

$$W_{\lessgtr} = \int_{L_{3\lessgtr}} \frac{\exp\{i[g_{\lessgtr}(\nu) + \nu\varphi]\}}{[(k\rho)^2 - \nu^2]^{1/4}}\, d\nu \tag{10.83}$$

and

$$g_<(\nu) = [(k\rho)^2 - \nu^2]^{1/2} - 2[(ka)^2 - \nu^2]^{1/2} - \nu\left[\arccos\frac{\nu}{k\rho} - 2\arccos\frac{\nu}{ka} + \frac{\pi}{2}\right] \tag{10.84a}$$

$$g_>(\nu) = [(k\rho)^2 - \nu^2]^{1/2} - \nu\left[\arccos\frac{\nu}{k\rho} + \frac{\pi}{2}\right]. \tag{10.84b}$$

The symbol $<$ denotes the portion of the integral on the left ($\nu < ka$) and $>$ denotes the portion of the integral on the right ($\nu > ka$).

The integrals $W_<$ and $W_>$ can now be evaluated, using the method of stationary phase. We find that the stationary point of $W_<$ is situated at

$$\nu_< \sim (ka)\cos(\pi - \varphi)/2, \qquad \rho \to \infty, \tag{10.85a}$$

while the stationary point of $W_>$ is given by

$$\nu_> = (k\rho)\cos(\varphi - \pi/2). \tag{10.85b}$$

We note

$$\begin{aligned}[(ka)^2 - \nu_<{}^2]^{1/2} &= (ka)\sin(\pi - \varphi)/2 \\ &= (ka)\cos\varphi/2, \qquad -\pi/2 < \varphi < \pi/2,\end{aligned} \tag{10.86a}$$

and

$$\begin{aligned}[(k\rho)^2 - \nu_>{}^2]^{1/2} &= k\rho\sin(\varphi - \pi/2) \\ &= -k\rho\cos\varphi, \qquad -\pi/2 < \varphi < \pi/2.\end{aligned} \tag{10.86b}$$

Now, we find

$$g_<(\nu_<) + \nu_<\varphi \sim k(\rho - 2a\cos\varphi/2), \tag{10.87a}$$

$$g_>(\nu_>) + \nu_>\varphi \sim -k\rho\cos\varphi, \tag{10.87b}$$

and

$$\frac{\partial^2}{\partial\nu^2}[g_<(\nu) + \nu\varphi]\bigg|_{\nu=\nu_<} \sim -\frac{2}{ka\cos\varphi/2} \tag{10.88a}$$

$$\frac{\partial^2}{\partial\nu^2}[g_>(\nu) + \nu\varphi]\bigg|_{\nu=\nu_>} \sim -(k\rho\cos\varphi)^{-1}. \tag{10.88b}$$

The path of integration can now be deformed to pass through the stationary points as shown in Fig. 10.8 and using the stationary phase formula (5.56), we find for $\rho \to \infty$

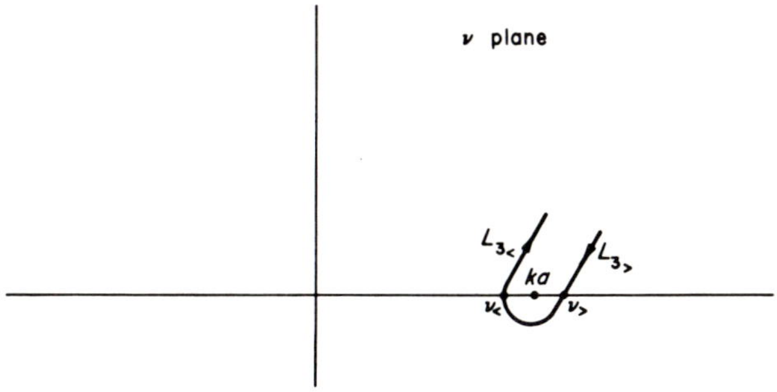

Fig. 10. 8. Deformed path of integration for evaluation of the geometrical optics field.

$$W_< \sim \left(\frac{\pi a \cos \varphi/2}{\rho}\right)^{1/2} \exp\left\{i\left[k\left(\rho - 2a\cos\frac{\varphi}{2}\right) - \frac{\pi}{4}\right]\right\} \tag{10.89a}$$

and

$$W_> \sim -i(2\pi)^{1/2} \exp\left[-i\left(k\rho\cos\varphi + \frac{\pi}{4}\right)\right]. \tag{10.89b}$$

Now substituting into (10.82a) and (10.82b), we obtain

$$E_z \sim E_0\left\{\exp(-ik\rho\cos\varphi) - \left(\frac{a\cos\varphi/2}{2\rho}\right)^{1/2} \exp\left[ik\left(\rho - 2a\cos\frac{\varphi}{2}\right)\right]\right\} \tag{10.90a}$$

and

$$H_z \sim H_0\left\{\exp(-ik\rho\cos\varphi) + \left(\frac{a\cos\varphi/2}{2\rho}\right)^{1/2} \exp\left[ik\left(\rho - 2a\cos\frac{\varphi}{2}\right)\right]\right\}, \tag{10.90b}$$

since the contribution from $I_2^{(e,m)}$ is negligible. It is clear that the first term in (10.90a) and (10.90b) represents the field incident on the cylinder since $k\rho\cos\varphi = x$. The second term represents the field that is reflected back in the region $-\pi/2 < \varphi < \pi/2$.

The formulas in (10.90) are in full agreement with the laws of geometrical optics. To show this, let us consider the diagram in Fig. 10.9. The incident ray of unity amplitude, which strikes the cylinder at point P and an angle φ' with respect to the local normal, has a phase factor $\exp(-ika\cos\varphi')$ at that

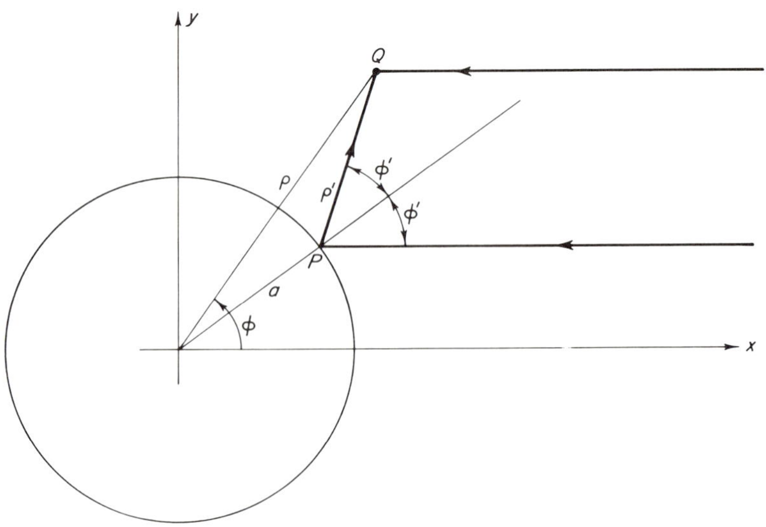

Fig. 10.9. Ray geometry in the illuminated region.

point. Upon reflection at the same angle φ', the ray acquires an amplitude A_r equal to the reflection coefficient (yet to be determined) at the cylinder surface. In traversing the distance P–Q, the ray acquires an additional phase factor $\exp(ik\rho')$. Hence the reflected ray at the point Q is of the form $A\ \exp[ik(\rho' - a\cos\varphi')]$. At a large distance from the cylinder, $\rho' \sim \rho - a \times \cos(\varphi - \varphi')$ and since also in this case $\varphi' \sim \varphi/2$, then $\rho' \sim \rho - a\cos\varphi/2$. Thus the reflected ray at the point Q has the form

$$\mp A_r \exp[ik(\rho - 2a\cos\varphi/2)], \tag{10.91}$$

where the upper sign is pertinent to the electric case, where reversal of polarization takes place upon reflection, and the lower sign is pertinent to the magnetic case, where no reversal of polarization takes place.

We must yet find the magnitude of the reflected ray, which can be obtained on the basis of the conservation of energy as follows. Consider the geometry shown in Fig. 10.10, where a thin rectangular beam of width $w \ll 1$ and of infinite extent in the direction parallel to the axis of the cylinder is incident on the cylinder with its upper edge making an angle φ' with respect to the local normal. By simple considerations it follows that the reflected beam is in the form of a wedge with its internal angle $2\beta \sim 2w/a\cos\varphi'$ when w is small. The apex of the wedge will be situated at the point O' a distance d from the point P. Again from simple considerations it follows that $d \sim (a/2)$

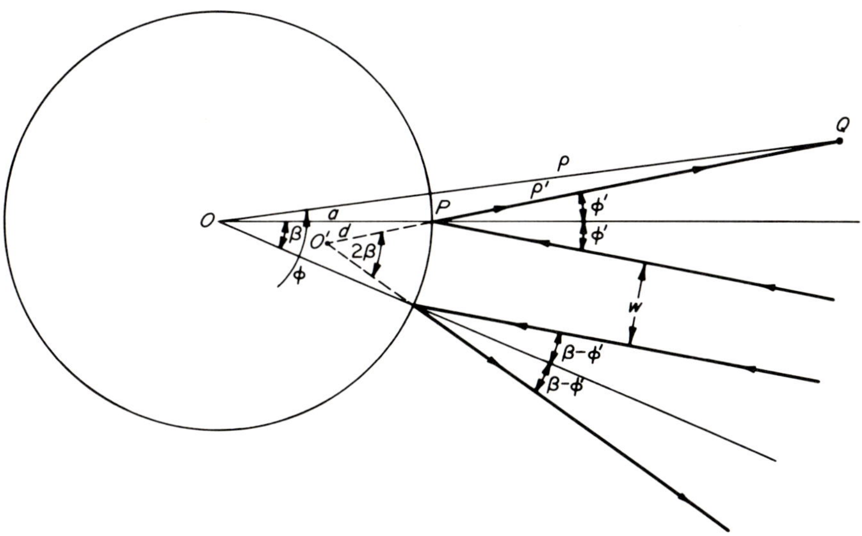

Fig. 10.10. Derivation of the reflection coefficient according to the laws of geometrical optics.

$\times \cos\varphi'$, and since the energies in the incident and the reflected beams must be equal, we must have

$$w = 2\beta(\rho' + d)A_r^2, \tag{10.92}$$

where A_r is the amplitude of the rays in the reflected beam while the amplitude of the rays in the incident beam is unity. Solving for A_r, with due consideration to our earlier remarks, we obtain

$$A_r = \left[\frac{(a/2)\cos\varphi'}{\rho + (a/2)\cos\varphi'}\right]^{1/2}. \tag{10.93}$$

Finally in the far zone $\rho' = \rho - a\cos(\varphi/2)$ and $\varphi' = \varphi/2$ and we obtain

$$A_r = \left(\frac{a\cos\varphi/2}{2\rho}\right)^{1/2}, \tag{10.94}$$

which together with (10.91) show that the formulas in (10.90) are identical to those obtained by means of geometrical optics.

10.2 Cylindrically Tipped Wedge

The Green function for a cylindrically tipped wedge was found in Chapter 8 to be

$$G_<^{(e)} = \frac{i\pi}{\varphi_0}\sum_{m=1}^{\infty}\left[J_\nu(k\rho) - \frac{J_\nu(ka)}{H_\nu^{(1)}(ka)}H_\nu^{(1)}(k\rho)\right]H_\nu^{(1)}(k\rho')\sin\nu\varphi\sin\nu\varphi' \tag{10.95}$$

in the electric case, and

$$G_<^{(m)} = \frac{i\pi}{2\varphi_0}\sum_{m=0}^{\infty}\varepsilon_m\left[J_\nu(k\rho) - \frac{J_\nu'(ka)}{H_\nu^{(1)\prime}(ka)}H_\nu^{(1)}(k\rho)\right] \times H_\nu^{(1)}(k\rho')\cos\nu\varphi\cos\nu\varphi' \tag{10.96}$$

for the magnetic excitation. If the source point is allowed to be moved to infinity, $k\rho' \to \infty$, then we can use the asymptotic formula

$$H_\nu^{(1)}(k\rho') \to \left(\frac{2}{\pi k\rho'}\right)^{1/2}(-i)^\nu \exp\left[i\left(k\rho' - \frac{\pi}{4}\right)\right] \tag{10.97}$$

to find the limiting form of the Green function pertinent to waves incident on the wedge from infinity. Since $E_z = i\omega\mu_0\mu I^{(e)}G^{(e)}$ and $H_z = i\omega\varepsilon_0 n^2 I^{(m)} \times G^{(m)}/\mu$, we obtain, using (10.97) and some trigonometric identities,

$$E_z = E_0\sum_{-\infty}^{\infty}(-i)^\nu\{\exp[i\nu(\varphi - \varphi')] - \exp[i\nu(\varphi + \varphi')]\} \times \left[J_\nu(k\rho) - \frac{J_\nu(ka)}{H_\nu^{(1)}(ka)}H_\nu^{(1)}(k\rho)\right], \tag{10.98a}$$

where

$$E_0 = -\frac{\omega\mu_0\mu}{4\varphi_0}\left(\frac{2\pi}{k\rho'}\right)^{1/2} \exp\left[i\left(k\rho' - \frac{\pi}{4}\right)\right] I^{(e)} \tag{10.98b}$$

and

$$H_z = H_0 \sum_{-\infty}^{\infty} (-i)^\nu \{\exp[i\nu(\varphi - \varphi')] + \exp[i\nu(\varphi + \varphi')]\}$$
$$\times \left[J_\nu(k\rho) - \frac{J_\nu'(ka)}{H_\nu^{(1)\prime}(ka)} H_\nu^{(1)}(k\rho)\right], \tag{10.99a}$$

where

$$H_0 = -\frac{\omega\varepsilon_0 n^2}{4\mu\varphi_0}\left(\frac{2\pi}{k\rho'}\right)^{1/2} \exp\left[i\left(k\rho' - \frac{\pi}{4}\right)\right] I^{(m)}. \tag{10.99b}$$

The formulas (10.98a) and (10.99a) give the total field (incident plus reflected) everywhere in the region $0 \le \varphi \le \varphi_0$, $\rho \ge a$ for a plane wave illumination incident at an angle $\varphi = \varphi'$on the structure. There is no simple way, however, to separate the reflected field from the total field for an arbitrary angle of the wedge φ_0. This can be appreciated by noting the expansion of the plane wave in terms of cylindrical waves

$$\exp[-ik\rho\cos(\varphi - \varphi')] = \sum_{-\infty}^{\infty} (-i)^n J_n(k\rho) \exp[in(\varphi - \varphi')]. \tag{10.100}$$

It can be seen that a term similar to the right side of (10.100) is indeed present in both (10.98a) and (10.99a) except that $\nu = m\pi/\varphi_0$ is not necessarily an integer unless $\varphi_0 = \pi$.

When $\varphi_0 = \pi$, the structure takes on the form shown in Fig. 10.11. Since

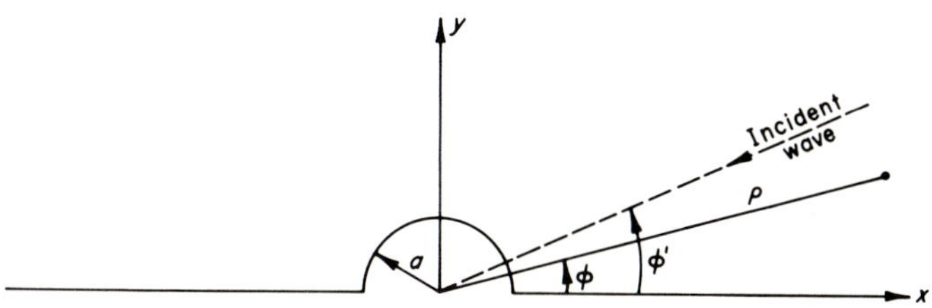

Fig. 10.11. Circular cylindrical boss on a plane.

in this case $\nu = m$, we can use (10.100) to rewrite (10.98a) and (10.99a) as

$$E_z = E_0\Big(\exp[-ik\rho\cos(\varphi - \varphi')] - \exp[-ik\rho\cos(\varphi + \varphi')] - \sum_{-\infty}^{\infty} (-i)^m$$
$$\times \{\exp[im(\varphi - \varphi')] - \exp[im(\varphi + \varphi')]\} \frac{J_m(ka)}{H_m^{(1)}(ka)} H_m^{(1)}(k\rho)\Big), \tag{10.101}$$

and

$$H_z = H_0\bigg(\exp[-ik\rho\cos(\varphi-\varphi')] + \exp[-ik\rho\cos(\varphi+\varphi')] - \sum_{-\infty}^{\infty}(-i)^m$$
$$\times \{\exp[im(\varphi-\varphi')] + \exp[im(\varphi+\varphi')]\}\frac{J_m'(ka)}{H_m^{(1)\prime}(ka)}H_m^{(1)}(k\rho)\bigg). \quad (10.102)$$

The first term in (10.101) and (10.102) represents the incident field. The second term represents the field reflected from the flat plate and it is negative in the electric case and positive in the magnetic case as it should be. Finally, the last term in (10.101) and (10.102) represents the portion of the field reflected from the cylindrical boss. In the far field we can replace $H_m^{(1)}(k\rho)$ by its asymptotic equivalent. Moreover, if the radius of the boss is small, so that $2\pi a < \lambda$, we can use the small argument approximation for $J_m(ka)$ and $H_m^{(1)}(ka)$ to obtain the approximate formulas

$$E_z \sim E_0\bigg\{\exp[-ik\rho\cos(\varphi-\varphi')] - \exp[-ik\rho\cos(\varphi+\varphi')]$$
$$+\left(\frac{\pi}{2k\rho}\right)^{1/2}(ka)^2\sin\varphi\sin\varphi'\exp\left[i\left(k\rho+\frac{\pi}{4}\right)\right]\bigg\} \quad (10.103)$$

and

$$H_z \sim H_0\bigg\{\exp[-ik\rho\cos(\varphi-\varphi')] + \exp[-ik\rho\cos(\varphi+\varphi')]$$
$$-\left(\frac{\pi}{2k\rho}\right)^{1/2}(ka)^2(1-2\cos\varphi\cos\varphi')\exp\left[i\left(k\rho+\frac{\pi}{4}\right)\right]\bigg\}. \quad (10.104)$$

valid when $2\pi a < \lambda$ and $2\pi\rho \gg \lambda$.

In the case when the radius of the boss is not small, so that $ka > 1$, the Watson transformation can be applied to convert the slow convergent harmonic series into the residue series. However, because of the presence of the terms $\exp[im(\varphi \pm \varphi')]$, the situation is considerably more complex than in the case, treated earlier, of an isolated cylinder. The regions of applicability of various representations (shadow region, illuminated region) will generally overlap and hence complicate the form of the final answer.

10.3 The Wedge

10.3.1 Method of Separation of Variables

The Green function for a perfectly conducting wedge was found in Chapter 8 to be

$$G_<^{(e)} = \frac{i\pi}{\varphi_0}\sum_{m=1}^{\infty} H_\nu^{(1)}(k\rho')J_\nu(k\rho)\sin\nu\varphi\sin\nu\varphi' \quad (10.105)$$

in the electric case and

$$G_<^{(m)} = \frac{i\pi}{2\varphi_0} \sum_{m=0}^{\infty} \varepsilon_m H_\nu^{(1)}(k\rho')J_\nu(k\rho) \cos \nu\varphi \cos \nu\varphi' \tag{10.106}$$

in the magnetic case. If the source point is allowed to be moved to infinity, $k\rho' \to \infty$, we can replace the Hankel function $H_\nu^{(1)}(k\rho')$ by its asymptotic equivalent to obtain

$$E_z = 2E_0 \sum_{m=1}^{\infty} (-i)^\nu [\cos \nu(\varphi - \varphi') - \cos \nu(\varphi + \varphi')] J_\nu(k\rho) \tag{10.107}$$

and

$$H_z = H_0 \sum_{m=0}^{\infty} (-i)^\nu \varepsilon_m [\cos \nu(\varphi - \varphi') + \cos \nu(\varphi + \varphi\ )] J_\nu(k\rho), \tag{10.108}$$

where E_0 and H_0 are given by (10.98b) and (10.99b), respectively.

In general it is not possible to separate the incident and the reflected waves in this case. In the case of an infinite plane ($\varphi_0 = \pi$, $\nu = m$), however, we readily obtain

$$E_z = E_0 \{\exp[-ik\rho \cos(\varphi - \varphi')] - \exp[-ik\rho \cos(\varphi + \varphi')]\} \tag{10.109}$$

and

$$H_z = H_0 \{\exp[-ik\rho \cos(\varphi - \varphi')] + \exp[-ik\rho \cos(\varphi + \varphi')]\}, \tag{10.110}$$

which express the well-known laws of reflection from a perfectly conducting plane.

In the case of a half plane along the positive x axis ($\varphi_0 = 2\pi$, $\nu = m/2$), we have

$$E_z = 2E_0 \sum_1^{\infty} (-i)^{m/2} \left[\cos \frac{m}{2} (\varphi - \varphi') - \cos \frac{m}{2} (\varphi + \varphi')\right] J_{m/2}(k\rho) \tag{10.111}$$

and

$$H_z = H_0 \sum_0^{\infty} \varepsilon_m (-i)^{m/2} \left[\cos \frac{m}{2} (\varphi - \varphi') + \cos \frac{m}{2} (\varphi + \varphi')\right] J_{m/2}(k\rho), \tag{10.112}$$

which, evidently, cannot be summed by any known means.

10.3.2 The Lebedev-Kontorovich Transform

In Chapter 8 we found an integral representation of the wedge solution as follows.

$$\Pi_z^{(e)} = -\frac{i\mu I^{(e)}}{16\omega\varepsilon_0 n^2 \varphi_0} (W_1 - W_2) \tag{10.113a}$$

and

$$\Pi_z^{(m)} = -\frac{iI^{(m)}}{16\omega\mu_0\,\mu\varphi_0}(W_1 + W_2), \tag{10.113b}$$

where

$$\begin{aligned} \underset{\varphi<\varphi'}{W_{1,2}} = -i4\varphi_0\Bigg(&\sum_{m=1}^{M_{1,2}} H_0^{(1)}\{k[\rho^2 + \rho'^2 - 2\rho\rho' \cos(2m\varphi_0 - \varphi_{1,2})]^{1/2}\} \\ &+ \sum_{n=0}^{N_{1,2}} H_0^{(1)}\{k[\rho^2 + \rho'^2 - 2\rho\rho' \cos(2n\varphi_0 + \varphi_{1,2})]^{1/2}\}\Bigg) \\ &+ \int_{-i\infty}^{i\infty} \frac{\sin 2v(\pi + x) H_0^{(1)}[k(\rho^2 + \rho'^2 + 2\rho\rho' \cos x)^{1/2}]}{\sin v(x + \pi + \varphi_{1,2}) \sin v(x + \pi - \varphi_{1,2})}\,dx \end{aligned} \tag{10.114}$$

where $v = \pi/2\varphi_0$, $\varphi_1 = \varphi' - \varphi$, and $\varphi_2 = \varphi' + \varphi$.

If the line source is allowed to be moved to infinity, $\rho' \to \infty$, we obtain

$$E_z = \frac{\varphi_0 E_0}{\pi}(V_1 - V_2), \tag{10.115}$$

$$H_z = \frac{\varphi_0 H_0}{\pi}(V_1 + V_2), \tag{10.116}$$

where E_0 and H_0 are defined in (10.98b) and (10.99b), respectively. Moreover,

$$\begin{aligned} \underset{\varphi<\varphi'}{V_{1,2}} &= \sum_{m=1}^{M_{1,2}} \exp[-ik\rho \cos(2m\varphi_0 - \varphi_{1,2})] \\ &\quad + \sum_{n=0}^{N_{1,2}} \exp[-ik\rho \cos(2n\varphi_0 + \varphi_{1,2})] + I_{1,2} \end{aligned} \tag{10.117}$$

and

$$I_{1,2} = \frac{i}{4\varphi_0}\int_{-i\infty}^{i\infty} \frac{\sin 2v(\pi + x)\exp(ik\rho \cos x)}{\sin v(x + \pi + \varphi_{1,2}) \sin v(x + \pi - \varphi_{1,2})}\,dx. \tag{10.118}$$

When the distance from the apex of the wedge is large, so that $k\rho \gg 1$, the integral $I_{1,2}$ can be evaluated by using the method of stationary phase. We obtain in this case

$$I_{1,2} \sim -\frac{1}{4\varphi_0}\left(\frac{2\pi}{k\rho}\right)^{1/2} \frac{\sin 2v\pi \exp[i(k\rho + \pi/4)]}{\sin v(\pi + \varphi_{1,2}) \sin v(\pi - \varphi_{1,2})}, \tag{10.119}$$

which, evidently, is valid for all φ except in the neighborhoods where $v(\pi \pm \varphi_{1,2})$ approaches a multiple of π.

In order to obtain a suitable representation valid in the regions where

$v(\pi \pm \varphi_{1,2})$ approaches a multiple of π, we proceed as follows.

Noting the identity

$$\frac{\sin 2v(\pi + x)}{\sin v(\pi + x + \varphi) \sin v(\pi + x - \varphi)} = \frac{\cos v(\pi + x)}{\cos v\varphi} \times \{[\sin v(\pi + x + \varphi)]^{-1} + [\sin v(\pi + x - \varphi)]^{-1}\}, \tag{10.120}$$

we can rewrite (10.118) as

$$I_{1,2} = \frac{i}{4\varphi_0 \cos v\varphi_{1,2}} [h(\varphi_{1,2}) + h(-\varphi_{1,2})], \tag{10.121}$$

where

$$h(\varphi) = \int_{-i\infty}^{i\infty} \frac{\cos v(\pi + x)}{\sin v(\pi + x + \varphi)} \exp(ik\rho \cos x)\, dx. \tag{10.122}$$

Since the main contribution to this integral comes from the neighborhood of the point $x = 0$ and the numerator is a slowly varying function, we can write

$$\begin{aligned} h(\varphi) &= \cos v\pi \int_{-i\infty}^{i\infty} \frac{\exp(ik\rho \cos x)}{\sin v(\pi + x + \varphi)}\, dx \\ &= \tfrac{1}{2} \cos v\pi \int_{-i\infty}^{i\infty} \left[\frac{\exp(ik\rho \cos x)}{\sin v(\pi + \varphi + x)} + \frac{\exp(ik\rho \cos x)}{\sin v(\pi + \varphi - x)}\right] dx \\ &= 2 \cos v\pi \sin v(\pi + \varphi) \int_{-i\infty}^{i\infty} \frac{\cos vx \exp(ik\rho \cos x)}{\cos 2vx - \cos 2v(\pi + \varphi)}\, dx. \end{aligned} \tag{10.123}$$

Now we set $\cos x = 1 + is^2$, so that $x \sim \sqrt{2} \exp(-i\pi/4)s$. Moreover, $\cos vx \sim 1$, $\cos 2vx \sim 1 + i4v^2s^2$, and putting

$$\eta = \frac{\sin v(\pi + \varphi)}{v\sqrt{2}}, \tag{10.124}$$

we obtain

$$h(\varphi) = \frac{2\eta \cos v\pi}{v} \exp\left[i\left(k\rho + \frac{\pi}{4}\right)\right] \int_0^{\infty} \frac{\exp(-k\rho s^2)}{s^2 - i\eta^2}\, ds. \tag{10.125}$$

The integral on the right in (10.125) can be reconciled with the complementary error function (Abramowitz and Stegun (1964), p. 297). Let us put

$$t = (k\rho)^{1/2} s, \tag{10.126a}$$

$$\zeta = \exp\left(\frac{i\pi}{4}\right)\eta(k\rho)^{1/2} = \exp\left(\frac{i\pi}{4}\right)\left(\frac{k\rho}{2}\right)^{1/2}\frac{\sin c(\pi + \varphi)}{v} \tag{10.126b}$$

and obtain

$$h(\varphi) = \frac{i\pi \cos v\pi}{v}\exp(ik\rho)[\exp(-\zeta^2)\operatorname{erfc}(-i\zeta)], \qquad \operatorname{Im}(\zeta) > 0, \tag{10.127a}$$

$$= -\frac{i\pi \cos v\pi}{v}\exp(ik\rho)[\exp(-\zeta^2)\operatorname{erfc}(-i\zeta)], \qquad \operatorname{Im}(\zeta) < 0, \tag{10.127b}$$

where we used the definition of the complementary error function

$$\zeta\int_0^\infty \frac{\exp(-t^2)\,dt}{t^2-\zeta^2} = \begin{cases} \dfrac{i\pi}{2}\exp(-\zeta^2)\operatorname{erfc}(-i\zeta), & \operatorname{Im}(\zeta) > 0, \\[2ex] -\dfrac{i\pi}{2}\exp(-\zeta^2)\operatorname{erfc}(i\zeta), & \operatorname{Im}(\zeta) < 0. \end{cases} \tag{10.128}$$

When the argument of the error function is large, i.e., when $v(\pi \pm \varphi_{1,2})$ is not a multiple of π, we can use the asymptotic formula

$$\operatorname{erfc}(\mp i\zeta) \sim \pm\frac{i}{\sqrt{\pi}}\frac{\exp(\zeta^2)}{\zeta} \tag{10.129}$$

valid for $\zeta \to \infty$, $|\arg(\mp i\zeta)| < 3\pi/4$, to obtain (10.119).

10.3.2.1 Diffraction by a Conducting Half Plane

We now apply the formalism of the Lebedev–Kontorovich representation to the classical problem of plane wave diffraction by a perfectly conducting half plane, the geometry of which is shown in Fig. 10.12. We note that in

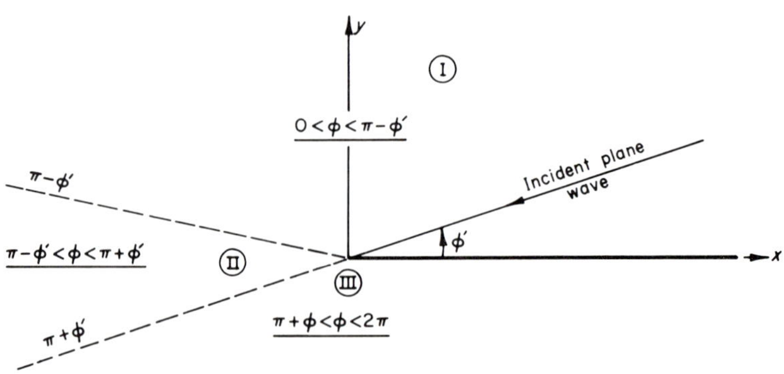

Fig. 10.12. Geometry of a half plane.

this case $\varphi_0 = 2\pi$ and $v = 1/4$. Without loss of generality we can assume that $\varphi' \leq \pi$ and it will follow from (8.191) that in this case M_1 and M_2 do not exist for either $\varphi < \varphi'$ or $\varphi > \varphi'$. Hence, the summations over m in (10.117) do not contribute. Moreover, it follows from (8.191) that for $\varphi < \varphi'$, $N_1 = 0$ and $N_2 = 0$ if $\varphi < \pi - \varphi' < \varphi'$. For $\varphi > \varphi'$ it follows that $N_1 = 0$ only in the region $\varphi' < \varphi < \pi + \varphi'$ and $N_2 = 0$ only in the region $\varphi' < \varphi < \pi - \varphi'$. It will prove convenient to divide the space about the half plane into regions I, II, and III with the corresponding bounds $0 < \varphi < \pi - \varphi'$, $\pi - \varphi' < \varphi < \pi + \varphi'$, and $\pi + \varphi' < \varphi < 2\pi$, as shown in Fig. 10.12. Then, using (10.117) we can write the following.

Region I

$$V_1 = \exp[-ik\rho \cos(\varphi - \varphi')] + I_1,$$
$$V_2 = \exp[-ik\rho \cos(\varphi + \varphi')] + I_2 . \tag{10.130a}$$

Region II

$$V_1 = \exp[-ik\rho \cos(\varphi - \varphi')] + I_1,$$
$$V_2 = I_2 , \tag{10.130b}$$

Region III

$$V_1 = I_1,$$
$$V_2 = I_2 . \tag{10.130c}$$

Except in the neighborhood of the points $\varphi = \pi - \varphi'$ and $\varphi = \pi + \varphi'$, I_1 and I_2 can be written, using (10.119),

$$I_1 \sim -\frac{\exp[i(k\rho + \pi/4)]}{2(2\pi k\rho)^{1/2}} \left[\cos\frac{(\varphi - \varphi')}{2}\right]^{-1}, \tag{10.131a}$$

$$I_2 \sim -\frac{\exp[i(k\rho + \pi/4)]}{2(2\pi k\rho)^{1/2}} \left[\cos\frac{(\varphi + \varphi')}{2}\right]^{-1}, \tag{10.131b}$$

so that

$$I_1 - I_2 = \left(\frac{2}{\pi k\rho}\right)^{1/2} \frac{\sin \varphi/2 \sin \varphi'/2}{\cos \varphi + \cos \varphi'} \exp\left[i\left(k\rho + \frac{\pi}{4}\right)\right]$$
$$= E_z^{(D)} \tag{10.132}$$

and

$$I_1 + I_2 = -\left(\frac{2}{\pi k\rho}\right)^{1/2} \frac{\cos \varphi/2 \cos \varphi'/2}{\cos \varphi + \cos \varphi'} \exp\left[i\left(k\rho + \frac{\pi}{4}\right)\right]$$
$$= H_z^{(D)}, \tag{10.133}$$

where $E_z^{(D)}$ and $H_z^{(D)}$ clearly represent the diffracted fields. Substituting the results above into (10.115) for the electric case and (10.116) for the magnetic case, we obtain the complete representation of the solution valid everywhere for $k\rho \to \infty$ except in the neighborhood of $\varphi = \pi \pm \varphi'$

$$\text{I:}\quad E_z \sim \exp[-ik\rho\cos(\varphi - \varphi')] - \exp[-ik\rho\cos(\varphi + \varphi')] + E_z^{(D)}, \tag{10.134a}$$

$$\text{II:}\quad E_z \sim \exp[-ik\rho\cos(\varphi - \varphi')] + E_z^{(D)}, \tag{10.134b}$$

$$\text{III:}\quad E_z \sim E_z^{(D)}, \tag{10.134c}$$

and

$$\text{I:}\quad H_z \sim \exp[-ik\rho\cos(\varphi - \varphi')] + \exp[-ik\rho\cos(\varphi + \varphi')] - H_z^{(D)} \tag{10.135a}$$

$$\text{II:}\quad H_z \sim \exp[-ik\rho\cos(\varphi - \varphi')] - H_z^{(D)}, \tag{10.135b}$$

$$\text{III:}\quad H_z \sim -H_z^{(D)}. \tag{10.135c}$$

The meaning of the various terms in (10.134) and (10.135) is clear from inspection of Fig. 10.12. The term $\exp[ik\rho\cos(\varphi - \varphi')]$ represents the incident plane wave and is present only in the illuminated regions, I and II. The term $\exp[-ik\rho\cos(\varphi + \varphi')]$ represents the reflected field from the half plane; it is negative in the electric case and positive in the magnetic case and is absent from region II, as expected. It is clear that the terms $\exp[-ik\rho \times \cos(\varphi \pm \varphi')]$ represent the geometrical optics field. The terms $E_z^{(D)}$ and $H_z^{(D)}$ denote the field that is diffracted by the edge and, as expected, this field is present in all regions of space about the half plane. This field decays as $\rho^{-1/2}$; hence, it can be thought of as one produced by a distribution of line sources at the edge of the half plane.

The remaining task is that of finding a proper representation for the fields in the neighborhood of the points $\varphi = \pi \mp \varphi'$ where the previous representation is not valid. Let us define

$$\zeta_1^{+} = 4(2k\rho)^{1/2}\sin\tfrac{1}{4}(\pi + \varphi - \varphi')\exp(i\pi/4), \tag{10.136a}$$

$$\zeta_2^{+} = 4(2k\rho)^{1/2}\sin\tfrac{1}{4}(\pi + \varphi + \varphi')\exp(i\pi/4), \tag{10.136b}$$

and

$$\zeta_1^{-} = 4(2k\rho)^{1/2}\sin\tfrac{1}{4}(\pi - \varphi + \varphi')\exp(i\pi/4), \tag{10.137a}$$

$$\zeta_2^{-} = 4(2k\rho)^{1/2}\sin\tfrac{1}{4}(\pi - \varphi - \varphi'))\exp(i\pi/4); \tag{10.137b}$$

then it follows that $\mathrm{Im}(\zeta_{1,2}^{+}) > 0$ in the entire region $0 < \varphi < 2\pi$, while $\mathrm{Im}(\zeta_1^{-}) \gtrless 0$ for $0 < \varphi \gtrless \pi + \varphi'$ and $\mathrm{Im}(\zeta_2^{-}) \gtrless 0$ for $0 < \varphi \gtrless \pi - \varphi'$. Hence, using (10.127) and (10.121), we obtain the following.

Region I $(0 < \varphi < \pi - \varphi')$

$$I_1 = -\frac{\exp(ik\rho)}{\sqrt{2}\cos(\varphi - \varphi')/4} \times \{\exp[-(\zeta_1^{+})^2]\operatorname{erfc}(-i\zeta_1^{+}) + \exp[-(\zeta_1^{-})^2]\operatorname{erfc}(-i\zeta_1^{-})\}, \quad (10.138a)$$

$$I_2 = -\frac{\exp(ik\rho)}{\sqrt{2}\cos(\varphi + \varphi')/4} \times \{\exp[-(\zeta_2^{+})^2]\operatorname{erfc}(-i\zeta_2^{+}) + \exp[-(\zeta_2^{-})^2]\operatorname{erfc}(-i\zeta_2^{-})\}. \quad (10.138b)$$

Region II $(\pi - \varphi' < \varphi < \pi + \varphi')$

$$I_1 = -\frac{\exp(ik\rho)}{\sqrt{2}\cos(\varphi - \varphi')/4} \times \{\exp[-(\zeta_1^{+})^2]\operatorname{erfc}(-i\zeta_1^{+}) + \exp[-(\zeta_1^{-})^2]\operatorname{erfc}(-i\zeta_1^{-})\}, \quad (10.139a)$$

$$I_2 = -\frac{\exp(ik\rho)}{\sqrt{2}\cos(\varphi + \varphi')/4} \times \{\exp[-(\zeta_2^{+})^2]\operatorname{erfc}(-i\zeta_2^{+}) - \exp[-(\zeta_2^{-})^2]\operatorname{erfc}(i\zeta_2^{-})\}. \quad (10.139b)$$

Region III $(\pi + \varphi' < \varphi < 2\pi)$

$$I_1 = -\frac{\exp(ik\rho)}{\sqrt{2}\cos(\varphi - \varphi')/4} \times \{\exp[-(\zeta_1^{+})^2]\operatorname{erfc}(-i\zeta_1^{+}) - \exp[-(\zeta_1^{-})^2]\operatorname{erfc}(i\zeta_1^{-})\}, \quad (10.140a)$$

$$I_2 = -\frac{\exp(ik\rho)}{\sqrt{2}\cos(\varphi + \varphi')/4} \times \{\exp[-(\zeta_2^{+})^2]\operatorname{erfc}(-i\zeta_2^{+}) - \exp[-(\zeta_2^{-})^2]\operatorname{erfc}(i\zeta_2^{-})\}. \quad (10.140b)$$

These formulas can be used to find the diffracted fields, $E_z^{(D)} = I_1 - I_2$, $H_z^{(D)} = I_1 + I_2$ in the respective regions.

It will be noted that as $k\rho \to \infty$, the arguments of all error functions are large and always positive in their respective regions. Consequently, we can use the asymptotic formula (10.129) to show that I_1 and I_2 become identical with the forms given in (10.131) derived by the method of stationary phase. For other ranges of $k\rho$ as well as in the vicinity of $\varphi = \pi \mp \varphi'$, the complete expression must be used. The computations are somewhat simplified by noting the relations

$$\operatorname{erfc}(\pm z) = 1 \mp \operatorname{erf}(z) \tag{10.141a}$$

and

$$\operatorname{erf}\left[\exp(-i\pi/4)x\left(\frac{\pi}{2}\right)^{1/2}\right] = \sqrt{2}\exp(-i\pi/4)[C(x) + iS(x)], \tag{10.141b}$$

where $C(x)$ and $S(x)$ are Fresnel integrals that have been extensively tabulated (Abramowitz and Stegun (1964), p. 321).

10.4 The Wiener-Hopf Technique

The Wiener–Hopf technique was first proposed in 1931 (Wiener and Hopf, 1931) to solve a certain type of integral equation. The technique was subsequently applied to solve problems arising in the theory of smoothing and prediction (Wiener, 1949), and to a class of problems of diffraction theory (Copson, 1946).

The technique is extremely useful in solving problems of diffraction when the boundary condition changes abruptly (e.g., conducting half plane, slit, half plane at the interface of two different media). The solution is formulated using the Fourier integral representation of the fields and the unknown coefficients are determined using analytic continuation. The technique will be illustrated by the following examples.

10.4.1 Diffraction by a Conducting Half Plane, Magnetic Case

Consider an infinite perfectly conducting half plane situated on the positive x axis as shown in Fig. 10.12. An incident plane wave has the magnetic field vector along the z axis and it is incident at an angle φ'. Hence, the incident field can be written

$$\begin{aligned} H_z^{(i)} &= \exp[-ik(x\cos\varphi' + y\sin\varphi')] \\ &= \exp[-ik\rho\cos(\varphi - \varphi')]. \end{aligned} \tag{10.142}$$

The total field around the half plane $H_z^{(t)}$ will consist of the incident field $H_z^{(i)}$ and scattered field $H_z^{(s)}$. Hence, we can write

$$H_z^{(t)} = H_z^{(i)} + H_z^{(s)} \tag{10.143}$$

and it is clear that the scattered field must satisfy the differential equation

$$\left(\frac{\partial^2}{\partial x^2} + \frac{\partial^2}{\partial y^2} + k^2\right)H_z^{(s)} = 0. \tag{10.144}$$

Moreover, the electric field components can be expressed in terms of H_z as

$$E_x = \frac{i\mu}{\omega\varepsilon_0 n^2}\frac{\partial H_z}{\partial y}, \tag{10.145a}$$

$$E_y = -\frac{i\mu}{\omega\varepsilon_0 n^2}\frac{\partial H_z}{\partial x}. \tag{10.145b}$$

The boundary conditions to be satisfied by the unknown field $H_z^{(s)}$ can be found as follows. Since $E_x^{(t)}$ must vanish for $y = 0$, $0 < x \leq \infty$, it will follow that

$$\frac{\partial H_z^{(s)}}{\partial y} = ik \sin\varphi' \exp(-ikx\cos\varphi'), \qquad y = 0, 0 < x < \infty. \tag{10.146a}$$

Moreover, since the tangential electric field is zero on both sides of the half plane, $y = 0$, $0 < x < \infty$, and must be continuous at $y = 0$, $-\infty \leq x \leq 0$, hence it must be continuous for all x at $y = 0$; thus

$$\frac{\partial H_{z<}^{(s)}}{\partial y} = \frac{\partial H_{z>}^{(s)}}{\partial y}, \qquad y = 0, -\infty < x < \infty. \tag{10.146b}$$

The tangential magnetic field must be continuous for $y = 0$, $-\infty \leq x < 0$, but since H_z does not vanish at the conducting half plane and it will assume different values on its top and bottom surface, we must write

$$H_{z<}^{(s)} = H_{z>}^{(s)}, \qquad y = 0, -\infty < x < 0. \tag{10.146c}$$

Now the integral representation for the unknown field $H_z^{(s)}$ can be written

$$H_{z>}^{(s)} = (2\pi)^{-1/2}\int_{-\infty}^{\infty} A_1(\alpha)\exp[i(\alpha x + \tau y)]\,d\alpha, \qquad y > 0, \tag{10.147a}$$

$$H_{z<}^{(s)} = (2\pi)^{-1/2}\int_{-\infty}^{\infty} A_2(\alpha)\exp[i(\alpha x - \tau y)]\,d\alpha, \qquad y < 0, \tag{10.147b}$$

where $\tau = (k^2 - \alpha^2)^{1/2}$ and $\mathrm{Im}(\tau) > 0$. It is clear that the $H_{z>}^{(s)}$ and $H_{z<}^{(s)}$ as stated in (10.147) satisfy the differential equation (10.144). By virtue of (10.146b) it follows that $A_2 = -A_1 = -A$ and we can write

$$H_{z>}^{(s)} = (2\pi)^{-1/2}\int_{-\infty}^{\infty} A(\alpha)\exp[i(\alpha x + \tau y)]\,d\alpha, \tag{10.148a}$$

$$H_{z<}^{(s)} = -(2\pi)^{-1/2}\int_{-\infty}^{\infty} A(\alpha)\exp[i(\alpha x - \tau y)]\,d\alpha. \tag{10.148b}$$

Now applying (10.146c) and (10.146a), we obtain

$$(2\pi)^{-1/2} \int_{-\infty}^{\infty} A(\alpha) \exp(i\alpha x)\, d\alpha = 0, \qquad x < 0, \tag{10.149a}$$

$$(2\pi)^{-1/2} \int_{-\infty}^{\infty} \tau A(\alpha) \exp(i\alpha x)\, d\alpha = k \sin \varphi' \exp(-ikx \cos \varphi'), \qquad x > 0. \tag{10.149b}$$

Equations (10.149a) and (10.149b) constitute a set of dual integral equations that can be solved for the unknown function $A(\alpha)$ by means of a procedure known as the Wiener–Hopf technique.

As the next step, we transform the integrals in (10.149a) and (10.149b) into contour integrals by adding infinite semicircles of which the contributions are to be vanishingly small.

In the integral of (10.149a), x is negative. Then by Jordan's lemma, if $A(\alpha) \to 0$ as $|\alpha| \to \infty$, we can close the path of integration with a large semicircle *below* the real axis without changing the value of the integral, providing that $A(\alpha)$ has no singularities in the lower half plane.,

In the integral of (10.149b), x is positive; consequently, we can close the path of integration in the upper half plane without affecting the value of the integral providing that $A(\alpha)(k^2 - \alpha^2)^{1/2} \to 0$ as $|\alpha| \to \infty$. Since, however, the right side of (10.149b) is different from zero, we must require that $A(\alpha)$ have an appropriate pole singularity, so that the result of this contour integration will yield a value equal to the right side of (10.149b).

Consider the contour as shown in Fig. 10.13. A simple consideration will reveal that the exponential function can be represented by a contour integral as follows.

$$\exp(-ikx \cos \varphi') = (2\pi i)^{-1} \int_{C+C_1} \frac{\exp(i\alpha x)}{\alpha + k \cos \varphi'}\, d\alpha \xrightarrow[R\to\infty]{} (2\pi i)^{-1} \times \int_C \frac{\exp(i\alpha x)}{\alpha + k \cos \varphi'}\, d\alpha. \tag{10.150}$$

The value of this integral will not change if we introduce in the integrand some function, say $\Phi_+(\alpha)$, that is regular in the upper half plane and for $|\alpha| \to \infty$ behaves in such a manner that the contribution from the semicircular arc still vanishes. Thus, we can write

$$\exp(-ikx \cos \varphi') = (2\pi i)^{-1} \int_C \frac{\Phi_+(\alpha) \exp(i\alpha x)\, d\alpha}{\Phi_+(-k \cos \varphi')(\alpha + k \cos \varphi')}. \tag{10.151}$$

Substituting (10.151) into (10.149b) and with the understanding that the path of integration from $-\infty$ to $+\infty$ shall be identical with path C in Fig. 10.13, we conclude that

$$\tau A(\alpha) = \frac{k \sin \varphi'}{i(2\pi)^{1/2}} \cdot \frac{\Phi_+(\alpha)}{\Phi_+(-k \cos \varphi')(\alpha + k \cos \varphi')}, \tag{10.152a}$$

which upon rearranging becomes

$$(k - \alpha)^{1/2}(\alpha + k \cos \varphi')A(\alpha) = \frac{k \sin \varphi'}{i(2\pi)^{1/2}} \cdot \frac{\Phi_+(\alpha)}{(k + \alpha)^{1/2}\Phi_+(-k \cos \varphi')}. \tag{10.152b}$$

The expression on the left in (10.152b) is regular in the lower half plane since $A(\alpha)$ has no singularities therein. The expression on the right in (10.152b) is free of singularities in the upper half plane since $\Phi_+(\alpha)$ is regular therein. Since both sides are to be equal, it will follow that either side will be regular in the entire α plane. Moreover, since $A(\alpha) \to 0$ as $|\alpha| \to \infty$, then the function represented by either side of (10.152b) is bounded. Now, Liouville's theorem states that a function that is regular for all finite values of the complex variable, say α, and is bounded at infinity must be a constant. Thus, we must have

$$C = -\frac{ik \sin \varphi'}{(2\pi)^{1/2}} \frac{\Phi_+(\alpha)}{(k + \alpha)^{1/2}\Phi_+(-k \cos \varphi')}. \tag{10.153a}$$

This constant may be easily evaluated by setting $\alpha = -k \cos \varphi'$. Hence

$$C = -\frac{ik \sin \varphi'}{[2\pi k(1 - \cos \varphi')]^{1/2}}, \tag{10.153b}$$

which upon substitution back into (10.152b) gives

$$A(\alpha) = -\frac{ik \sin \varphi'}{[2\pi k(1 - \cos \varphi')]^{1/2}} \frac{1}{(k - \alpha)^{1/2}(\alpha + k \cos \varphi')}. \tag{10.154}$$

Finally, substituting back into (10.148), we obtain

$$H_{z \gtrless}^{(s)} = \mp \frac{ik \sin \varphi'}{2\pi[k(1 - \cos \varphi')]^{1/2}} \int_C \frac{\exp[i(\alpha x + \tau |y|)]}{(k - \alpha)^{1/2}(\alpha + k \cos \varphi')} d\alpha, \tag{10.155}$$

where C is the path running along the real axis with indentations as shown in Fig. 10.13. Equation (10.155) constitutes a formal answer to the problem we have set forth to solve.

In order to recast (10.155) into a form suitable for numerical calculations, we proceed as follows. Making the transformations

$$\alpha = k \cos \beta, \quad x = \rho \cos \varphi, \quad y = \rho \sin \varphi, \tag{10.156}$$

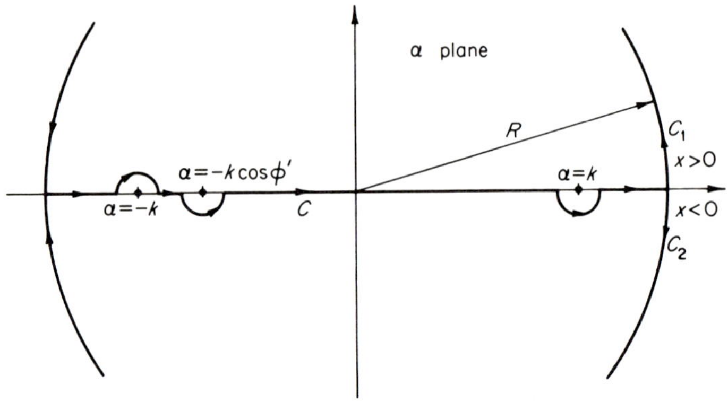

Fig. 10.13. Contour for evaluation of the dual integral equations.

and using some trigonometrical identities, we obtain in place of (10.155) for $H_{z>}^{(s)}$

$$H_{z>}^{(s)} = -\frac{i}{\pi}\int_{\Gamma_2} \frac{\cos \varphi'/2 \cos \beta/2}{\cos \beta + \cos \varphi'} \exp[ik\rho \cos(\beta - \varphi)]\, d\beta, \qquad (10.157)$$

where Γ_2 is a contour in the complex β plane, as shown in Fig. 10.14. Next, we attempt to deform the original path of integration Γ_2 to the path of steepest descents Γ. Here we note that the integrand in (10.157) has a pole where the denominator vanishes, i.e., where $\beta_P = \pi - \varphi'$. The integral

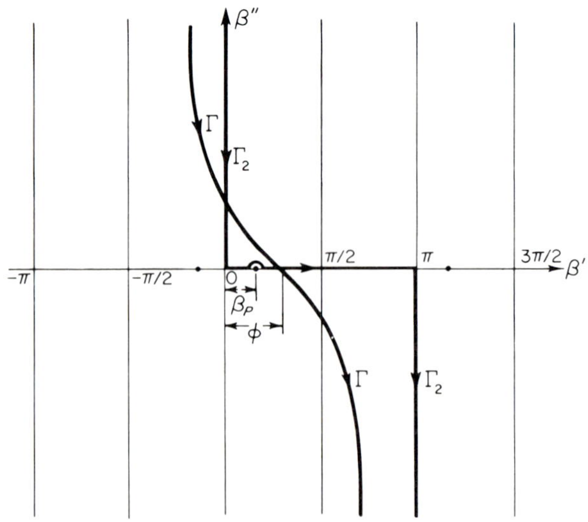

Fig. 10.14. The complex β plane for evaluation of the field integrals ($\beta_P = \pi - \varphi'$).

(10.157) evaluated at this pole gives $\exp[-ik\rho\cos(\varphi+\varphi')]$ and it is clear that this residue will be captured only if $\varphi<\pi-\varphi'$. Hence, we obtain

$$\begin{aligned} H_{z>}^{(s)} &= \exp[-ik\rho\cos(\varphi+\varphi')] - H_{z>}^{(D)}, \qquad 0<\varphi<\pi-\varphi', \\ H_{z>}^{(s)} &= -H_{z>}^{(D)}, \qquad \pi-\varphi'<\varphi<\pi, \end{aligned} \tag{10.158}$$

where

$$H_{z>}^{(D)} = \frac{i}{\pi}\int_\Gamma \frac{\cos\varphi'/2\cos\beta/2}{\cos\beta+\cos\varphi'}\exp[ik\rho\cos(\beta-\varphi)]\,d\beta, \qquad 0<\varphi<\pi, \tag{10.159a}$$

$$H_{z>}^{(D)} \sim \left(\frac{2}{\pi k\rho}\right)^{1/2}\cdot\frac{\cos\varphi'/2\cos\varphi/2}{\cos\varphi+\cos\varphi'}\exp\left[i\left(k\rho+\frac{\pi}{4}\right)\right], \qquad k\rho\to\infty. \tag{10.159b}$$

The integral $H_{z<}^{(s)}$ can be evaluated in a similar manner by putting, say, $\theta=2\pi-\varphi$, so that $0<\theta<\pi$ when $\pi<\varphi<2\pi$ and using the results obtained for $H_{z>}^{(s)}$. Finally, substituting back for θ, we obtain

$$H_{z<}^{(s)} = -H_{z<}^{(D)}, \qquad \pi<\varphi<\pi+\varphi', \tag{10.160a}$$

$$H_{z<}^{(s)} = -\exp[ik\rho\cos(\varphi-\varphi')] - H_{z<}^{(D)}, \tag{10.160b}$$

where

$$H_{z<}^{(D)} = -\frac{i}{\pi}\int_\Gamma \frac{\cos\varphi'/2\cos\beta/2}{\cos\beta+\cos\varphi'}\exp[ik\rho\cos(\beta-2\pi+\varphi)]\,d\beta, \tag{10.161a}$$

$$H_{z<}^{D} \sim \left(\frac{2}{\pi k\rho}\right)^{1/2}\frac{\cos\varphi'/2\cos\varphi/2}{\cos\varphi+\cos\varphi'}\exp\left[i\left(k\rho+\frac{\pi}{4}\right)\right], \qquad k\rho\to\infty. \tag{10.161b}$$

Combining the foregoing results with the incident field of (10.142), we obtain, with reference to the regions defined in Fig. 10.12, results that are in complete agreement with those of (10.135), obtained by means of the Lebedev–Kontorovich transform.

The approximate representations (10.159b) and (10.161b) are, of course, not valid in the neighborhood where $\varphi=\pi\pm\varphi'$ and the denominator vanishes. It will be shown that in this case the integrals (10.159a) and (10.161a) can be expressed in terms of the complementary error function, as was done earlier in the case of the Lebedev–Kontorovich transform. Let us rewrite (10.159a) as

$$H_{z>}^{(D)} = \frac{i}{4\pi}[h(\varphi')+h(-\varphi')], \tag{10.162}$$

where

$$h(\varphi') = \int_\Gamma \frac{\exp[ik\rho\cos(\beta-\varphi)]}{\cos(\beta-\varphi')/2}\,d\beta. \tag{10.163}$$

With the substitution $\beta - \varphi = \alpha$, we write

$$h(\varphi') = \int_{\Gamma(0)} \frac{\exp(ik\rho\cos\alpha)}{\cos(\alpha + \varphi - \varphi')/2}\,d\alpha$$

$$= \frac{1}{2}\int_{\Gamma(0)} \{[\cos(\alpha + \varphi - \varphi')/2]^{-1} + [\cos(\alpha - \varphi + \varphi')/2]^{-1}\}$$

$$\times \exp(ik\rho\cos\alpha)\,d\alpha$$

$$= 2\int_{\Gamma(0)} \frac{\cos(\varphi - \varphi')/2 \cdot \cos\alpha/2}{\cos\alpha + \cos(\varphi - \varphi')} \exp(ik\rho\cos\alpha)\,d\alpha, \tag{10.164}$$

where the second integral in the foregoing was obtained by putting $\alpha = -\alpha$. On the path of steepest descents passing through the saddle point $\alpha = 0$, we set $\cos\alpha = 1 + is^2$ and obtain

$$h(\varphi') = -4\eta \exp\left[i\left(k\rho + \frac{\pi}{4}\right)\right] \int_0^\infty \frac{\exp(-k\rho s^2)}{s^2 - i\eta^2}\,ds \tag{10.165}$$

where $\eta = [1 + \cos(\varphi - \varphi')]^{1/2}$ in this case. The integral in (10.165) was expressed in terms of the complex Fresnel integral by Clemmow (1951), but it can also be expressed in terms of the complementary error function, as was done in (10.125). We obtain in this case

$$h(\varphi') = \mp i2\pi \exp[-ik\rho\cos(\varphi - \varphi')]$$
$$\times \operatorname{erfc}[\mp i\exp(i\pi/4)(2k\rho)^{1/2}\cos(\varphi - \varphi')/2], \tag{10.166}$$

where the $\mp$ go according to $\cos(\varphi - \varphi') \gtrless 0$. The integral $h(-\varphi')$ of (10.162) can be obtained from the preceding formula by changing the sign of φ'. Finally, the integral $H_{z<}^{(D)}$ in (10.161a) can be obtained by using the result of (10.166) and substituting $\varphi - 2\pi$ in place of φ. As expected, the final results in this case are in complete agreement with those obtained by means of the Lebedev–Kontorovich transform.

10.4.2 Diffraction by a Two-Part Plane Surface, Magnetic Case

Consider the problem of plane wave diffraction by a two-part plane surface as shown in Fig. 10.15. To the right, $x > 0$, the surface is a plane conductor of infinite conductivity and to the left, $x < 0$, the surface is an imperfect conductor specified by a surface impedance $Z_s = Z_0/n$ where n is large. Such configuration is relevant to investigation of plane wave diffraction at very low frequencies by a linear shoreline in a planar land–sea boundary.

As before, the incident field can be written

$$H_z^{(i)} = \exp[-ik_0(x\cos\varphi' + y\sin\varphi')]$$
$$= \exp[-ik_0\rho\cos(\varphi - \varphi')], \tag{10.167}$$

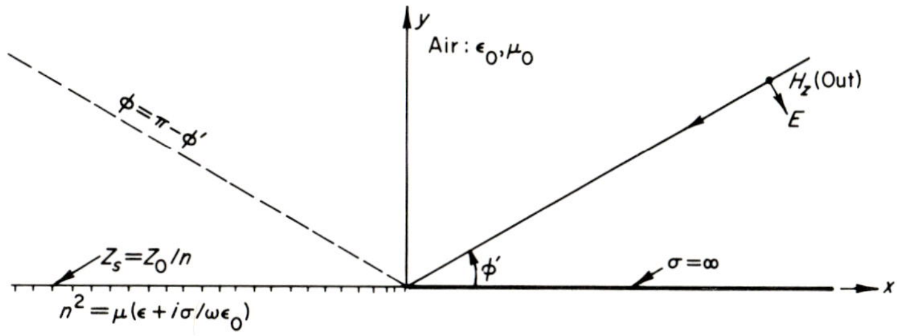

Fig. 10.15. Two-part plane surface.

and the total field in the region $y > 0$ will consist of the incident field $H_z^{(i)}$ plus an unknown field H_z; hence

$$H_z^{(t)} = H_z^{(i)} + H_z. \tag{10.168}$$

Our task, as before, is to determine the nature of the unknown field H_z, which must satisfy the differential equation

$$\left(\frac{\partial^2}{\partial x^2} + \frac{\partial^2}{\partial y^2} + k_0^2\right)H_z = 0. \tag{10.169}$$

In what follows it will be convenient to split the unknown field into the field that would be reflected from a surface impedance sheet Z_s, $y = 0$, $-\infty \leq x < 0$, say $H_z^{(R)}$, plus the scattered field $H_z^{(s)}$. Hence, $H_z = H_z^{(R)} + H_z^{(s)}$ and we write in place of (10.168)

$$H_z^{(t)} = H_z^{(i)} + H_z^{(R)} + H_z^{(s)}. \tag{10.170}$$

For the reflected field we write immediately

$$H_z^{(R)} = R\exp[-ik_0(x\cos\varphi' - y\sin\varphi')] = R\exp[-ik_0\,\rho\cos(\varphi + \varphi')], \tag{10.171}$$

where R is the plane wave reflection coefficient

$$R = \frac{n\sin\varphi' - 1}{n\sin\varphi' + 1}. \tag{10.172}$$

For the scattered field it is appropriate to use the integral representation

$$H_z^{(s)} = (2\pi)^{-1/2}\int_{-\infty}^{\infty} A(\alpha)\exp[i(\alpha x + \tau y)]\,d\alpha, \tag{10.173}$$

where in this case $\tau = (k_0^2 - \alpha^2)^{1/2}$ and $\text{Im}(\tau) > 0$, as before.

The boundary conditions to be satisfied by the unknown field $H_z^{(s)}$ can be found as follows. Since $E_x^{(t)}$ must vanish for $y = 0$, $0 < x \leq \infty$, we must have

$$\frac{\partial H_z^{(s)}}{\partial y} = \frac{i2k_0 \sin \varphi'}{n \sin \varphi' + 1} \exp(-ik_0 x \cos \varphi'), \qquad y = 0, \quad 0 < x \leq \infty. \quad (10.174)$$

At the boundary $y = 0$, $x < 0$ the ratio $E_x^{(t)}/H_z^{(t)} = Z_s$, which implies that

$$H_z^{(s)} - \frac{in}{k_0} \frac{\partial H_z^{(s)}}{\partial y} = 0, \qquad y = 0, \quad -\infty \leq x < 0. \quad (10.175)$$

These boundary conditions can now be applied to (10.173) to obtain

$$(2\pi)^{-1/2} \int_{-\infty}^{\infty} \tau A(\alpha) \exp(i\alpha x)\, d\alpha = \frac{2k_0 \sin \varphi'}{n \sin \varphi' + 1} \exp(-ik_0 x \cos \varphi'), \qquad x > 0, \quad (10.176a)$$

and

$$(2\pi)^{-1/2} \int_{-\infty}^{\infty} \left(1 + \frac{n\tau}{k_0}\right) A(\alpha) \exp(i\alpha x)\, d\alpha = 0, \qquad x < 0. \quad (10.176b)$$

In the integral (10.176b), x is negative. Then if the integrand is of algebraic growth as $|\alpha| \to \infty$ and has no singularities in the lower half plane, the contour can be closed by adding a large semicircle that adds nothing to the value of the integral. Along the real axis, the contour must be indented above the branch point $\alpha = -k_0$ and below the branch point $\alpha = k_0$ to satisfy the radiation condition as shown in Fig. 10.13. Thus, we write

$$(2\pi)^{-1/2} \left[1 + \frac{n}{k_0} (k_0{}^2 - \alpha^2)^{1/2}\right] A(\alpha) = \Phi_-(\alpha), \quad (10.177)$$

where $\Phi_-(\alpha)$ represents a function free of singularities in the lower half plane and of algebraic growth at infinity.

In the integral (10.176a) x is positive and, consequently, we can close the contour in the upper half plane provided that $A(\alpha)$ has an algebraic growth at infinity and a proper pole in the upper half plane to account for the right side of (10.176a). We recall that

$$\exp(-ik_0 \cos \varphi') = (2\pi i)^{-1} \int_C \frac{\Phi_+(\alpha) \exp(i\alpha x)\, d\alpha}{\Phi_+(-k_0 \cos \varphi')(\alpha + k_0 \cos \varphi')}, \quad (10.178)$$

where C is the contour shown in Fig. 10.13 and the function $\Phi_+(\alpha)$ is regular in the upper half plane and for $|\alpha| \to \infty$ it behaves in such a manner that the contribution from the large semicircular arc vanishes. Comparison of (10.176a) and (10.178) gives

$$A(\alpha) = -i\left(\frac{2}{\pi}\right)^{1/2} \frac{k_0 \sin \varphi'}{(n \sin \varphi' + 1)\Phi_+(-k_0 \cos \varphi')} \frac{\Phi_+(\alpha)}{(k_0{}^2 - \alpha^2)^{1/2}(\alpha + k_0 \cos \varphi')}. \quad (10.179)$$

Eliminating $A(\alpha)$ between (10.177) and (10.179), we obtain

$$\frac{\Phi_-(\alpha)}{1 + k_0/n \cdot (k_0{}^2 - \alpha^2)^{1/2}} = \frac{n \sin \varphi'}{i\pi(n \sin \varphi' + 1)\Phi_+(-k_0 \cos \varphi')} \frac{\Phi_+(\alpha)}{(\alpha + k_0 \cos \varphi')}. \tag{10.180}$$

It can be seen that any further success in finding the functions $\Phi_\pm(\alpha)$ and hence $A(\alpha)$ is dependent on our being able to factorize the function

$$K(\alpha) = 1 + k_0/n \cdot (k_0{}^2 - \alpha^2)^{1/2} \tag{10.181}$$

into two functions $K_+(\alpha)$ and $K_-(\alpha)$ regular and zeroless in the upper and lower half planes, respectively, so that

$$K(\alpha) = K_+(\alpha)/K_-(\alpha). \tag{10.182}$$

A general procedure for factorization of a large class of functions has been established in the original work of Wiener and Hopf (1931); hence, the existence of the functions $K_+(\alpha)$ and $K_-(\alpha)$ is thus assured. The formulas of Wiener and Hopf are, however, cumbersome to use, inasmuch as the behavior of the factors in the particular regions of interest is not clear. Thus, even though the possibility of factorization is assured, the problem of expressing the factors in a tractable form is not simple.

The factorization of the function $K(\alpha)$ defined in (10.181) and (10.182) has been accomplished by several authors (Bazer and Karp, 1962; Heins and Feshbach, 1954; Noble, 1962; Senior, 1952). The analysis leading to this result is not simple and we state only the final result

$$K_+(\alpha) = K_+(0) \exp\left[(2\pi i)^{-1} \int_0^\alpha L(\alpha)\, d\zeta\right], \tag{10.183}$$

where

$$K_-(\alpha) = [K_+(-\alpha)]^{-1}, \tag{10.184a}$$

$$L(\alpha) = -i\left(\frac{\pi}{\alpha + k_0} + \frac{k_0}{n}\left\{\frac{f(\alpha) - f[k_0(1 - n^{-2})^{1/2}]}{\alpha - k_0(-n^{-2})^{1/2}} + \frac{f(\alpha) - f[-k_0(1 - n^{-2})^{1/2}]}{\alpha + k_0(1 - n^{-2})^{1/2}}\right\}\right), \tag{10.184b}$$

$$f(\alpha) = \frac{\operatorname{arc\,cos}(\alpha/k_0)}{(k_0{}^2 - \alpha^2)^{1/2}}, \tag{10.184c}$$

$$K_+(0) = (i + n^{-1})^{1/2}. \tag{10.184d}$$

Substituting (10.182) into (10.180) and rearranging, we obtain

$$\Phi_-(\alpha)K_-(\alpha)(\alpha + k_0 \cos \varphi') = \frac{n \sin \varphi' \Phi_+(\alpha)K_+(\alpha)}{i\pi(n \sin \varphi' + 1)\Phi_+(-k_0 \cos \varphi')}. \tag{10.185}$$

The expression on the left in (10.185) is regular in the lower half plane since $\Phi_-(\alpha)$ and $K_-(\alpha)$ have no singularities therein. The expression on the right is free of singularities in the upper half plane since $\Phi_+(\alpha)$ and $K_+(\alpha)$ are regular therein. Since both sides are equal, it follows that either side will be regular in the entire α plane. Moreover, since $\Phi_-(\alpha) \to 0$ as $|\alpha| \to \infty$, then either side is bounded and, hence, by Liouville's theorem it must be a constant. Thus,

$$C = \frac{n \sin \varphi' \Phi_+(\alpha) K_+(\alpha)}{i\pi(n \sin \varphi' + 1)\Phi_+(-k_0 \cos \varphi')}$$

$$= \frac{n \sin \varphi' K_+(-k_0 \cos \varphi')}{i\pi(n \sin \varphi' + 1)}$$

$$= \frac{n \sin \varphi'}{i\pi(n \sin \varphi' + 1)K_-(k_0 \cos \varphi')}. \qquad (10.186)$$

Now referring to the left side of (10.185), we solve for $\Phi_-(\alpha)$ and then substitute into (10.177) to obtain

$$A(\alpha) = \frac{(2\pi)^{1/2} n \sin \varphi'}{i\pi(n \sin \varphi' + 1)K_-(k_0 \cos \varphi')}$$

$$\times \left\{\left[1 + \frac{n}{k_0}(k_0{}^2 - \alpha^2)^{1/2}\right](\alpha + k_0 \cos \varphi')K_-(\alpha)\right\}^{-1}, \qquad (10.187\text{a})$$

$$= \frac{(2\pi)^{1/2} k_0}{i\pi n K_+(k_0 \cos \varphi')}[(k_0{}^2 - \alpha^2)^{1/2}(\alpha + k_0 \cos \varphi')K_+(\alpha)]^{-1}, \qquad (10.187\text{b})$$

where in the last expression $K_-(\alpha)$ has been written in terms of $K_+(\alpha)$ by using (10.182).

Substituting (10.187b) into (10.173) and using the transformations of (10.156), we obtain as the formal solution

$$H_z^{(s)} = [i\pi n K_+(k_0 \cos \varphi')]^{-1} \int_{\Gamma_2} \frac{\exp[ik_0 \rho \cos(\beta - \varphi)]}{(\cos \beta + \cos \varphi')K_+(k_0 \cos \beta)}\, d\beta, \qquad (10.188)$$

where the path Γ_2 is shown in Fig. 10.14. The path Γ_2 can be deformed to the path of steepest descents Γ with due regard to the pole at $\beta = \pi - \varphi'$, the residue of which must be added when $\varphi < \pi - \varphi'$. Hence, using (10.184a) and (10.182), we can write

$$H_z^{(s)} = H_z^{(D)} + \left(\frac{2}{n \sin \varphi' + 1}\right) \exp[-ik_0 \rho \cos(\varphi + \varphi')], \qquad \varphi < \pi < \varphi', \qquad (10.189\text{a})$$

$$H_z^{(s)} = H_z^{(D)}, \qquad \varphi > \pi - \varphi', \tag{10.189b}$$

where

$$H_z^{(D)} = [i\pi n K_+(k_0 \cos \varphi')]^{-1} \int_\Gamma \frac{\exp[ik_0 \rho \cos(\beta - \varphi)]}{(\cos \beta + \cos \varphi')K_+(k_0 \cos \beta)} \, d\beta, \tag{10.190a}$$

$$\xrightarrow[k_0\rho \to \infty]{} -\left(\frac{2}{\pi k_0 \rho}\right)^{1/2} \frac{\exp[i(k_0 \rho + \pi/4)]}{nK_+(k_0 \cos \varphi')(\cos \varphi + \cos \varphi')K_+(k_0 \cos \varphi)}. \tag{10.190b}$$

Finally, substituting the foregoing results into (10.170) and also (10.171), we obtain in the region $0 < \varphi < \pi - \varphi'$

$$H_z^{(t)} = \exp[-ik_0 \rho \cos(\varphi - \varphi')] + \exp[-ik_0 \rho \cos(\varphi + \varphi')] + H_z^{(D)} \tag{10.191a}$$

and in the region $\pi - \varphi' < \varphi < \pi$

$$H_z^{(t)} = \exp[-ik_0 \rho \cos(\varphi - \varphi')] + \left(\frac{n \sin \varphi' - 1}{n \sin \varphi' + 1}\right) \times \exp[-ik_0 \rho \cos(\varphi + \varphi')] + H_z^{(D)}. \tag{10.191b}$$

The meaning of the various terms in (10.191a) and (10.191b) is clear from inspection of Fig. 10.15. In the region $0 < \varphi < \pi - \varphi'$ we have the incident wave, the wave reflected from the perfectly conducting surface (unity reflection coefficient), and the wave diffracted from the edge $H_z^{(D)}$. In the region $\pi - \varphi' < \varphi < \pi$, we again have the incident wave, the wave reflected from the impedance half plane (reflection coefficient equal to $(n \sin \varphi' - 1)/(n \sin \varphi' + 1)$), and the wave diffracted from the edge $H_z^{(D)}$.

The functions $K_+(k_0 \cos \varphi)$ and $K_+(k_0 \cos \varphi')$ can, supposedly, be calculated from (10.183), but in general the result is intractable. When, however, n is large, we can expand the exponential term in (10.183) as a power series in inverse powers of n to show that

$$K_+(k_0 \cos \varphi) = \exp\left(\frac{i\pi}{4}\right)\left[1 - \frac{1}{n}\left(\frac{1+i}{2} - \frac{\varphi}{\pi \sin \varphi}\right) + \cdots\right]. \tag{10.192}$$

Finally, it must be mentioned that the asymptotic formula (10.190b) is not valid in the neighborhood of $\varphi = \pi - \varphi'$ since the denominator vanishes there. We can, in this case, obtain a representation in terms of the complementary error function, as was done in the case of a perfectly conducting half plane in this chapter.

Problems

10.1 Show that adding a semicircle of infinite radius in the upper half plane to the integrals of (10.24) and (10.25) does not affect the values of these integrals. (*Hint*: Use the results in (10.60).)

10.2 Starting with (10.75e), verify (10.78a) and (10.78b).

10.3 Starting with (10.78a, b), verify (10.90a, b).

10.4 Let the cylinder in Fig. 10.1 be imperfectly conducting ($Z_s = Z_0/n$) and the incident plane wave axially polarized (E_z only). Find the diffracted field when the radius of the cylinder is small and large. (Carry out the analysis as far as possible and state all of the necessary assumptions.)

10.5 Use the asymptotic form of the complementary error function as given in (10.129) to show that as $\varphi_{1,2} \neq \pi$, then (10.121) becomes identical with (10.119).

10.6 Prove (10.192).

10.7 Carry out an analysis similar to that of Section 10.4.1 for the electric case (E_z only).

10.8 Carry out an analysis similar to that of Section 10.4.2 for the electric case (E_z only).

References

Abramowitz, M., and Stegun, I. A., (eds.) (1964). "Handbook of Mathematical Functions." U.S. Gov't. Printing Office, Washington, D.C. (*Natl. Bur. Std., U.S. Appl. Math. Ser.* **55**.)

Bazer, J., and Karp S, N. (1962). Propagation of plane electromagnetic waves past a shoreline, *J. Res.* (*Natl. Bur. Std.*) **D66**, 319–334.

Beckmann, P., and Franz, W. (1957). Asymptotisches Verhalten der Zylinderfunktionen in Abhangigkeit von Komplexen Index, *Z. Angew. Math. Mech.* **37**, 17–27.

Carrier, G. F., Krook, M., and Pearson, C. E. (1966). "Functions of a Complex Variable." McGraw-Hill, New York.

Clemmow, P. C. (1951). A method for the exact solution of two-dimensional diffraction problems, *Proc. Roy. Soc.* (*London*) *Ser. A* **205**, 286–308.

Clemmow, P. C. (1966). "The Plane Wave Spectrum Representation of Electromagnetic Fields." Pergamon, Oxford.

Copson, E. T. (1946). On an integral equation arising in the theory of diffraction, *Quart. J. Math.* **17**, 19–34.

Dettman, J. W. (1962). "Mathematical Methods in Physics and Engineering." McGraw-Hill, New York.

Franz, W. (1957). "Theorie der Beugung elektromagnetischer Wellen." Springer, Berlin.

Heins, A. E., and Feshbach, H. (1954). On the coupling of two half planes, *Proc. Symp. Appl. Math.* **75**; McGraw-Hill, New York.

Jones, D. S. (1964). "The Theory of Electromagnetism." Macmillan, New York.

Noble, B. (1958). "Methods Based on the Wiener–Hopf Technique." Pergamon, Oxford.

Senior, T. B. A. (1952). Diffraction by a semi-infinite metallic sheet, *Proc. Roy. Soc. (London) Ser. A* **213**, 436–458.

Watson, G. N. (1962). "Theory of Bessel Functions." Cambridge Univ. Press, London and New York.

Wiener, N. (1949). "The Extrapolation, Interpolation, and Smoothing of Stationary Time Series." Wiley, New York.

Wiener, N., and Hopf, E. (1931). Ueber eine Klasse singulärer Integralgleichungen, *Sitzber. Preuss. Akad. Wiss.* **1931**, 696–706.

CHAPTER 11

APERTURES ON CYLINDRICAL STRUCTURES

In this chapter we are concerned with the finding of electromagnetic fields around structures containing arbitrarily shaped apertures at which the tangential components of the electric field are specified. The simplest problem in this category is that of an aperture on an infinite perfectly conducting plane, which will be treated first. This will be followed by the treatment of an aperture on an infinite circular cylinder and an infinite wedge.

11.1 Apertures on Planes

Consider the geometry shown in Fig. 11.1. The tangential electric fields $\mathscr{E}_x$ and $\mathscr{E}_y$ at the aperture are considered to be known. Our task is to find all the field components at some point (x, y, z) away from the aperture.

It is clear that everywhere away from the aperture the electric and magnetic fields must satisfy the homogeneous differential equations

$$(\nabla^2 + k^2)\mathbf{E} = 0, \tag{11.1a}$$

$$(\nabla^2 + k^2)\mathbf{H} = 0, \tag{11.1b}$$

whereas at $z = 0$, $E_x(x, y, 0) = \mathscr{E}_x$ and $E_y(x, y, 0) = \mathscr{E}_y$. It will be more convenient, however, to formulate the problem in terms of two components of the Hertzian vector of the electric and magnetic type, each satisfying the scalar wave equation

$$(\nabla^2 + k^2)\Pi_z^{(e,m)} = 0, \tag{11.2}$$

in which case the electric and magnetic field components are found from

$$E_x = i\omega\mu_0\,\mu\,\frac{\partial\Pi_z^{(m)}}{\partial y} + \frac{\partial^2\Pi_z^{(e)}}{\partial x\,\partial z}, \tag{11.3a}$$

$$E_y = -\,i\omega\mu_0\,\mu\,\frac{\partial\Pi_z^{(m)}}{\partial x} + \frac{\partial^2\Pi_z^{(e)}}{\partial y\,\partial z}, \tag{11.3b}$$

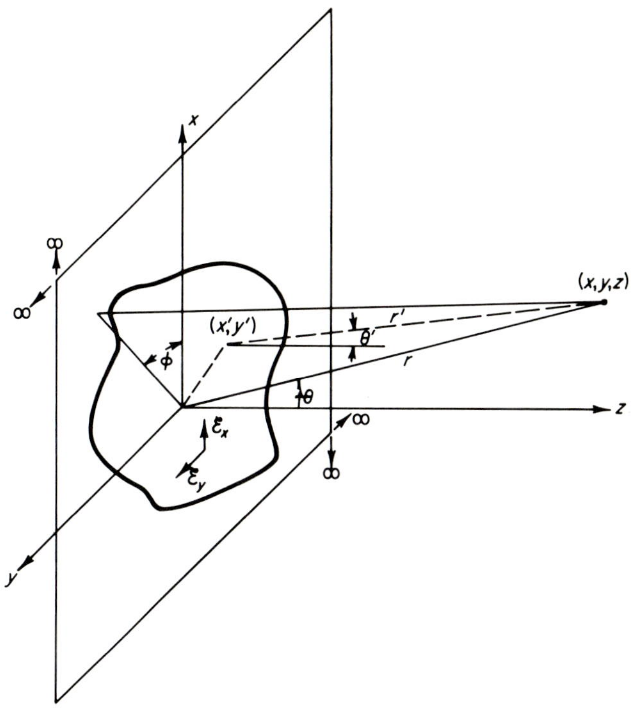

Fig. 11.1. Aperture on an infinite plane.

$$E_z = \left(k^2 + \frac{\partial^2}{\partial z^2}\right)\Pi_z^{(e)}, \tag{11.3c}$$

and

$$H_x = -\frac{i\omega\varepsilon_0 n^2}{\mu}\frac{\partial \Pi_z^{(e)}}{\partial y} + \frac{\partial^2 \Pi_z^{(m)}}{\partial x\, \partial z}, \tag{11.4a}$$

$$H_y = \frac{i\omega\varepsilon_0 n^2}{\mu}\frac{\partial \Pi_z^{(e)}}{\partial x} + \frac{\partial^2 \Pi_z^{(m)}}{\partial y\, \partial z}, \tag{11.4b}$$

$$H_z = \left(k^2 + \frac{\partial^2}{\partial z^2}\right)\Pi_z^{(m)} . \tag{11.4c}$$

Suitable integral representation for $\Pi_z^{(e)}$ and $\Pi_z^{(m)}$ can be obtained by using a two-dimensional Fourier transform on the x and y coordinates as follows.

$$\tilde{\Pi}(\alpha_1, \alpha_2, z) = (2\pi)^{-1} \iint_{-\infty}^{\infty} \Pi(x, y, z) \exp[-i(\alpha_1 x + \alpha_2 y)]\, dx\, dy, \tag{11.5a}$$

$$\Pi(x, y, z) = (2\pi)^{-1} \iint_{-\infty}^{\infty} \tilde{\Pi}(\alpha_1, \alpha_2, z) \exp[i(\alpha_1 x + \alpha_2 y)]\, d\alpha_1\, d\alpha_2, \tag{11.5b}$$

from which we obtain

$$\Pi_z^{(e)} = (2\pi)^{-1} \iint_{-\infty}^{\infty} A(\alpha_1, \alpha_2) \exp[i(\alpha_1 x + \alpha_2 y + \tau z)]\, d\alpha_1\, d\alpha_2, \tag{11.6a}$$

$$\Pi_z^{(m)} = (2\pi)^{-1} \iint_{-\infty}^{\infty} B(\alpha_1, \alpha_2) \exp[i(\alpha_1 x + \alpha_2 y + \tau z)]\, d\alpha_1\, d\alpha_2, \tag{11.6b}$$

where $\tau = (k^2 - \alpha_1{}^2 - \alpha_2{}^2)^{1/2}$ and $\mathrm{Im}(\tau) > 0$. Since at $z = 0$, $E_x(x, y, 0) = \mathscr{E}_x$ and $E_y(x, y, 0) = \mathscr{E}_y$, we obtain, using (11.3) and (11.6),

$$\mathscr{E}_x = -(2\pi)^{-1} \iint_{-\infty}^{\infty} [\alpha_1 \tau A + \omega\mu_0 \mu\alpha_2 B] \exp[i(\alpha_1 x + \alpha_2 y)]\, d\alpha_1\, d\alpha_2, \tag{11.7a}$$

$$\mathscr{E}_y = -(2\pi)^{-1} \iint_{-\infty}^{\infty} [\alpha_2 \tau A - \omega\mu_0 \mu\alpha_1 B] \exp[i(\alpha_1 x + \alpha_2 y)]\, d\alpha_1\, d\alpha_2. \tag{11.7b}$$

It will be noted that the equations above express a Fourier transform relation. Hence, the integrands of (11.7) are Fourier transforms of $\mathscr{E}_x$ and $\mathscr{E}_y$, respectively. Thus, we find

$$\begin{pmatrix} \alpha_1 \tau & \omega\mu_0 \mu\alpha_2 \\ \alpha_2 \tau & -\omega\mu_0 \mu\alpha_1 \end{pmatrix} \begin{pmatrix} A \\ B \end{pmatrix} = \begin{pmatrix} -\tilde{\tilde{\mathscr{E}}}_x \\ -\tilde{\tilde{\mathscr{E}}}_y \end{pmatrix}, \tag{11.8}$$

where

$$\tilde{\tilde{\mathscr{E}}} = (2\pi)^{-1} \iint_{-\infty}^{\infty} \mathscr{E}(x', y') \exp[-i(\alpha_1 x' + \alpha_2 y')]\, dx'\, dy'. \tag{11.9}$$

The coefficients A and B can now be expressed in terms of the known transforms $\tilde{\tilde{\mathscr{E}}}_x$ and $\tilde{\tilde{\mathscr{E}}}_y$ and substituted back into (11.6) to obtain

$$\Pi_z^{(e)} = \frac{i}{2\pi}\left(\frac{\partial I_x}{\partial x} + \frac{\partial I_y}{\partial y}\right), \tag{11.10a}$$

$$\Pi_z^{(m)} = (2\pi\omega\mu_0 \mu)^{-1} \frac{\partial}{\partial z}\left(\frac{\partial I_x}{\partial y} - \frac{\partial I_y}{\partial x}\right), \tag{11.10b}$$

where

$$I_x = \iint_{-\infty}^{\infty} \frac{\tilde{\tilde{\mathscr{E}}}_x}{\tau(\alpha_1{}^2 + \alpha_2{}^2)} \exp[i(\alpha_1 x + \alpha_2 y + \tau z)]\, d\alpha_1\, d\alpha_2, \tag{11.11a}$$

$$I_y = \iint_{-\infty}^{\infty} \frac{\tilde{\tilde{\mathscr{E}}}_y}{\tau(\alpha_1{}^2 + \alpha_2{}^2)} \exp[i(\alpha_1 x + \alpha_2 y + \tau z)]\, d\alpha_1\, d\alpha_2 . \qquad (11.11b)$$

Note that even though the integration in (11.9) extends for minus infinity, to plus infinity, since the tangential electric field exists only at the aperture, the contribution to the integral will arise only from the integration over the aperture.

Alternately, the integrals I_x and I_y can be written

$$I_x = (2\pi)^{-1} \iint_{\text{ap}} \mathscr{E}_x(x', y') W(x, x'; y, y'; z)\, dx'\, dy', \qquad (11.12a)$$

$$I_y = (2\pi)^{-1} \iint_{\text{ap}} \mathscr{E}_y(x', y') W(x, x'; y, y'; z)\, dx'\, dy', \qquad (11.12b)$$

where

$$W = \iint_{-\infty}^{\infty} \frac{\exp\{i[\alpha_1(x - x') + \alpha_2(y - y') + \tau z]\}}{\tau(\alpha_1{}^2 + \alpha_2{}^2)}\, d\alpha_1\, d\alpha_2 . \qquad (11.13)$$

When the distance

$$r' = [(x - x')^2 + (y - y')^2 + z^2]^{1/2} \qquad (11.14)$$

from a point on the aperture (x', y') to the field point (x, y, z) is large with respect to the wavelength, so that $kr \gg 1$, then the integral W in (11.13) can be evaluated by the method of steepest descents, using (5.94). Hence we obtain in this case

$$W \sim -\frac{i2\pi \exp(ikr')}{k^2 \sin^2 \theta' r'}, \qquad (11.15)$$

where $\theta' = \arccos(z/r')$. Moreover, if the aperture dimensions are small, so that $(x'^2 + y'^2)^{1/2} < r$, we can approximate as follows.

$$r' \sim r - \sin\theta \cos\varphi x' - \sin\theta \sin\varphi y' \qquad (11.16)$$

in the exponent of (11.5), whereas in the denominator we put $r' \sim r$ and $\theta' \sim \theta$. Hence we obtain

$$I_x \sim -\frac{i \exp(ikr)}{k^2 \sin^2 \theta r} \iint_{\text{ap}} \mathscr{E}_x(x', y') \exp[-ik \sin\theta(x' \cos\varphi + y' \sin\varphi)]\, dx'\, dy', \qquad (11.17a)$$

$$I_y \sim -\frac{i \exp(ikr)}{k^2 \sin^2 \theta r} \iint_{\text{ap}} \mathscr{E}_y(x', y') \exp[-ik \sin\theta(x' \cos\varphi + y' \sin\varphi)]\, dx'\, dy'. \qquad (11.17b)$$

We can also express the result in spherical coordinates by noting that in the far zone $\partial/\partial x \sim ik \cos\varphi \sin\theta$, $\partial/\partial y \sim ik \sin\varphi \sin\theta$, and $\partial/\partial z \sim ik \cos\theta$. Hence, using (11.10a) and (11.10b), we obtain

$$\Pi_z^{(e)} \sim \frac{i \exp(ikr)}{2\pi k \sin\theta r} \iint_{\mathrm{ap}} [\mathscr{E}_x \cos\varphi + \mathscr{E}_y \sin\varphi] \times \exp[-ik \sin\theta(x' \cos\varphi + y' \sin\varphi)]\, dx'\, dy' \tag{11.18a}$$

and

$$\Pi_z^{(m)} \sim \frac{i \cos\theta \exp(ikr)}{2\pi\omega\mu_0 \mu \sin\theta r} \iint_{\mathrm{ap}} [\mathscr{E}_x \sin\varphi - \mathscr{E}_y \cos\varphi] \times \exp[-ik \sin\theta(x' \cos\theta + y' \sin\theta)]\, dx'\, dy'. \tag{11.18b}$$

Moreover, using (11.3) and (11.4), the electric and magnetic field components can be written and

$$E_x \sim -k \sin\theta[\omega\mu_0 \mu \text{ s n } \varphi\Pi_z^{(m)} + k \cos\varphi \cos\theta\Pi_z^{(e)}], \tag{11.19a}$$

$$E_y \sim k \sin\theta[\omega\mu_0 \mu \cos\varphi\Pi_z^{(m)} - k \sin\varphi \cos\theta\Pi_z^{(e)}], \tag{11.19b}$$

$$E_z \sim k^2 \sin^2\theta\Pi_z^{(e)}, \tag{11.19c}$$

and

$$H_x \sim k \sin\theta[(\omega\varepsilon_0 n^2/\mu) \sin\varphi\Pi_z^{(e)} - k \cos\varphi \cos\theta\Pi_z^{(m)}], \tag{11.20a}$$

$$H_y \sim -k \sin\theta[(\omega\varepsilon_0 n^2/\mu) \cos\varphi\Pi_z^{(e)} + k \sin\varphi \cos\theta\Pi_z^{(m)}], \tag{11.20b}$$

$$H_z \sim k^2 \sin^2\theta\Pi_z^{(m)}. \tag{11.20c}$$

The field components can also be expressed in spherical coordinates, using the transformation

$$\begin{pmatrix} A_\theta \\ A_\varphi \\ A_r \end{pmatrix} = \begin{pmatrix} \cos\theta\cos\varphi & \cos\theta\sin\varphi & -\sin\theta \\ -\sin\varphi & \cos\varphi & 0 \\ \sin\theta\cos\varphi & \sin\theta\sin\varphi & \cos\theta \end{pmatrix} \begin{pmatrix} A_x \\ A_y \\ A_z \end{pmatrix}, \tag{11.21}$$

to obtain for the radiation field, $kr \to \infty$,

$$E_\theta = \frac{in \exp(ikr)}{\lambda_0 r} \iint_{\mathrm{ap}} [\mathscr{E}_x(x', y') \cos\varphi + \mathscr{E}_y(x', y') \sin\varphi] \times \exp[-ik \sin\theta(x' \cos\varphi + y' \sin\varphi)]\, dx'\, dy', \tag{11.22a}$$

$$E_\varphi = \frac{in \cos\theta \exp(ikr)}{\lambda_0 r} \iint_{\mathrm{ap}} [\mathscr{E}_x(x', y') \sin\varphi - \mathscr{E}_y(x', y') \cos\varphi] \times \exp[-ik \sin\theta(x' \cos\varphi + y' \sin\varphi)]\, dx'\, dy', \tag{11.22b}$$

$$E_r = 0 \tag{11.22c}$$

and

$$H_\theta = -\left(\frac{\varepsilon_0 \varepsilon + i\sigma/\omega}{\mu_0 \mu}\right)^{1/2} E_\varphi, \tag{11.23a}$$

$$H_\varphi = \left(\frac{\varepsilon_0 \varepsilon + i\sigma/\omega}{\mu_0 \mu}\right)^{1/2} E_\theta, \tag{11.23b}$$

$$H_r = 0. \tag{11.23c}$$

Let us now consider two specific examples of apertures on planes.

11.1.1 Rectangular Aperture

Consider a linearly polarized rectangular aperture, as shown in Fig. 11.2. The aperture field $\mathscr{E}_x(x', y') = \mathscr{E}_x$ is constant in the aperture $-a \leq x' \leq a$ and $-b \leq y' \leq b$. Using (11.22), the fields at a distance $r \to \infty$ can be written

$$E_\theta = \frac{in \cos\varphi \mathscr{E}_x \exp(ikr)}{\lambda_0 r} \int_{-a}^{a} \exp(-ik \sin\theta \cos\varphi x')\, dx' \times \int_{-b}^{b} \exp(-ik \sin\theta \sin\varphi y')\, dy', \tag{11.24a}$$

$$E_\varphi = \frac{in \cos\theta \sin\varphi \mathscr{E}_x \exp(ikr)}{\lambda_0 r} \int_{-a}^{a} \exp(-ik \sin\theta \cos\varphi x')\, dx' \times \int_{-b}^{b} \exp(-ik \sin\theta \sin\varphi y')\, dy'. \tag{11.24b}$$

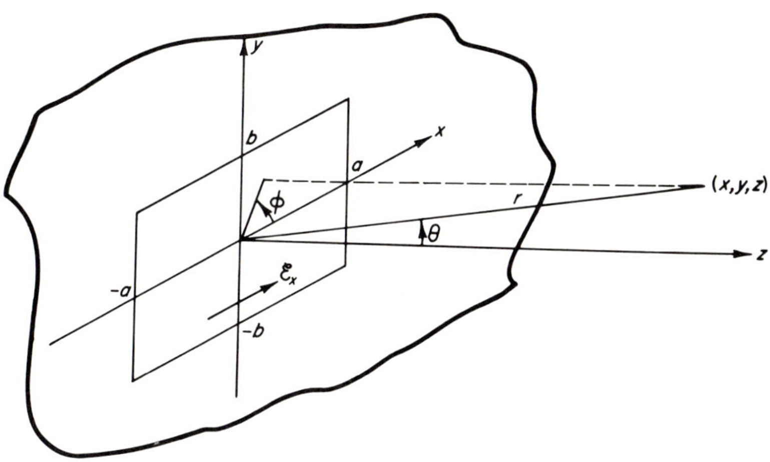

Fig. 11.2. Rectangular aperture on an infinite plane.

The indicated integration can be readily carried out, and we obtain

$$E_\theta = \frac{i4abn \cos\varphi \mathscr{E}_x e^{ikr}}{\lambda_0 r}\left[\frac{\sin(ka \sin\theta \cos\varphi)}{ka \sin\theta \cos\varphi}\right]\left[\frac{\sin(kb \sin\theta \sin\varphi)}{kb \sin\theta \sin\varphi}\right] \tag{11.25a}$$

$$E_\varphi = \frac{i4abn \cos\theta \sin\varphi \,\mathscr{E}_x e^{ikr}}{\lambda_0 r}\left[\frac{\sin(ka \sin\theta \cos\varphi)}{ka \sin\theta \cos\varphi}\right]\left[\frac{\sin(kb \sin\theta \sin\varphi)}{kb \sin\theta \sin\varphi}\right]. \tag{11.25b}$$

It is seen that the field components are of the form $\sin x/x$ in the $\varphi = 0$ and $\varphi = \pi/2$ planes.

11.1.2 Circular Aperture

The geometry of a circular aperture in a plane is shown in Fig. 11.3. If the aperture is linearly polarized and of constant excitation $\mathscr{E}_x(x', y') = \mathscr{E}_x$, then using the transformation

$$x' = \rho' \cos\varphi', \qquad y' = \rho' \sin\varphi', \tag{11.26}$$

we obtain

$$E_\theta = \frac{in \cos\varphi \mathscr{E}_x \exp(ikr)}{\lambda_0 r} I, \tag{11.27a}$$

$$E_\varphi = \frac{in \cos\theta \sin\varphi \mathscr{E}_x \exp(ikr)}{\lambda_0 r} I, \tag{11.27b}$$

where

$$I = \int_0^a d\rho' \,\rho' \int_{-\pi}^{\pi} \exp[-ik\rho' \sin\theta \cos(\varphi - \varphi')]\, d\varphi'. \tag{11.28}$$

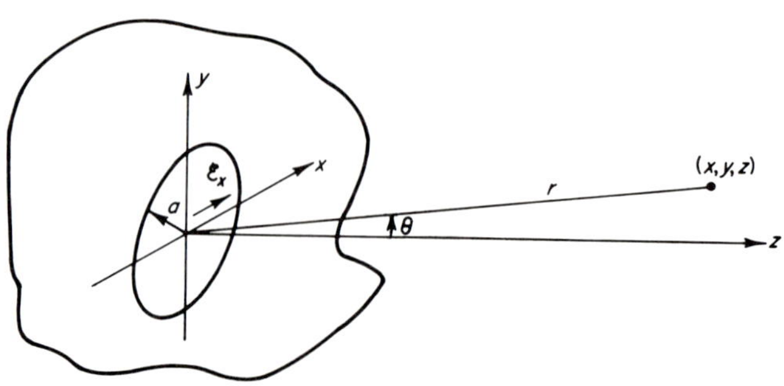

Fig. 11.3. Circular aperture on an infinite plane.

The last integral in (11.28) is equal to $2\pi J_0(k\rho' \sin\theta)$. Moreover, since $xJ_0(x) = [xJ_1(x)]'$ we readily obtain

$$I = 2\pi a^2 \left[\frac{J_1(ka\sin\theta)}{ka\sin\theta}\right]; \tag{11.29}$$

hence, the field components are

$$E_\theta = \frac{i2\pi a^2 n \cos\varphi \mathscr{E}_x e^{ikr}}{\lambda_0 r}\left[\frac{J_1(ka\sin\theta)}{ka\sin\theta}\right] \tag{11.30a}$$

$$E_\varphi = \frac{i2\pi a^2 n \cos\theta \sin\varphi \mathscr{E}_x e^{ikr}}{\lambda_0 r}\left[\frac{J_1(ka\sin\theta)}{ka\sin\theta}\right]. \tag{11.30b}$$

It will be noted that the function $J_1(x)/x$ has a behavior similar to $\sin x/x$.

11.2 Apertures on Circular Cylinders

Consider a circular cylinder of radius a and infinite in the z direction as shown in Fig. 11.4. On the cylinder there is an aperture of an arbitrary shape

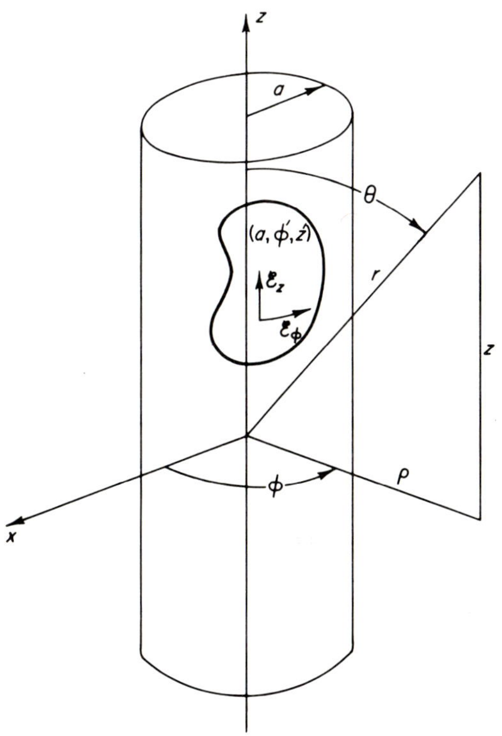

Fig. 11.4. Arbitrary aperture on a circular cylinder.

at which the tangential components of the electric field are specified, i.e.,

$$E_\varphi(a, \varphi, z) = \mathscr{E}_\varphi(\varphi, z), \tag{11.31a}$$

$$E_z(a, \varphi, z) = \mathscr{E}_z(\varphi, z), \tag{11.31b}$$

and they are zero everywhere else on the surface of the cylinder.

We shall formulate the present problem in terms of the axial components of the Hertzian vector of the electric and magnetic type satisfying the differential equation

$$\left(\frac{1}{\rho}\frac{\partial}{\partial\rho}\rho\frac{\partial}{\partial\rho} + \frac{1}{\rho^2}\frac{\partial^2}{\partial\varphi^2} + \frac{\partial^2}{\partial z^2} + k^2\right)\Pi_z^{(e,m)} = 0 \tag{11.32}$$

where the electric and magnetic field components are found from

$$E_\rho = \frac{\partial^2 \Pi_z^{(e)}}{\partial\rho\,\partial z} + \frac{i\omega\mu_0\,\mu}{\rho}\frac{\partial\Pi_z^{(m)}}{\partial\varphi}, \tag{11.33a}$$

$$E_\varphi = \frac{1}{\rho}\frac{\partial^2\Pi_z^{(e)}}{\partial\varphi\,\partial z} - i\omega\mu_0\,\mu\frac{\partial\Pi_z^{(m)}}{\partial\rho}, \tag{11.33b}$$

$$E_z = \left(k^2 + \frac{\partial^2}{\partial z^2}\right)\Pi_z^{(e)}, \tag{11.33c}$$

and

$$H_\rho = \frac{\partial^2\Pi_z^{(m)}}{\partial\rho\,\partial z} - \frac{i\omega\varepsilon_0\,n^2}{\mu\rho}\frac{\partial\Pi_z^{(e)}}{\partial\varphi}, \tag{11.34a}$$

$$H_\varphi = \frac{1}{\rho}\frac{\partial^2\Pi_z^{(m)}}{\partial\varphi\,\partial z} + \frac{i\omega\varepsilon_0\,n^2}{\mu}\frac{\partial\Pi_z^{(e)}}{\partial\rho}, \tag{11.34b}$$

$$H_z = \left(k^2 + \frac{\partial^2}{\partial z^2}\right)\Pi_z^{(m)}. \tag{11.34c}$$

Since the cylinder is of infinite extent, it will be convenient to use the Fourier transform pair

$$\tilde{\Pi}(\rho, \varphi, \alpha) = (2\pi)^{-1/2}\int_{-\infty}^{\infty}\Pi(\rho, \varphi, z)\exp(-i\alpha z)\,dz, \tag{11.35a}$$

$$\Pi(\rho, \varphi, z) = (2\pi)^{-1/2}\int_{-\infty}^{\infty}\tilde{\Pi}(\rho, \varphi, \alpha)\exp(i\alpha z)\,d\alpha, \tag{11.35b}$$

whereupon (11.32) can be transformed into

$$\left(\frac{1}{\rho}\frac{\partial}{\partial\rho}\rho\frac{\partial}{\partial\rho} + \frac{1}{\rho^2}\frac{\partial^2}{\partial\varphi^2} + \tau^2\right)\tilde{\Pi} = 0 \tag{11.36}$$

with $\tau = (k^2 - \alpha^2)^{1/2}$.

It is clear from the geometry of the problem that the fields must be periodic with the period of 2π. This suggests the Fourier expansion

$$\tilde{\Pi}(\rho, \varphi, \alpha) = \sum_{-\infty}^{\infty} F_m(\rho, \alpha) \exp(im\varphi), \tag{11.37a}$$

$$F_m(\rho, \alpha) = (2\pi)^{-1} \int_0^{2\pi} \tilde{\Pi}(\rho, \varphi, \alpha) \exp(-im\varphi)\, d\varphi, \tag{11.37b}$$

and it will follow that each Fourier coefficient will satisfy the Bessel differential equation

$$\left[\frac{1}{\rho}\frac{d}{d\rho}\rho\frac{d}{d\rho} + \left(\tau^2 - \frac{m^2}{\rho^2}\right)\right] F_m = 0. \tag{11.38}$$

Since the fields must be finite as $\rho \to \infty$ and the origin $\rho = 0$ is not included, we choose $H_m^{(1)}(\tau\rho)$, $\mathrm{Im}(\tau) > 0$, as the appropriate solution. Transforming back according to (11.35b), we obtain the representations

$$\Pi_z^{(e)} = (2\pi)^{-1/2} \sum_{-\infty}^{\infty} \exp(im\varphi) \int_{-\infty}^{\infty} A_m(\alpha) H_m^{(1)}(\tau\rho) \exp(i\alpha z)\, d\alpha, \tag{11.39a}$$

$$\Pi_z^{(m)} = (2\pi)^{-1/2} \sum_{-\infty}^{\infty} \exp(im\varphi) \int_{-\infty}^{\infty} B_m(\alpha) H_m^{(1)}(\tau\rho) \exp(i\alpha z)\, d\alpha. \tag{11.39b}$$

In order to apply the boundary conditions (11.31a) and (11.31b), we must determine the representation of the tangential electric fields. Using (11 33), we find

$$E_z = (2\pi)^{-1/2} \sum_{-\infty}^{\infty} e^{im\varphi} \int_{-\infty}^{\infty} \tau^2 A_m H_m^{(1)}(\tau\rho) e^{i\alpha z}\, d\alpha, \tag{11.40a}$$

$$E_\varphi = (2\pi)^{-1/2} \sum_{-\infty}^{\infty} e^{im\varphi} \int_{-\infty}^{\infty} \left[\frac{m\alpha}{\rho} A_m H_m^{(1)}(\tau\rho) + i\omega\mu_0 \mu B_m H_m^{(1)\prime}(\tau\rho)\right] e^{i\alpha z}\, d\alpha. \tag{11.40b}$$

At $\rho = a$, $E_z = \mathscr{E}_z(\varphi', z')$, and we obtain

$$\mathscr{E}_z(\varphi', z') = \sum_{-\infty}^{\infty} \left[(2\pi)^{-1/2} \int_{-\infty}^{\infty} \tau^2 A_m H_m^{(1)}(\tau a) \exp(i\alpha z')\, d\alpha\right] \exp(im\varphi'). \tag{11.41}$$

The expression in (11.41) can be looked upon as a Fourier series expansion of the function $\mathscr{E}_z(\varphi', z')$. The Fourier coefficients (quantity enclosed in the square brackets) can be found to give

$$(2\pi)^{-1/2} \int_{-\infty}^{\infty} \tau^2 A_m H_m^{(1)}(\tau a) \exp(i\alpha z')\, d\alpha = (2\pi)^{-1} \int_0^{2\pi} \mathscr{E}_z(\varphi', z') \exp(im\varphi')\, d\varphi'. \tag{11.42}$$

Now, since the integral on the left is a Fourier transform of the function $\tau^2 A_m H_m^{(1)}(\tau a)$, then the integral on the right must be its inverse. Thus, we obtain

$$A_m = [2\pi\tau^2 H_m^{(1)}(\tau a)]^{-1} \int_0^{2\pi} \tilde{\mathscr{E}}_z(\varphi', \alpha) \exp(-im\varphi')\, d\varphi', \tag{11.43}$$

where

$$\tilde{\mathscr{E}}_z = (2\pi)^{-1/2} \int_{-\infty}^{\infty} \mathscr{E}_z(\varphi', z') \exp(-i\alpha z')\, dz'. \tag{11.44}$$

Substituting (11.43) into (11.39a) gives

$$\Pi_z^{(e)} = (4\pi^2)^{-1} \sum_{-\infty}^{\infty} \exp(im\varphi) \iint_{\text{ap}} d\varphi'\, dz'\, I_1(\rho, z, z')\mathscr{E}_z(\varphi', z') \exp(-im\varphi'), \tag{11.45}$$

$$I_1(\rho, z, z') = \int_{-\infty}^{\infty} \frac{H_m^{(1)}(\tau\rho) \exp[i\alpha(z - z')]}{\tau^2 H_m^{(1)}(\tau a)}\, d\alpha. \tag{11.46}$$

Also at $\rho = a$, $E_\varphi = \mathscr{E}_\varphi(\varphi', z')$, which gives

$$(m\alpha/\rho)A_m H_m^{(1)}(\tau a) + i\omega\mu_0 \mu\tau B_m H_m^{(1)\prime}(\tau a)$$
$$= -(2\pi)^{-1} \int_0^{2\pi} \tilde{\mathscr{E}}_\varphi(\varphi', \alpha) \exp(-im\varphi')\, d\varphi', \tag{11.47}$$

where $\tilde{\mathscr{E}}_\varphi$ is the Fourier transform of $\mathscr{E}_\varphi$, similar to (11.44). Hence, using (11.43), we obtain

$$B_m = \frac{i}{2\pi\omega\mu_0 \mu\tau H_m^{(1)\prime}(\tau a)} \int_0^{2\pi} \left[\frac{m\alpha}{a\tau^2} \tilde{\mathscr{E}}_z + \tilde{\mathscr{E}}_\varphi\right] \exp(-im\varphi')\, d\varphi'. \tag{11.48}$$

Now, substituting back into (11.39b), we obtain

$$\Pi_z^{(m)} = (4\pi^2\omega\mu_0 \mu)^{-1} \sum_{-\infty}^{\infty} e^{im\varphi} \iint_{\text{ap}} d\varphi'\, dz'\, e^{-im\varphi'} \left[\frac{m}{a} \mathscr{E}_z \frac{\partial}{\partial z} + i\mathscr{E}_\varphi\left(k^2 + \frac{\partial^2}{\partial z^2}\right)\right] I_2, \tag{11.49}$$

where

$$I_2 = \int_{-\infty}^{\infty} \frac{H_m^{(1)}(\tau\rho)}{\tau^3 H_m^{(1)\prime}(\tau a)} \exp[i\alpha(z - z')]\, d\alpha. \tag{11.50}$$

The foregoing results constitute a formal solution for the electromagnetic fields due to an arbitrary aperture on a circular cylinder.

11.2.1 Far-Zone Fields

We now attempt to evaluate the integrals I_1 and I_2 as given in (11.46) and (11.50), respectively, for distances far away from the aperture, using the method of steepest descents. To this end we must examine the singularities of the integrals as functions of the variable of integration α.

The points where $\tau = 0$, $\alpha = \pm k$, are branch points and it is clear from what has been said earlier that the integration must be carried out on the Riemann sheet on which $\mathrm{Im}(\tau) > 0$. Turning to the possibility of poles where either $H_m^{(1)}(\tau a)$ or $H_m^{(1)\prime}(\tau a)$ may vanish, we note (Abramowitz and Stegun (1964), p. 373) that this can occur only if $\mathrm{Im}(\tau a) < 0$. Hence, inasmuch as the integration is to be performed on the sheet $\mathrm{Im}(\tau) > 0$, these poles must be disregarded.

Now we make the transformations

$$\alpha = k \cos \theta, \quad \rho = r \sin \theta, \quad z = r \cos \theta, \tag{11.51}$$

and obtain

$$I_1 = \frac{(-i)^m \exp(-i\pi/4)}{k} \int_{\Gamma_2} \frac{\breve{H}_m^{(1)}(k\rho \sin \beta) \exp(-ik \cos \beta z')}{\sin \beta H_m^{(1)}(ka \sin \beta)} \times \exp[ik\rho \cos(\beta - \theta)]\, d\beta \tag{11.52a}$$

and

$$I_2 = \frac{(-i)^m \exp(-i\pi/4)}{k^2} \int_{\Gamma_2} \frac{\breve{H}_m(k\rho \sin \beta) \exp(-ik \cos \beta z')}{\sin^2 \beta H_m^{(1)\prime}(ka \sin \beta)} \times \exp[ik\rho \cos(\beta - \theta)]\, d\beta. \tag{11.52b}$$

Since all of the zeros of $H_m^{(1)}(ka \sin \beta)$ and $H_m^{(1)\prime}(ka \sin \beta)$ lie on the lower Riemann sheet, they must be disregarded in integration and, consequently, the original path of integration Γ_2 can be deformed freely into the path of steepest descents Γ. Also note that the pole at $\beta = 0$ does not lie between the paths Γ_2 and Γ and the pole $\beta = \pi$ is never reached since $0 < \theta < \pi$ does not include the end points $\theta = 0, \pi$. Thus, we obtain for the leading terms

$$I_1 = \frac{2(-i)^{m+1} \exp(-ikz' \cos \theta)}{k^2 \sin^2 \theta H_m^{(1)}(ka \sin \theta)} \cdot \frac{\exp(ikr)}{r}, \tag{11.53a}$$

$$I_2 = \frac{2(-i)^{m+1} \exp(-ikz' \cos \theta)}{k^3 \sin^3 \theta H_m^{(1)\prime}(ka \sin \theta)} \cdot \frac{\exp(ikr)}{r}. \tag{11.53b}$$

Substituting (11.53a) and (11.53b) in (11.45) and (11.49), respectively, we obtain in the far field

$$\Pi_z^{(e)} = \frac{e^{ikr}}{2\pi^2 k^2 \sin^2 \theta r} \sum_{-\infty}^{\infty} \frac{(-i)^{m+1} e^{im\varphi}}{H_m^{(1)}(ka \sin \theta)} F_z(m, \theta), \tag{11.54a}$$

$$\Pi_z^{(m)} = \frac{e^{ikr}}{2\pi^2 k\omega\mu_0 \mu \sin \theta r} \sum_{-\infty}^{\infty} \frac{(-i)^{m+1} e^{im\varphi}}{H_m^{(1)\prime}(ka \sin \theta)} \left[\frac{m \cos \theta}{ka \sin^2 \theta} F_z(m, \theta) + F_\varphi(m, \theta)\right], \tag{11.54b}$$

where

$$F_z(m, \theta) = \iint_{\text{ap}} \mathscr{E}_z(\varphi', z') \exp[-i(m\varphi' + k \cos \theta z')]\, d\varphi'\, dz', \tag{11.55a}$$

$$F_\varphi(m, \theta) = \iint_{\text{ap}} \mathscr{E}_\varphi(\varphi', z') \exp[-i(m\varphi' + k \cos \theta z')]\, d\varphi'\, dz'. \tag{11.55b}$$

To find the far-zone field components in spherical coordinates, we note the transformations

$$\frac{\partial}{\partial z} \sim \cos \theta \frac{\partial}{\partial r} = ik \cos \theta, \tag{11.56a}$$

$$\frac{\partial}{\partial \rho} \sim \sin \theta \frac{\partial}{\partial r} = ik \sin \theta, \tag{11.56b}$$

and

$$\begin{pmatrix} A_\theta \\ A_r \end{pmatrix} = \begin{pmatrix} \cos \theta & -\sin \theta \\ \sin \theta & \cos \theta \end{pmatrix} \begin{pmatrix} A_\rho \\ A_z \end{pmatrix}. \tag{11.57}$$

Hence, we obtain in the far zone

$$E_\theta = -k^2 \sin \theta \Pi_z^{(e)}, \quad E_\varphi = k\omega\mu_0 \mu \sin \theta \Pi_z^{(m)}, \quad E_r = 0 \tag{11.58}$$

and

$$H_\theta = -(n\mu/Z_0)E_\phi, \quad H_\phi = (n\mu/Z_0)E_\theta, \quad H_r = 0. \tag{11.59}$$

11.2.2 Field of an Axial Slot

Consider an axial slot on a circular cylinder, as shown in Fig. 11.5. We assume that the slot is excited by an electric field of the form

$$\mathscr{E}_\varphi(\varphi', z') = \frac{V}{\varphi_0 a} \cos \frac{\pi z'}{L},$$

$$\mathscr{E}_z(\varphi', z') = 0 \tag{11.60}$$

in the slot $-L/2 \leq z' \leq L/2$, $-\varphi_0/2 \leq \varphi' \leq \varphi_0/2$. We find, using (11.55),

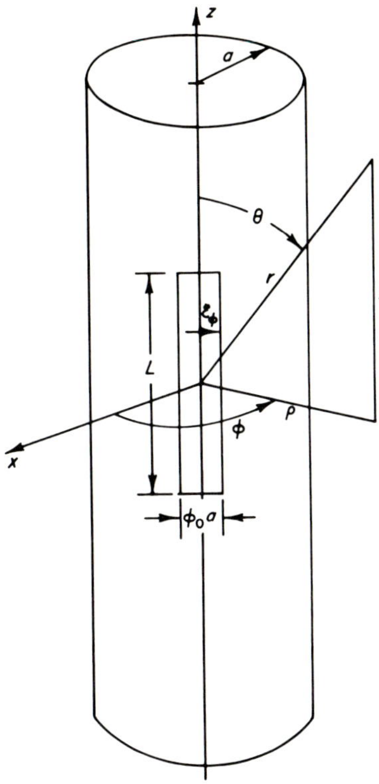

Fig. 11.5. Axial slot on a circular cylinder.

$$F_\varphi = \frac{2VL}{\pi a}\frac{\sin m\varphi_0/2}{m\varphi_0/2}\left\{\frac{\cos[(\pi L/\lambda)\cos\theta]}{1-[(2L/\lambda)\cos\theta]^2}\right\}, \quad (11.61a)$$

$$F_z = 0. \quad (11.61b)$$

Hence, the only nonvanishing components of the far fields are

$$E_\varphi = \frac{VLe^{ikr}}{\pi^3 ar}\left\{\frac{\cos[(L/\lambda)\cos\theta]}{1-[(2L/\lambda)\cos\theta]^2}\right\}\sum_0^\infty \frac{\varepsilon_m(-i)^m\cos m\varphi}{H_m^{(1)\prime}(ka\sin\theta)}\frac{\sin m\varphi_0/2}{m\varphi_0/2} \quad (11.62a)$$

$$H_\theta = -\frac{n\mu}{Z_0}E_\varphi. \quad (11.62b)$$

If the radius of the cylinder is small, so that $2\pi a/\lambda < 1$, then the Hankel function can be replaced by its small argument approximation and the resulting series will converge fast. In this case the first two terms of the series are found to be

$$E_\varphi = -\frac{ikVL\sin\theta e^{ikr}}{2\pi^2 r}\left\{\frac{\cos[(L/\lambda)\cos\theta]}{1-[(2L/\lambda)\cos\theta]^2}\right\}$$

$$\times\left[1 - i2ka\left(\frac{\sin\varphi_0/2}{\varphi_0/2}\right)\sin\theta\cos\varphi\right]. \quad (11.63)$$

When the slot is very long (infinite) and narrow with the excitation $\mathscr{E}_\varphi = (V/a)\delta(\varphi')$, then it follows that the only nonvanishing component of the magnetic field will be

$$H_z = \frac{iVn}{2\pi a Z_0 \mu}\sum_{-\infty}^{\infty} e^{im\varphi}\frac{H_m^{(1)}(k\rho)}{H_m^{(1)\prime}(ka)}, \quad (11.64)$$

which can be used for numerical calculations when the radius of the cylinder is small, $ka < 1$.

11.2.3 Watson Transformation for an Axial Slot

In the case of a circular cylinder of small radius, the harmonic series representation of (11.64) will converge rapidly enough so that only a few terms will be needed for a good degree of accuracy. In the case of large cylinders, $ka \gg 1$, the harmonic series converges very slowly and hence is not suitable for calculations. We can, in this case, transform the series into a contour integral, as was done in Section 10.1.2 for the analogous case of diffraction by a circular cylinder. We write, using the result of (10.19),

$$H_z = \frac{iA}{2}\int_C \frac{H_\nu^{(1)}(k\rho)}{H_\nu^{(1)\prime}(ka)}\frac{\exp[i\nu(\varphi-\pi)]}{\sin\nu\pi}\,d\nu \quad (11.65)$$

since $H^{(1)\prime}(ka)$ has no zeros on the real axis; C is the contour shown in Fig. 10.2 and

$$A = \frac{iV}{2\pi a}\left(\frac{\varepsilon_0\varepsilon}{\mu_0\mu}\right)^{1/2}, \qquad \sigma = 0. \quad (11.66)$$

Since $H_{-\nu}(k\rho)/H_{-\nu}^{(1)\prime}(ka) = H_\nu^{(1)}(k\rho)/H_\nu^{(1)\prime}(ka)$, we can write (11.65) as

$$H_z = iA\int_{-\infty}^{\infty}\frac{H_\nu^{(1)}(k\rho)}{H_\nu^{(1)\prime}(ka)}\frac{\cos[\nu(\varphi-\pi)]}{\sin\nu\pi}\,d\nu, \quad (11.67)$$

where the path of integration is understood to run slightly above the real axis.

The contour can now be closed in the upper half plane with a large semicircle, the contribution of which adds nothing to the value of the integral. Integrating over the poles produced by the zeros of $H_{\nu_n}^{(1)\prime}(ka) = 0$ (see Fig. 10.6), we obtain

$$H_z = -2\pi A \sum_{n=1}^{\infty} \frac{\cos \nu_n'(\varphi - \pi)}{\sin \nu_n' \pi} \frac{H_{\nu_n'}^{(1)}(k\rho)}{\partial/\partial\nu \, H_\nu^{(1)\prime}(ka)|_{\nu=\nu_n'}}. \tag{11.68}$$

Moreover, using (10.67) and substituting back from (11.68), we obtain

$$H_z = -\frac{iV\omega\varepsilon_0\,\varepsilon}{4} \sum_{n=1}^{\infty} \frac{\cos \nu_n'(\varphi - \pi)}{\sin \nu_n' \pi} \frac{H_{\nu_n'}^{(1)}(k\rho)}{a_n' \, Ai(a_n')}, \tag{11.69}$$

where the a_n' are the zeros of the derivative of the Airy function (see Table 10.1).

In the radiation zone, $k\rho \to \infty$, and we can use the asymptotic equivalent of $H_{\nu_n'}^{(1)}(k\rho)$. Moreover, noting the relation

$$\frac{\cos \nu_n'(\varphi - \pi)}{\sin \nu_n' \pi} \exp\left(-i\nu_n' \frac{\pi}{2}\right) = -i \frac{\exp[i\nu_n'(3\pi/2 - \varphi)] + \exp[i\nu_n'(\varphi - \pi/2)]}{1 - \exp(i2\pi\nu_n')}$$
$$\sim -i\left\{\exp\left[i\nu_n'\left(\frac{3\pi}{2} - \varphi\right)\right] + \exp\left[i\nu_n'\left(\varphi - \frac{\pi}{2}\right)\right]\right\}. \tag{11.70}$$

where the last step is justified inasmuch as ν_n' has an appreciable positive imaginary part, we obtain for the radiation field

$$H_z = -\frac{V\omega\varepsilon_0\,\varepsilon}{2(2\pi k\rho)^{1/2}} \exp\left[i\left(k\rho - \frac{\pi}{4}\right)\right] \times \sum_{n=1}^{\infty} \frac{\exp[i\nu_n'(3\pi/2 - \varphi)] + \exp[i\nu_n'(\varphi - \pi/2)]}{a_n' \, Ai(a_n')}. \tag{11.71}$$

The series in (11.71) is well suited for numerical calculations of the fields in the "shadow" region of the slot, $\pi/2 < \varphi < 3\pi/2$. Usually, only a few terms of this series will suffice for a high degree of accuracy since the imaginary part of ν_n' is positive and increases with n. This series is, however, not suitable for calculation of the fields in the "illuminated" region, $-\pi/2 < \varphi < \pi/2$, and we must yet obtain a suitable expression for that region. Before leaving this section, however, it will be fruitful to point out the creeping wave interpretation of the fields in the shadow region.

To this end we must return to (11.69) and employ an asymptotic form of $H_{\nu_n'}^{(1)}(k\rho)$ when $k\rho$ is large, $k\rho > \nu_n'$, but not infinite. Using (10.35a), we can write

$$H_{\nu_n'}^{(1)}(k\rho) \sim \left[\frac{2}{\pi k(\rho^2 - a^2)^{1/2}}\right]^{1/2} \exp\{i[k(\rho^2 - a^2)^{1/2} - \nu_n' \arccos(a/\rho) - \pi/4]\}, \tag{11.72}$$

where we approximated $[(k\rho)^2 - v_n'^2]^{1/2} \sim k(\rho^2 - a^2)^{1/2}$ and $\arccos(v_n'/k\rho) \sim \arccos(a/\rho)$. Moreover, since $\operatorname{Im}(v_n') > 0$, we can write

$$(\sin v_n'\pi)^{-1} = -2i\exp(iv_n'\pi)\sum_{m=0}^{\infty}\exp(i2mv_n'\pi) \tag{11.73}$$

and hence

$$\frac{\cos v_n'(\varphi - \pi)}{\sin v_n'\pi} = -i\sum_{m=0}^{\infty}(\exp\{iv_n'[2\pi(1+m) - \varphi]\} + \exp[iv_n'(2m\pi + \varphi)]). \tag{11.74}$$

Now substituting (11.72) and (11.74) into (11.69), we obtain

$$H_z = \sum_{n=1}^{\infty} D_n \sum_{m=0}^{\infty} \exp(i\{k(\rho^2 - a^2)^{1/2} + v_n'[2\pi(1+m) - \varphi - \arccos(a/\rho)]\}) + \exp(i\{k(\rho^2 - a^2)^{1/2} + v_n'[2m\pi + \varphi - \arccos(a/\rho)]\}) \tag{11.75}$$

where

$$D_n = -\frac{V\omega\varepsilon_0\,\varepsilon\exp(-i\pi/4)}{2[2\pi k(\rho^2 - a^2)^{1/2}]^{1/2} a_n' Ai(a_n')}. \tag{11.76}$$

Equation (11.75) is useful in explaining the presence of the fields in the shadow region of the slot, as shown in Fig. 11.6, where a geometrical interpretation of the $m = 0$ waves is depicted. As it appears, these waves travel along the surface of the cylinder with a complex propagation constant $v_n' = ka - a_n'(ka/2)^{1/3}\exp(-i\pi/3)$ in both directions away from the slot.

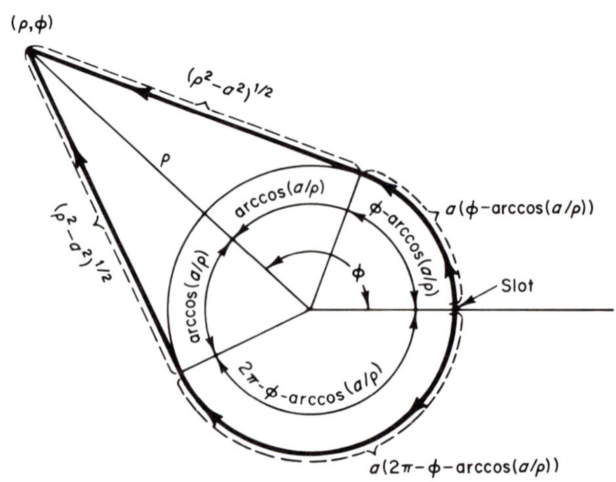

Fig. 11.6. Geometry of the creeping waves produced by a radiating slot.

Inasmuch as the propagation constant is complex, these creeping waves evidently lose energy by radiation as they progress along the surface of the cylinder. After leaving the surface of the cylinder by radiation, they travel through a distance $(\rho^2 - a^2)^{1/2}$ in free space (propagation constant k) to recombine at some point (ρ, φ) in the shadow region of the slot. The higher-order creeping waves, $m > 0$, are those that make one or more complete circuits around the cylinder before recombining at the point (ρ, φ). It is clear that these higher-order waves will be progressively damped, which follows from (11.75) as well as from physical considerations.

To complete the evaluation of the fields for large cylinders, we must still obtain a suitable representation in the region directly illuminated by the slot, where the residue series representation fails. To this end, we note the relation

$$\frac{\cos \nu(\varphi - \pi)}{\sin \nu\pi} = \frac{\cos \nu\varphi}{\sin \nu\pi} \exp(i\nu\pi) - i \exp(i\nu\varphi) \tag{11.77}$$

and rewrite (11.67) as

$$H_z = A(I_1 + iI_2), \tag{11.78}$$

where

$$I_1 = \int_{-\infty}^{\infty} \frac{H_\nu^{(1)}(k\rho)}{H_\nu^{(1)\prime}(ka)} \exp(i\nu\varphi)\, d\nu \tag{11.79}$$

and

$$I_2 = \int_{-\infty}^{\infty} \frac{\cos \nu\varphi}{\sin \nu\pi} \cdot \frac{H_\nu^{(1)}(k\rho)}{H_\nu^{(1)\prime}(ka)} \exp(i\nu\pi)\, d\nu, \tag{11.80}$$

where the line of integration is slightly above the real axis.

First we dispose of the integral I_2 in (11.80). It is clear that this integral can be evaluated by closing the contour with a large semicircle in the upper half plane and using the residue theorem. It follows that the contribution from the semicircular arc vanishes as its radius approaches infinity and, consequently,

$$I_2 = 2\pi i \sum_{n=1}^{\infty} \frac{\cos \nu_n' \varphi \exp(i\nu_n' \pi) H_{\nu_n'}^{(1)}(k\rho)}{\sin \nu_n' \pi \; \partial/\partial\nu \; H_\nu^{(1)\prime}(ka)\big|_{\nu=\nu_n'}}. \tag{11.81}$$

Now, since

$$\frac{\cos \nu_n' \varphi}{\sin \nu_n' \pi} \exp(i\nu_n' \pi) = -i \frac{\exp[i\nu_n'(2\pi + \varphi)] + \exp[i\nu_n'(2\pi - \varphi)]}{1 - \exp(i2\pi\nu_n')}, \tag{11.82}$$

it follows that the contribution of I_2 will be vanishingly small in the illuminated region $-\pi/2 < \varphi < \pi/2$ since ν_n' has a positive imaginary part.

We can now concentrate on the evaluation of the integral I_1 in (11.79) by means of the method of stationary phase. To this end we must investigate the existence of stationary points of the integrand in the complex v plane when $k\rho \to \infty$ and $ka \gg 1$.

Noting the asymptotic forms of the Hankel function and its derivative in (10.35) and (10.36), we obtain

$$\frac{H_v^{(1)}(k\rho)}{H_v^{(1)\prime}(ka)}\exp(iv\varphi) \sim -\frac{ika}{\{k\rho[(ka)^2 - v^2]^{1/2}\}^{1/2}}\exp[i(g_< + v\varphi)],$$

$$v < ka < k\rho, \qquad (11.83a)$$

and

$$\frac{H_v^{(1)}(k\rho)}{H_v^{(1)\prime}(ka)}\exp(iv\varphi) \sim -\frac{ka\exp(i\pi/4)}{\{k\varphi[v^2 - (ka)^2]^{1/2}\}^{1/2}}\exp[i(g_> + v\varphi)],$$

$$ka < v < k\rho, \qquad (11.83b)$$

where

$$\begin{aligned} g_< &= [(k\rho)^2 - v^2]^{1/2} - [(ka)^2 - v^2]^{1/2} \\ &\quad + v[\operatorname{arc\,cos}(v/ka) - \operatorname{arc\,cos}(v/k\rho)], \end{aligned} \qquad (11.84a)$$

$$\begin{aligned} g_> &= [(k\rho)^2 - v^2]^{1/2} - i[v^2 - (ka)^2]^{1/2} \\ &\quad + v\left\{i\ln\left(\frac{v + [v^2 - (ka)^2]^{1/2}}{ka}\right) - \operatorname{arc\,cos}\left(\frac{v}{k\rho}\right)\right\}. \end{aligned} \qquad (11.84b)$$

Now the stationary points are located where the derivative of the exponent in (11.83) vanishes, i.e., where $\partial g_{\lessgtr}/\partial v = -\varphi$. But since

$$\begin{aligned} \frac{\partial g_<}{\partial v} &= \operatorname{arc\,cos}\left(\frac{v}{ka}\right) - \operatorname{arc\,cos}\left(\frac{v}{k\rho}\right) \\ &\xrightarrow[k\rho\to\infty]{} \operatorname{arc\,cos}\left(\frac{v}{ka}\right) - \frac{\pi}{2}, \end{aligned} \qquad (11.85)$$

then a stationary point will be situated at $v = v_<$ where

$$v_< = ka\sin\varphi. \qquad (11.86)$$

In the case $ka < v < k\rho$, we find

$$\frac{\partial g_>}{\partial v} \xrightarrow[k\rho\to\infty]{} \pm\frac{\pi}{2} + i\ln\left\{\frac{v}{ka} + \left[\left(\frac{v}{ka}\right)^2 - 1\right]^{1/2}\right\} \qquad (11.87)$$

which, on account of being a complex number, cannot be equal to a real angle φ. Hence, we conclude that these are no stationary points other than the one in (11.86).

Now we find

$$g_< + v\varphi\,|_{v=v_<} \sim k(\rho - a\cos\varphi),$$
$$\frac{\partial^2}{\partial v^2}[g_< + v\varphi]_{v=v_<} \sim -(ka\cos\varphi)^{-1} \tag{11.88}$$

and using the formula (5.56), we obtain

$$I_1 \sim -ika\left(\frac{2\pi}{k\rho}\right)^{1/2} \exp\left\{i\left[k(\rho - a\cos\varphi) - \frac{\pi}{4}\right]\right\}. \tag{11.89}$$

Thus, the magnetic field in the illuminated region becomes

$$H_z \sim \frac{V}{(\mu_0\mu/\varepsilon_0\varepsilon)^{1/2}}\left(\frac{k}{2\pi\rho}\right)^{1/2} \exp\left\{i\left[k(\rho - a\cos\varphi) - \frac{\pi}{4}\right]\right\}. \tag{11.90}$$

It is worthwhile to note that the field represented by (11.90) is similar to the field of a narrow slot on a perfectly conducting plane. Using the substitution $\rho \sim a\cos\varphi + \rho'$ where ρ' is the distance from the slot to the field point, we obtain

$$H_z \sim \frac{V}{(\mu_0\mu/\varepsilon_0\varepsilon)^{1/2}}\left(\frac{k}{2\pi\rho'}\right)^{1/2} \exp\left[i\left(k\rho' - \frac{\pi}{4}\right)\right], \tag{11.91}$$

which is the correct expression describing the far field from an infinitely long narrow slot on a ground plane.

11.3 Apertures on Wedges

Consider an infinite perfectly conducting wedge of internal angle φ_0 as shown in Fig. 11.7. One side of the wedge contains an aperture of an arbitrary shape on which the tangential components of the electric field are specified, i.e.,

$$E_\rho(\rho, 0, z) = \mathscr{E}_\rho(\rho, z), \tag{11.92a}$$
$$E_z(\rho, 0, z) = \mathscr{E}_z(\rho, z), \tag{11.92b}$$

and they are zero elsewhere on the wedge.

As in the case of the circular cylinder, we formulate the present problem in terms of the axial components of the Hertzian vector of the electric and magnetic type whose Fourier transforms must satisfy the wave equation

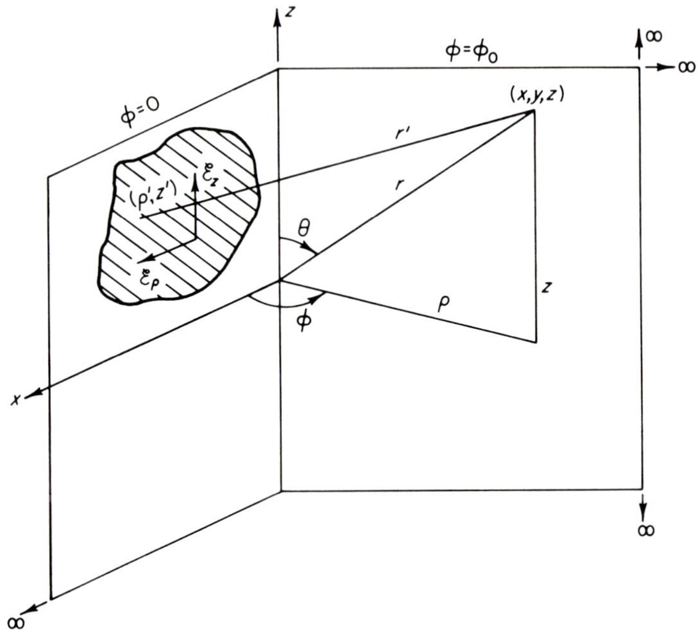

Fig. 11.7. Arbitrary aperture on an infinite wedge.

$$\left(\frac{1}{\rho}\frac{\partial}{\partial\rho}\rho\frac{\partial}{\partial\rho}+\frac{1}{\rho^2}\frac{\partial^2}{\partial\varphi^2}+\tau^2\right)\tilde{\Pi}^{(e,m)}=0 \tag{11.93a}$$

where $\tau=(k^2-\alpha^2)^{1/2}$, $\text{Im}(\tau)>0$, and

$$\Pi_z^{(e,m)}=(2\pi)^{-1/2}\int_{-\infty}^{\infty}\tilde{\Pi}^{(e,m)}e^{i\alpha z}\,d\alpha. \tag{11.93b}$$

The electric and magnetic field components are found from

$$E_\rho=\frac{\partial^2\Pi_z^{(e)}}{\partial\rho\,\partial z}+\frac{i\omega\mu_0\,\mu}{\rho}\frac{\partial\Pi_z^{(m)}}{\partial\varphi}, \tag{11.94a}$$

$$E_\varphi=\frac{1}{\rho}\frac{\partial^2\Pi_z^{(e)}}{\partial\varphi\,\partial z}+i\omega\mu_0\,\mu\frac{\partial\Pi_z^{(m)}}{\partial\rho}, \tag{11.94b}$$

$$E_z=\left(k^2+\frac{\partial^2}{\partial z^2}\right)\Pi_z^{(e)}, \tag{11.94c}$$

and

$$H_\rho = \frac{\partial^2 \Pi_z^{(m)}}{\partial \rho\, \partial z} - \frac{i\omega\varepsilon_0 n^2}{\mu\rho} \frac{\partial \Pi_z^{(e)}}{\partial \varphi}, \tag{11.95a}$$

$$H_\varphi = \frac{1}{\rho} \frac{\partial^2 \Pi_z^{(m)}}{\partial \varphi\, \partial z} + \frac{i\omega\varepsilon_0 n^2}{\mu} \frac{\partial \Pi_z^{(e)}}{\partial \rho}, \tag{11.95b}$$

$$H_z = \left(k^2 + \frac{\partial^2}{\partial z^2}\right)\Pi_z^{(m)}. \tag{11.95c}$$

Next, we make use of the Lebedev–Kontorovich transform pair

$$P(u) = \int_0^\infty \tilde{\Pi} K_u(-i\tau\rho)\, \frac{d\rho}{\rho}, \tag{11.96a}$$

$$\tilde{\Pi}(\rho) = (\pi i)^{-1} \int_{-i\infty}^{i\infty} uP(u) I_u(-i\tau\rho)\, du, \tag{11.96b}$$

where $I_u(\;)$ and $(K_u(\;)$ are the modified Bessel functions of the first and second kind satisfying the differential equation

$$\left[\frac{1}{z}\frac{d}{dz} z \frac{d}{dz} - \left(1 + \frac{u^2}{z^2}\right)\right] \begin{matrix} I_u(z) \\ K_u(z) \end{matrix} = 0$$

and related to the Bessel functions as follows.

$$I_u(z) = \exp(-iu\pi/2) J_u(iz),$$
$$K_u(z) = i\pi/2 \exp(iu\pi/2) H_u^{(1)}(iz), \qquad -\pi < \arg z \le \pi/2. \tag{11.97}$$

However, as concluded in Section 8.4.7, the integral representation of P(u) in (11.96a) has a meaning only if $\Pi(\rho) \to \rho^\beta$, $\beta > 0$, as $\rho \to 0$. In the case of $\tilde{\Pi}^{(e)}$ this condition is automatically satisfied since $\tilde{\Pi}^{(e)} \sim E_z$ and E_z must vanish at the edge. In the case of $\Pi^{(m)}$, however, since $\tilde{\Pi}^{(m)} \sim H_z$, which does not have to vanish at the edge, the transform (11.96a) cannot be applied directly. As in Section 8.7.7, we can in this case define a new function

$$\tilde{M}(\rho, \varphi) = \tilde{\Pi}^{(m)}(\rho, \varphi) - \tilde{\Pi}^{(m)}(0) e^{i\tau\rho}, \tag{11.98}$$

so that $\tilde{M}(0)$ vanishes at the edge. Substituting from (11.98) into (11.93a), we obtain

$$\left(\frac{1}{\rho}\frac{\partial}{\partial\rho}\rho\frac{\partial}{\partial\rho} + \frac{1}{\rho^2}\frac{\partial^2}{\partial\varphi^2} + \tau^2\right)\tilde{M} = -\frac{i\tau}{\rho}\tilde{\Pi}^{(m)}(0) e^{i\tau\rho}, \tag{11.99}$$

to which the transform (11.96a) can now be applied. Hence, we obtain, using (8.15a),

$$\left(\frac{d^2}{d\varphi^2} + u^2\right)P^{(m)} = \frac{u\pi}{\sin u\pi}\,\tilde{\Pi}^{(m)}(0) \tag{11.100}$$

and

$$\left(\frac{d^2}{d\varphi^2} + u^2\right)P^{(e)} = 0. \tag{1.101}$$

Equation (11.101) can be solved without difficulty to give $P^{(e)} = A \cos u\varphi + B \sin u\varphi$. The solution to (11.100) can be written

$$P^{(m)} = C \cos u\varphi + D \sin u\varphi + \frac{\pi}{u \sin u\pi}\,\tilde{\Pi}^{(m)}(0). \tag{11.102}$$

Hence, inverting with respect to the u and α transform variables, we obtain

$$\Pi_z^{(e)}(\mathbf{r}) = \int_{-\infty}^{\infty} d\alpha\, e^{i\alpha z} \int_{-}^{i\infty} u[A(u, \alpha) \cos u\varphi + B(u, \alpha) \sin u\varphi] I_u(-i\tau\rho)\, du, \tag{11.103a}$$

and

$$\Pi_z^{(m)}(\mathbf{r}) = \int_{-\infty}^{\infty} d\alpha\, e^{i\alpha z} \int_{-\infty i}^{i\infty} u[C(u, \alpha) \cos u\varphi + D(u, \alpha) \sin u\varphi] I_u(-i\tau\rho)\, du, \tag{11.103b}$$

where we used the result of (8.168). It can be readily verified (problem) that the integral representations of $\Pi_z^{(e)}$ and $\Pi_z^{(m)}$ in (11.103) satisfy the wave equation in (11.93a). Hence they can be used to formulate a solution to the problem we have set forth to solve.

The plane $\varphi = \varphi_0$ contains no aperture and, hence, the tangential electric field components E_ρ and E_z must vanish identically there for all ρ and z. It follows from (11.94a) and (11.94c) that this boundary condition can be satisfied only if

$$\begin{aligned} \Pi_z^{(e)}(\rho, \varphi_0, z) &= 0, \\ \partial\Pi_z^{(m)}(\rho, \varphi_0, z)/\partial\varphi &= 0; \end{aligned} \tag{11.104}$$

consequently, we must have $B = -\cot u\varphi_0\, A$, and $C = \cot u\varphi_0\, D$. Thus we can rewrite (11.103) as

$$\Pi_z^{(e)}(\mathbf{r}) = \int_{-\infty}^{\infty} d\alpha\, e^{i\alpha z} \int_{-i\infty}^{i\infty} \frac{u \sin u(\varphi_0 - \varphi)}{\sin u\varphi_0} A(u, \alpha) I_u(-i\tau\rho)\, du \tag{11.105a}$$

and

$$\Pi_z^{(m)}(\mathbf{r}) = \int_{-\infty}^{\infty} d\alpha\, e^{i\alpha z} \int_{-i\infty}^{i\infty} \frac{u \cos u(\varphi_0 - \varphi)}{\sin u\varphi_0} D(u, \alpha) I_u(-i\tau\rho)\, du. \tag{11.105b}$$

We are now ready to apply the boundary conditions at the plane $\varphi = 0$. Using (11.92b), (11.94c), and (11.105a), we write

$$(2\pi)^{-1/2}\mathscr{E}_z(\rho, z) = (2\pi)^{-1/2} \int_{-\infty}^{\infty} \left[\tau^2 \int_{-i\infty}^{i\infty} uA(u, \alpha)I_u(-i\tau\rho)\, du\right] e^{i\alpha z}\, d\alpha. \tag{11.106}$$

It is seen that the quantity in square brackets on the right is simply a Fourier transform of the quantity on the left. Hence we obtian, upon inversion,

$$(\pi i)^{-1} \int_{-i\infty}^{i\infty} uA(u, \alpha)I_u(-i\tau\rho)\, du = -\frac{i}{2\pi^2} \int_{-\infty}^{\infty} \frac{\mathscr{E}_z(\rho, z)}{\tau^2} e^{-i\alpha z}\, dz. \tag{11.107}$$

Now the right side of (11.107) is identified as the Lebedev–Kontorovich transform of the unknown function $A(u, \alpha)$. Hence we obtain, by comparison with (11.96),

$$A(u, \alpha) = -\frac{i}{2\pi^2} \iint_{\mathrm{ap}} \frac{\mathscr{E}_z(\rho', z')K_u(-i\tau\rho')}{\tau^2\rho'} e^{-i\alpha z'}\, d\rho'\, dz', \tag{11.108}$$

where the integration is performed over the aperture ap only since $\mathscr{E}_z$ is zero elsewhere. Applying a similar procedure to the boundary condition (11.92a), we obtain, upon inversion with respect to the Fourier transform variable α,

$$-\frac{i}{2\pi^2} \int_{-\infty}^{\infty} \mathscr{E}_\rho(\rho, z)e^{-i\alpha z}\, dz = (i\pi)^{-1}$$

$$\times \int_{-i\infty}^{i\infty} \left[u\tau\alpha A(u, \alpha)I_u'(-i\tau\rho) + \frac{i\omega\mu_0\mu}{\rho} u^2 D(u, \alpha)I_u(-i\tau\rho)\right] du. \tag{11.109}$$

The inversion with respect to u cannot be performed immediately because of the first term on the right in (11.109). However, we note that the differentiation of (11.107) with respect to ρ gives

$$(i\pi)^{-1} \int_{-i\infty}^{i\infty} \alpha\tau u A(u, \alpha)I_u'(-i\tau\rho)\, du = (2\pi^2)^{-1} \int_{-\infty}^{\infty} \frac{\alpha}{\tau^2} \frac{\partial\mathscr{E}_z(\rho, z)}{\partial\rho} e^{-i\alpha z}\, dz, \tag{11.110}$$

where, in addition, we multiplied both sides by α. Note that the left side of (11.110) is identical with the first term on the right in (11.109). Hence we rewrite (11.109) as

$$(i\pi)^{-1} \int_{-i\infty}^{i\infty} u[uD(u, \alpha)]I_u(-i\tau\rho)\, du$$

$$= -(2\pi^2\omega\mu_0\mu)^{-1} \int_{\infty}^{\infty} \rho\left[\mathscr{E}_\rho(\rho, z) - \frac{\alpha}{\tau^2}\frac{\partial\mathscr{E}_z(\rho, z)}{\partial\rho}\right] e^{-i\alpha z}\, dz, \tag{11.111}$$

which can be readily inverted according to (11.96) to give

$$uD(u, \alpha) = -(2\pi^2\omega\mu_0\mu)^{-1} \iint_{\text{ap}} \left[\mathscr{E}_\rho(\rho', z') - \frac{i\alpha}{\tau^2}\frac{\partial\mathscr{E}_z(\rho', z')}{\partial\rho'}\right] \times K_u(-i\tau\rho')e^{-i\alpha z'}\,d\rho'\,dz'. \tag{11.112}$$

The results of (11.108) and (11.112) together with (11.105) constitute a formal solution to the problem. In order to express this solution in a more useful form, we substitute (11.108) and (11.112) into (11.105) and write

$$\Pi_z^{(e)}(\mathbf{r}) = -(2\pi)^{-1}\frac{\partial}{\partial\varphi}\int_{-\infty}^{\infty} d\alpha \iint_{\text{ap}} \frac{\mathscr{E}_z(\rho', z')}{\tau^2\rho'}\, g(\boldsymbol{\rho}, \rho', \alpha)\exp[i\alpha(z - z')]\,d\rho'\,dz' \tag{11.113}$$

and

$$\Pi_z^{(m)}(\mathbf{r}) = \frac{i}{2\pi\omega\mu_0\mu}\int_{-\infty}^{\infty} d\alpha \iint_{\text{ap}} \left[\mathscr{E}_\rho(\rho', z') - \frac{i\alpha}{\tau^2}\frac{\partial\mathscr{E}_z(\rho', z')}{\partial\rho'}\right] \times g(\boldsymbol{\rho}, \rho', \alpha)\exp[i\alpha(z - z')]\,d\rho'\,dz', \tag{11.114}$$

where

$$g(\boldsymbol{\rho}, \rho', \alpha) = -\frac{1}{2}\int_{-i\infty}^{i\infty} \frac{\cos u(\varphi_0 - \varphi)}{\sin u\varphi_0} J_u(\tau\rho)H_u^{(1)}(\tau\rho')\,du \tag{11.115}$$

and in order to arrive at (11.115) we also used (11.97). It will follow that $g(\ )$ is symmetrical in ρ and ρ'; i.e., $g(\rho, \varphi, \rho', \alpha) = g(\rho', \varphi, \rho, \alpha)$.

It will be noted that $g(\boldsymbol{\rho}, \rho', \alpha)$ in (11.115) is identical in form to $G^{(m)}(\boldsymbol{\rho}, \boldsymbol{\rho}')$ in (8.169b) when $\varphi' = 0$ and k is replaced by τ in the latter. Hence, using (8.176), we can write for the present case

$$g_{\rho>\rho'}(\boldsymbol{\rho}, \rho', \alpha) = \frac{i\pi}{2\varphi_0}\sum_{m=0}^{\infty} \varepsilon_m \cos v\varphi J_v(\tau\rho')H_v^{(1)}(\tau\rho) \tag{11.116}$$

where $v = m\pi/\varphi_0$. Alternately, using (8.189) and (8.191) with appropriate changes, we obtain an alternate representation

$$\begin{aligned} \underset{\rho>\rho}{g}(\boldsymbol{\rho}, \rho', \alpha) = \frac{i}{2}\Bigg(&\sum_{m=1}^{M} H_0^{(1)}\{\tau[\rho^2 + \rho'^2 - 2\rho\rho'\cos(2m\varphi_0 - \varphi)]^{1/2}\} \\ &+ \sum_{n=0}^{N} H_0^{(1)}\{\tau[\rho^2 + \rho'^2 - 2\rho\rho'\cos(2n\varphi_0 + \varphi)]^{1/2}\}\Bigg) \\ &- \frac{1}{8\varphi_0}\int_{-i\infty}^{i\infty} \frac{\sin 2v(x + \pi)H_0^{(1)}[\tau(\rho^2 + \rho'^2 + 2\rho\rho'\cos x)^{1/2}]}{\sin v(x + \pi + \varphi)\sin v(x + \pi - \varphi)}\,dx, \end{aligned} \tag{11.117}$$

where $v = \pi/2\varphi_0$ and

$$0 < 2M\varphi_0 - \varphi < \pi, \qquad M = 1, 2, 3, \ldots, \tag{11.118a}$$

$$0 < 2N\varphi_0 + \varphi < \pi, \qquad N = 0, 1, 2, 3, \ldots. \tag{11.118b}$$

11.3.1 SERIES REPRESENTATION OF THE SOLUTION

Substituting (11.116) into (11.113) and (11.114) and rearranging, we obtain

$$\Pi_z^{(m)}(\mathbf{r}) = \frac{i}{2\varphi_0} \sum_{m=1}^{\infty} v \sin v\varphi \iint_{\text{ap}} \frac{1}{\rho'} \mathscr{E}_z(\rho', z') U(\rho, z, \rho', z')\, d\rho'\, dz' \tag{11.119}$$

and

$$\Pi_z^{(m)}(\mathbf{r}) = (4\varphi_0 \omega\mu_0 \mu)^{-1} \sum_{m=0}^{\infty} \varepsilon_m \cos v\varphi$$
$$\times \left[\frac{\partial}{\partial z} \iint_{\text{ap}} \frac{\partial \mathscr{E}_z(\rho', z')}{\partial \rho'} U(\rho, z, \rho', z')\, d\rho'\, dz'\right.$$
$$\left. - \left(k^2 + \frac{\partial^2}{\partial z^2}\right) \iint_{\text{ap}} \mathscr{E}_\rho(\rho', z') U(\rho, z, \rho', z')\, d\rho'\, dz'\right], \tag{11.120}$$

where

$$U(\rho, z, \rho', z') = \int_{-\infty}^{\infty} \tau^{-2} J_v(\tau\rho') H_v^{(1)}(\tau\rho) \exp[i\alpha(z - z')]\, d\alpha. \tag{11.121}$$

The above integral is of the form of that in (9.12) with $F(\alpha) = \tau^{-2}$. Hence, when the distance from the apex to the field point is large, $k\rho \to \infty$, we use (9.22) to obtain

$$U(\rho, z, \rho', z') = -\frac{2i \exp(ikr)}{k^2 \sin^2 \theta r} (-i)^v J_v(k\rho' \sin\theta) \exp(-ik\cos\theta z'). \tag{11.122}$$

Thus, in the far field, $k\rho \to \infty$, we have

$$\Pi_z^{(e)}(\mathbf{r}) = \frac{e^{ikr}}{\varphi_0 k^2 \sin^2 \theta r} \sum_{m=1}^{\infty} (-i)^v v \sin v\varphi Q_v^{(e)}(\theta) \tag{11.123}$$

and

$$\Pi_z^{(m)}(\mathbf{r}) = \frac{i e^{ikr}}{2\varphi_0 \omega\mu_0 \mu r} \sum_{m=0}^{\infty} (-i)^v \varepsilon_m \cos v\varphi Q_v^{(m)}(\theta), \tag{11.124}$$

where

$$Q_v^{(e)}(\theta) = \iint_{\text{ap}} \frac{1}{\rho'} \mathscr{E}_z(\rho', z') J_v(k\rho' \sin\theta) \exp(-ik\cos\theta z')\, d\rho\, dz' \tag{11.25}$$

and

$$Q_\nu^{(m)}(\theta) = \iint_{\text{ap}} \mathscr{E}_\rho(\rho', z') J_\nu(k\rho' \sin\theta) \exp(-ik\cos\theta z')\, d\rho'\, dz'$$

$$+ i\cot\theta \iint_{\text{ap}} \mathscr{E}_z(\rho', z') J_\nu'(k\rho' \sin\theta) \exp(-ik\cos\theta z')\, d\rho'\, dz'. \tag{11.126}$$

To arrive at (11.126) we have integrated by parts once with respect to ρ' the integral containing the term $\partial\mathscr{E}_z(\rho', z')/\partial\rho'$.

The resulting field components in spherical coordinates can be found by using (11.58) and (11.59). We find for the radiation zone

$$E_\theta = -\frac{e^{ikr}}{\varphi_0 r \sin\theta} \sum_{m=1}^{\infty} (-i)^\nu \nu \sin \nu\varphi Q_\nu^{(e)}(\theta) \tag{11.127}$$

and

$$E_\varphi = \frac{ik \sin e^{ikr}}{2\varphi_0 r} \sum_{m=0}^{\infty} \varepsilon_m (-i)^\nu \cos \nu\varphi Q_\nu^{(e)}(\theta). \tag{11.128}$$

Moreover, $H_\varphi = (n/\mu)(\varepsilon_0/\mu_0)^{1/2} E_\theta$ and $H_\theta = -(n/\mu)(\varepsilon_0/\mu_0)^{1/2} E_\varphi$, as expected.

If the aperture is infinite in the axial direction, $\mathscr{E}_z = 0$, and the exciting field $\mathscr{E}_\rho$ does not depend on the axial coordinate, then $\Pi_z^{(e)}(\mathbf{r}) = 0$ and by (11.114)

$$\Pi_z^{(m)}(\boldsymbol{\rho}) = \frac{i}{\omega\mu_0\mu} \int_{\text{ap}} \mathscr{E}_\rho(\rho') g(\boldsymbol{\rho}, \rho', 0)\, d\rho', \tag{11.129}$$

where we used the integral representation of the delta function

$$2\pi\,\delta(\alpha) = \int_{\infty}^{\infty} \exp(-i\alpha z')\, dz'. \tag{11.130}$$

As an example of the application of the formulas derived in this section, consider the problem of a thin axial slot on an infinite wedge as shown in Fig. 11.8. Let the exciting field at the aperture be constant and linearly polarized in the radial direction, i.e.,

$$\begin{aligned} &\mathscr{E}_z(\rho', z') = 0, \\ &\mathscr{E}_\rho(\rho', z') = -V\,\delta(\rho' - a), \end{aligned} \tag{11.131}$$

where V is the voltage across the slot. In this case $Q_\nu^{(e)}(\theta) = 0$ and we obtain

$$E_\varphi = -\frac{iV \sin\theta e^{ikr}}{\varphi_0 r \cos\theta} \sin\left(\frac{kL}{2}\cos\theta\right) \sum_{m=0}^{\infty} \varepsilon_m (-i)^\nu \cos\nu\varphi J_\nu(ka\sin\theta). \tag{11.132}$$

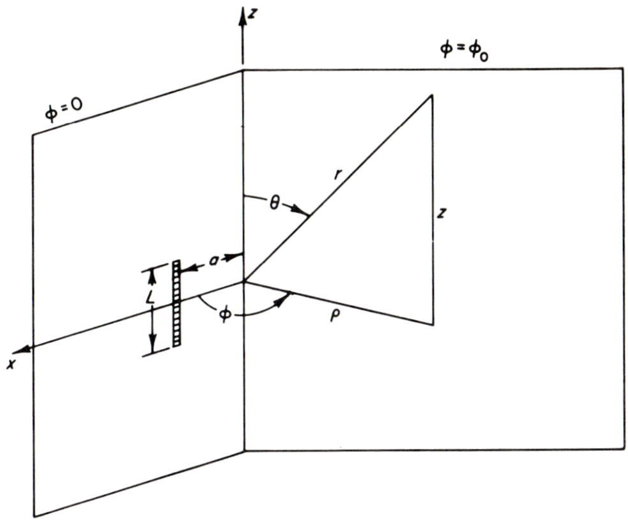

Fig. 11.8. Axial slot on an infinite wedge.

If the distance from the apex to the slot is small, so that $ka < 1$, the Bessel function can be replaced by its small argument approximation, i.e., $J_\nu(x) \sim (x/2)^\nu/\Gamma(1+\nu)$, to simplify numerical calculations. If, however, $ka \gg 1$, then many terms of the series will be needed for a given degree of accuracy, especially in the vicinity of $\theta = \pi/2$.

If the slot in Fig. 11.8 is of infinite extent in the axial direction, $L \to \infty$, then we obtain, using (11.129),

$$\Pi_z^{(m)}(\boldsymbol{\rho}) = \frac{V\pi}{2\varphi_0\,\omega\mu_0\,\mu} \sum_{m=0}^{\infty} \varepsilon_m \cos \nu\varphi J_\nu(ka) H_\nu^{(1)}(k\rho), \qquad (11.133)$$

where, again, many terms of the series will be needed if ka is large.

11.3.2 Alternative Representation of the Solution

When the aperture exciting field is radial ($\mathscr{E}_z = 0$), a simpler expression for the radiation fields can be obtained as follows.

Since in this case $\Pi_z^{(e)} = 0$, all of the field components can be derived from $\Pi_z^{(m)}$, which, using (11.114), can be written in the form

$$\Pi_z^{(m)}(\mathbf{r}) = \frac{i}{2\pi\omega\mu_0\,\mu} \iint_{\text{ap}} f(\mathbf{r}, \rho', z')\mathscr{E}_\rho(\rho', z')\, d\rho'\, dz', \qquad (11.134)$$

where

$$f(\mathbf{r}, \rho', z') = \int_{-\infty}^{\infty} g(\boldsymbol{\rho}, \rho', \alpha) \exp[i\alpha(z - z')]\, d\alpha, \tag{11.135}$$

and $g(\boldsymbol{\rho}, \rho', \alpha)$ is given by (11.117). In order to perform the indicated integration in (11.135) we note that the integral

$$\int_{-\infty}^{\infty} H_0^{(1)}(\tau s) \exp[i\alpha(z - z')]\, d\alpha = -2i \frac{\exp\{ik[s^2 + (z - z')^2]^{1/2}\}}{[s^2 + (z - z')^2]^{1/2}} \tag{11.136}$$

and, hence, we can write in place of (11.135)

$$f(\mathbf{r}, \rho', z') = \sum_{m=1}^{M} \frac{\exp(ikR_1)}{R_1} + \sum_{n=0}^{N} \frac{\exp(ikR_2)}{R_2}$$

$$+ \frac{i}{4\varphi_0} \int_{-i\infty}^{i\infty} \frac{\sin 2v(x + \pi) \exp(ikR_3)}{\sin v(x + \pi + \varphi) \sin v(x + \pi - \varphi)}\, dx, \tag{11.137}$$

where

$$R_1 = \{r^2 + \rho'^2 + z'^2 - 2[\rho\rho' \cos(2m\varphi_0 - \varphi) + zz']\}^{1/2}, \tag{11.138a}$$

$$R_2 = \{r^2 + \rho'^2 + z'^2 - 2[\rho\rho' \cos(2n\varphi_0 + \varphi) + zz']\}^{1/2}, \tag{11.138b}$$

$$R_3 = [r^2 + \rho'^2 + z'^2 + 2(\rho\rho' \cos x - zz')]^{1/2}. \tag{11.138c}$$

In the radiation zone $r^2 \gg \rho'^2 + z'^2$ and, moreover, since $\rho/r = \sin\theta$ and $z/r = \cos\theta$ we can approximate as follows.

$$R_1 \sim r - z' \cos\theta - \rho' \sin\theta \cos(\varphi - 2m\varphi_0), \tag{11.139a}$$

$$R_2 \sim r - z' \cos\theta - \rho' \sin\theta \cos(\varphi + 2n\varphi_0), \tag{11.139b}$$

$$R_3 \sim r - z' \cos\theta + \rho' \sin\theta \cos x, \tag{11.139c}$$

in the exponents of (11.137), whereas in the numerators we set $R_1 \sim R_2 \sim R_3 \sim r$. Thus, (11.137) becomes

$$f(\mathbf{r}, \rho', z') = \frac{\exp[ik(r - z'\cos\theta)]}{r}$$

$$\times \Big\{ I(\rho', \varphi, \theta) + \sum_{m=1}^{M} \exp[-ik\rho' \sin\theta \cos(\varphi - 2m\varphi_0)]$$

$$+ \sum_{n=0}^{N} \exp[-ik\rho' \sin\theta \cos(\varphi + 2n\varphi_0)] \Big\}, \tag{11.140}$$

where

$$I(\rho', \varphi, \theta) = \frac{i}{4\varphi_0} \int_{-i\infty}^{i\infty} \frac{\sin 2v(x + \pi) \exp(ik\rho' \sin\theta \cos x)}{\sin v(x + \pi + \varphi) \sin v(x + \pi - \varphi)} dx. \quad (11.141)$$

When $k\rho' \sin\theta \gg 0$, the integral in (11.141) can be evaluated approximately by using the method of stationary phase. We obtain in this case

$$I(\rho', \varphi, \theta) \sim -\frac{1}{2\varphi_0}\left(\frac{\pi}{2k\rho' \sin}\right)^{1/2} \cdot \frac{\sin 2v\pi \exp[i(k\rho' \sin\theta + \pi/4)]}{\sin v(\pi + \varphi) \sin v(\pi - \varphi)}, \quad (11.142)$$

which, evidently, is valid for all $0 < \varphi < \varphi_0$ except the neighborhoods where $v(\pi + \varphi)$ approaches zero or a multiple of π. To obtain suitable representation in these critical regions, we can use the results of Section 10.3.2 with appropriate changes and write

$$I(\rho', \varphi, \theta) = -\frac{\cos v\pi}{2 \cos v\varphi} [Q(\varphi) + Q(-\varphi)] \exp(ik\rho' \sin\theta), \quad (11.143)$$

where

$$Q(\varphi) = \pm\exp(-\zeta_+^2) \operatorname{erfc}(\mp i\zeta_+), \qquad \operatorname{Im}(\zeta_+) \gtrless 0, \quad (11.144a)$$

$$Q(-\varphi) = \pm\exp(-\zeta_-^2) \operatorname{erfc}(\mp i\zeta_-), \qquad \operatorname{Im}(\zeta_-) \gtrless 0, \quad (11.144b)$$

and

$$\zeta_+ = (\varphi_0/\pi) \sin v(\pi + \varphi)(2k\rho' \sin\theta)^{1/2} \exp(i\pi/4), \quad (11.145a)$$

$$\zeta_- = (\varphi_0/\pi) \sin v(\pi - \varphi)(2k\rho' \sin\theta)^{1/2} \exp(i\pi/4). \quad (11.145b)$$

Away from the critical point $v(\pi \pm \varphi) = m\pi$ where the arguments of the error functions are large, $\zeta_+ \to \infty$, we can use the asymptotic formula $\operatorname{erfc}(\mp i\zeta) \sim \pm i\zeta^{-1}\pi^{-1/2} \exp(\zeta^2)$ to show that

$$Q(\pm\varphi) \to \frac{\exp(i\pi/4)}{\varphi_0}\left(\frac{\pi}{2k\rho' \sin\theta}\right)^{1/2} \cdot [\sin v(\pi \pm \varphi)]^{-1}, \quad (11.146)$$

which upon substitution into (11.143) gives (11.142).

It will be noted that the integral $I(\rho', \varphi, \theta)$ vanishes whenever $\varphi_0 = \pi/l$, where l is a positive integer other than zero. Moreover, it will be noted that the critical points where the denominator of (11.142) vanishes occur only at $\varphi = \pi$ if $\varphi_0 > \pi$, and at $\varphi = 2m\varphi_0 - \pi$ and $\varphi = \pi - 2n\varphi_0$, where m and n are positive integers other than zero, if $\varphi_0 < \pi$.

To illustrate the usefulness of the present representation, let us return to the problem of the axial slot on an infinite wedge shown in Fig. 11.8. In this case $\mathscr{E}_z = 0$, $\mathscr{E}_\rho = -V\,\delta(\rho' - a)$, and since $E_\theta = 0$, $E_\varphi = k\omega\mu_0\,\mu \sin\theta\Pi^{(m)}$, we have

$$E_\varphi = \frac{iV \sin\theta \exp(ikr)}{\pi r \cos\theta} \sin\left(\frac{kL}{2}\cos\theta\right) \times \left\{ I(a, \varphi, \theta) + \sum_{m=1}^{M} \exp[-ika \sin\theta \cos(\varphi - 2m\varphi_0)] + \sum_{n=0}^{N} \exp[-ika \sin\theta \cos(\varphi + 2n\varphi_0)] \right\}. \tag{11.147}$$

Let us now consider the following wedge configurations.

Plane

In this case $\varphi_0 = \pi$, $v = 1/2$, and the integral $I(a, \varphi, \theta)$ vanishes. Examination of (11.118) reveals that M does not exist and $N = 0$. Hence, using (11.147), we obtain immediately

$$E_\varphi = -\frac{iV \sin\theta}{\pi r \cos\theta} \sin\left(\frac{kL}{2}\cos\theta\right) \exp[ik(r - a \sin\theta \cos\varphi)]. \tag{11.148}$$

It can be shown (problem) that (11.148) is in agreement with the results of Section 11.1.1.

45-*Degree Wedge*

In this case $\varphi_0 = \pi/4$, $v = 2$, and again, the integral vanishes. Examination of (11.118) reveals that $M = 2$ and $N = 1$. Hence we obtain, using (11.147),

$$E_\varphi = -\frac{i2V \sin\theta e^{ikr}}{\pi r \cos\theta} \sin\left(\frac{kL}{2}\cos\theta\right) \times [\cos(ka \sin\theta \cos\varphi) + \cos(ka \sin\theta \sin\varphi)]. \tag{11.149}$$

Half Plane

In this case $\varphi_0 = 2\pi$ and $v = \frac{1}{4}$, so that the integral $I(a, \varphi, \theta)$ does not vanish. Examination of (11.118) reveals that M does not exist and $N = 0$ exists only if $\varphi < \pi$. Hence we obtain

$$E_\varphi = -\frac{iV \sin\theta \exp(ikr)}{\pi r \cos\theta} \sin\left(\frac{kL}{2}\cos\theta\right) [I(a, \varphi, \theta) + \exp(-ika \sin\theta \cos\varphi)] \tag{11.150a}$$

for the illuminated region, $\varphi < \pi$, and

$$E_\varphi = -\frac{iV \sin\theta e^{ikr}}{\pi r \cos\theta} \sin\left(\frac{kL}{2}\cos\theta\right) I(a, \varphi, \theta) \tag{11.150b}$$

for the shadow region, $\varphi > \pi$, where in this case

$$I(a, \varphi, \theta) = -\frac{\exp[i(ka \sin\theta + \pi/4)]}{2\cos(\varphi/2)(2ka \sin\theta)^{1/2}}, \qquad \varphi \neq \pi. \tag{11.151}$$

It is clear that in this case $I(a, \varphi, \theta)$ represents the portion of the field that is diffracted by the edge of the half plane. In fact, it can be seen that as the distance from the edge to the slot becomes large ($a \to \infty$), then $I(a, \varphi, \theta) \to 0$, the field in the shadow region vanishes, and the field in the illuminated region becomes identical to that of a slot on an infinite half plane as found in (11.148).

In the vicinity of $\varphi = \pi$ the formula in (11.151) is not appropriate because $\cos(\varphi/2)$ vanishes there. In this case we must use the representation in (11.143). Since $Q(\varphi)$ is finite at $\varphi = \pi$, we can write, using (11.146),

$$Q(\varphi) = \frac{\exp(i\pi/4)}{2 \sin \frac{1}{4}(\pi + \varphi)(2\pi ka \sin \theta)^{1/2}}, \tag{11.152}$$

whereas for $Q(-\varphi)$ we use (11.144b) to obtain

$$Q(-\varphi) = \pm\exp(-\zeta_-{}^2)\,\mathrm{erfc}(\pm i\zeta_-), \tag{11.153}$$

where the upper signs are appropriate to $\varphi < \pi$ and the lower signs to $\varphi > \pi$ and

$$\zeta_- = 2 \sin \tfrac{1}{4}(\pi - \varphi)(2ka \sin \theta)^{1/2} \exp(i\pi/4). \tag{11.154}$$

When φ is very close to π, then we can use the small argument approximation for the complementary error function

$$\mathrm{erfc}(\pm z) \sim 1 \mp (2/\sqrt{\pi})z, \qquad |z| \to 0, \tag{11.155}$$

and obtain in this case

$$Q(-\varphi) \sim \pm 1 \mp \frac{i2\zeta_-}{\sqrt{\pi}}, \tag{11.156}$$

where again the upper signs are appropriate to $\varphi < \pi$ and the lower signs to $\varphi > \pi$. Hence, using (11.143), we write for $\varphi \sim \pi$

$$I(a, \varphi, \theta) = -\exp(ika \sin \theta) \times \left[\frac{\exp(i\pi/4)}{4(2\pi ka \sin \theta)^{1/2}} \pm \frac{1}{2} \pm \frac{\exp(-i\pi/4)}{2\sqrt{\pi}} \sin \tfrac{1}{4}(\pi - \varphi)(2ka \sin \theta)^{1/2}\right]. \tag{11.157}$$

Substitution of (11.157) into either (11.150a) or (11.150b) and letting $\varphi = \pi$ gives the value of E_φ at $\varphi = \pi$ as

$$E_\varphi = -\frac{iV \sin \theta \exp[ik(r + a \sin \theta)]}{2\pi r \cos \theta} \sin\left(\frac{kL}{2} \cos \theta\right)\left[1 - \frac{1 + i}{4(\pi ka \sin \theta)^{1/2}}\right]. \tag{11.158}$$

Thus we see that the fields are continuous everywhere in the region $0 < \varphi < 2\pi$, as expected.

Problems

11.1 Show that the integral representations of $\Pi_z^{(e)}(\mathbf{r})$ and $\Pi_z^{(m)}(\mathbf{r})$ in (11.105) satisfy the differential equation in (11.93a).

11.2 Show that $g(\boldsymbol{\rho}, \rho', \alpha)$ in (11.115) is symmetrical in ρ and ρ'.

11.3 Prove (11.117).

11.4 Prove (11.143).

11.5 Show that (11.148) is in agreement with the results of Section 11.1.1.

11.6 Prove (11.149) and show graphically the image representation of the solution.

References

Abramowitz, M., and Stegun, I. A. (eds.) (1964). "Handbook of Mathematical Functions." U.S. Gov't. Printing Office, Washington, D.C. (*Natl. Bur. Std., U.S. Appl. Math. Ser.* **55**.)

Harrington, R. F. (1961). "Time Harmonic Electromagnetic Fields." McGraw-Hill, New York.

Wait, J. R. (1959). "Electromagnetic Radiation from Cylindrical Structures." Pergamon, Oxford.

INDEX

Page entries followed by the letter n refer to entries to be found in the footnotes on that page.